Lecture Notes in Artificial Intelli

Subseries of Lecture Notes in Computer Science

Sheela Ramanna   Pawan Lingras
Chattrakul Sombattheera   Aneesh Krishna (Eds.)

Multi-disciplinary Trends in Artificial Intelligence

7th International Workshop, MIWAI 2013
Krabi, Thailand, December 9-11, 2013
Proceedings

 Springer

Volume Editors

Sheela Ramanna
University of Winnipeg, Applied Computer Science Department
Winnipeg, MB, Canada
E-mail: s.ramanna@uwinnipeg.ca

Pawan Lingras
Saint Mary's University, Mathematics and Computing Science
Halifax, NS, Canada
E-mail: pawan@cs.smu.ca

Chattrakul Sombattheera
Mahasarakham University, Faculty of Informatics
Khamreang, Kantarawichai, Mahasarakham, Thailand
E-mail: chattrakul.s@msu.ac.th

Aneesh Krishna
Curtin University, Department of Computing, Perth, WA, Australia
E-mail: a.krishna@curtin.edu.au

ISSN 0302-9743   e-ISSN 1611-3349
ISBN 978-3-642-44948-2   e-ISBN 978-3-642-44949-9
DOI 10.1007/978-3-642-44949-9
Springer Heidelberg New York Dordrecht London

Library of Congress Control Number: 2013952451

CR Subject Classification (1998): I.2, H.2.8, I.4, H.3, H.4, F.2

LNCS Sublibrary: SL 7 – Artificial Intelligence

Typesetting: Camera-ready by author, data conversion by Scientific Publishing Services, Chennai, India

Printed on acid-free paper

Springer is part of Springer Science+Business Media (www.springer.com)

Preface

This volume contains papers selected for presentation at the 7th Multi-Disciplinary International Workshop on Artificial Intelligence (MIWAI) held during December 9–11, 2013, at Krabi, Thailand.

The MIWAI workshop series started in 2007 in Thailand as the Mahasarakham International Workshop on Artificial Intelligence and has been held every year since then. It has emerged as an international workshop with participants from around the world. MIWAI 2011 was held in Hyderabad, India and MIWAI 2012 was held in Ho Chi Minh City, Vietnam.

The MIWAI series of workshops serves as a forum for artificial intelligence (AI) researchers and practitioners to discuss and deliberate cutting-edge AI research. It also aims to elevate the standards of AI research by providing researchers and students with feedback from an internationally renowned Program Committee.

The theory, methods, and tools that encompass AI research include such areas as: cognitive science, computational intelligence, computational philosophy, game theory, machine learning, multi-agent systems, natural language, representation and reasoning, speech, vision and the Web. The above methods have broad applications in big data, bioinformatics, biometrics, decision support, e-commerce, image processing, analysis and retrieval, industrial applications, knowledge management, privacy, recommender systems, security, software engineering, spam filtering, surveillance, telecommunications and Web services. Submissions received by MIWAI 2013 were wide ranging and covered both theory as well as applications.

MIWAI 2013 received 65 full papers from 21 countries and regions including Austria, Bangladesh, Canada, Chile, China, France, India, Japan, Korea, Malaysia, Morocco, Pakistan, Poland, Russia, Serbia, Taiwan, Thailand, Tunisia, UK, USA, and Vietnam. Following the success of previous MIWAI conferences, MIWAI 2013 continued the tradition of a rigorous review process. Submissions received were subject to a brief review by the program and general chairs to ensure a blind review. Every submission was reviewed by at least two Program Committee members and domain experts. Additional reviews were sought when necessary. Papers with conditional acceptances were reviewed by the program and general chairs before acceptance. A total of 30 papers were accepted with an acceptance rate of 46%. Some of the papers that were excluded from the proceedings showed promise, but had to be rejected to maintain the quality of the proceedings. We would like to thank all authors for their submissions. Without their contribution, this workshop would not have been possible.

We are grateful to Nick Cercone for accepting our invitation to deliver the keynote talk. Special thanks to Vivek Singh and Ashwin Srinivasan for organizing a special session on Machine Learning and Text Analytics. We wish to thank the members of the Steering Committee for their support. We are indebted to the Program Committee members and external reviewers for their effort in ensuring a rich scientific program.

We acknowledge the use of the EasyChair Conference System for the paper submission, review, and compilation of the proceedings. We are thankful to Alfred Hofmann, Anna Kramer, and the excellent LNCS team at Springer for their support and co-operation in publishing the proceedings as a volume of the *Lecture Notes in Computer Science*.

September 2013

Sheela Ramanna
Pawan Lingras
Chattrakul Sombattheera
Aneesh Krishna

Organization

Committees

Steering Committee

Arun Agarwal	University of Hyderabad, India
Rajkumar Buyya	University of Melbourne, Australia
Patrick Doherty	University of Linkoping, Sweden
Jerome Lang	University of Paris-Dauphine, France
James Peters	University of Manitoba, Canada
Srinivasan Ramani	IIIT Bangalore, India
C. Raghavendra Rao	University of Hyderabad, India
Leon Van Der Torre	University of Luxembourg, Luxembourg

Conveners

Richard Booth	University of Luxembourg, Luxembourg
Chattrakul Sombattheera	Mahasarakham University, Thailand

General Co-chairs

Sheela Ramanna	University of Winnipeg, Canada
Chattrakul Sombattheera	Mahasarakham University, Thailand

Program Co-chairs

Pawan Lingras	Saint Mary's University, Canada
Aneesh Krishna	Curtin University, Australia

Program Committee

Samir Aknine	Claude Bernard University of Lyon 1, France
Ricardo Aler	Universidad Carlos III, Spain
Dan Ames	Idaho State University, USA
Grigoris Antoniou	University of Huddersfield, UK
Costin Badica	University of Craiova, Romania
Chaitan Baru	UC San Diego, USA
Philippe Besnard	Université Paul Sabatier, France
Raj Bhatnagar	University of Cincinnati, USA
Antonis Bikakis	University College London, UK

K.K. Biswas   Indian Institute of Technology Delhi, India
Laor Boongasame   Bangkok University, Thailand
Veera Boonjing   KMITL, Thailand
Darko Brodic   University of Belgrade, Serbia
Patrice Caire   University of Luxembourg (SnT), Luxembourg
David Camacho   Universidad Autonoma de Madrid, Spain
Maria do Carmo Nicoletti   FACCAMP-SP and UFSCar-SP, Brazil
Narendra S. Chaudhari   Indian Institute of Technology Indore, India
Broderick Crawford   Pontificia Universidad Catolica de Valparaiso,
                                 Chile
Maria D. R-Moreno   Universidad de Alcala, Spain
Tiago De Lima   University of Artois and CNRS, France
Jurgen Dix   TU Clausthal, Germany
Nguyen Duy Binh   Hanoi University of Agriculture, Vietnam
David F. Barrero   Universidad de Alcala, Spain
Giorgos Flouris   FORTH-ICS, Greece
Christopher Henry   University of Winnipeg, Canada
Julio Hernandez   School of Computing, France
Andreas Herzig   Université Paul Sabatier, France
Prakash Hiremath   Gulbarga University, India
Sachio Hirokawa   Kyushu University, Japan
Estevam Rafael Hruschka   Federal University of Sao Carlos, Brazil
Nguyen Huynh   Ho Chi Minh City University of Technology,
                                 Vietnam
Sarun Intakosum   KMITL, Thailand
Julian Jang   CSIRO, Australia
Jason Jung   Yeungnam University, Korea
Mohan Kankanhalli   National University, Singapore
Satish Kolhe   North Maharashtra University, India
Emiliano Lorini   Université Paul Sabatier, France
B.M. Mehtre   IDRBT Hyderabad, India
Jerome Mengin   Université Paul Sabatier, France
Sebastian Moreno   Purdue University, USA
Narasimha Murty Musti   Indian Institute of Science, India
Ekawit Nantajeewarawat   Thammasat University, Thailand
Sven Naumann   University of Trier, Germany
Atul Negi   University of Hyderabad, India
Theodore Patkos   University of Crete, Greece
Heiko Paulheim   TU Darmstadt, Germany
Laurent Perrussel   University of Toulouse, France
Vu Pham-Tran   Ho Chi Minh City University of Technology,
                                 Vietnam
Guilin Qi   Southeast University, China
Ag Ramakrishnan   Indian Institute of Science, India
O.B.V. Ramanaiah   JNTU Hyderabad, India

Swarupa Rani   University of Hyderabad, India
V. Ravi   Sri Venkateswara University, India
Phattanapon Rhienmora   United Nations University - IIST, Macau
Harvey Rosas   Universidad de Valparaso, Chile
Andre Rossi   Université de Bretagne-Sud, France
Carolina Saavedra   Université de Lorraine, France
Samrat Sabat   University of Hyderabad, India
José Hiroki Saito   FACCAMP-SP, and UFSCar-SP, Brazil
Rodrigo Salas   Universidad de Valparaso, Chile
Jun Shen   University of Wollongong, Australia
Alok Singh   University of Hyderabad, India
Virach Sornlertlamvanich   National Electronics and Computer Technology
                               Center, Thailand
Kannan Srinathan   International Institute of Information
                               Technology, India
Siriwan Suebnukarn   Thammasat University, Thailand
Boontawee Suntisrivaraporn   Sirindhorn International Institute of
                               Technology, Thailand
Quan Thanh Tho   Ho Chi MInh City University of Technology,
                               Vietnam
Pham Thien   Nong Lam University, Vietnam
Romina Torres   Universidad Andres Bello, Chile
Nitin Tripathi   Asian Institute of Technology, Thailand
A.Y. Turhan   TU Dresden, Germany
Siba Kumar Udgata   University of Hyderabad, India
Jose Valls   Universidad Carlos III, Spain
Sergio Velastin   Kingston University, UK
Alejandro Veloz   Universidad de Valparaiso, Chile
Vo Thi Ngoc Chau   Ho Chi Minh City University of Technology,
                               Vietnam
Rajeev Wankar   University of Hyderabad, India
Alejandro Weinstein   Universidad de Valparaiso, Chile
Paul Weng   Université Paris 6, France

Publicity Co-chairs

Panich Sudkhot   Mahasarakham University, Thailand
Surasak La-ongkham   Mahasarakham University, Thailand

Local Organizing Committee

Benjawan Intara   Mahasarakham University, Thailand
Suwicha Natawong   Mahasarakham University, Thailand
Pichet Wayalun   Mahasarakham University, Thailand

Additional Reviewers

Baghel, Anurag
Biswas, Bhaskar
Henry, Christopher
Ouivirach, Kan
Pinto, David
Piryani, Rajesh

Siddiqui, Tanveer
Singh, Manoj Kumar
Singh, Sanjay
Singhal, Achintya
Vijaykumar, T.V
Waila, Pranav

Table of Contents

An ET-Based Low-Level Solution
for Query-Answering Problems

Kiyoshi Akama[1] and Ekawit Nantajeewarawat[2]

[1] Information Initiative Center, Hokkaido University, Hokkaido, Japan
akama@iic.hokudai.ac.jp
[2] Computer Science Program, Sirindhorn International Institute of Technology
Thammasat University, Pathumthani, Thailand
ekawit@siit.tu.ac.th

Abstract. Query-answering (QA) problems have attracted wider attention in recent years. Methods for solving QA problems based on the equivalent transformation (ET) principle have been recently developed. Meanwhile efficient satisfiability solvers (SAT solvers) have been invented and successfully applied to many kinds of problems. In this paper, we propose an ET-based low-level solution for QA problems. By slightly modifying it, we also propose a low-level solution using an all-solution SAT solver. We show that the obtained SAT-solver-based solution can also be seen as another ET-based low-level solution. Our findings clarify that the ET principle supports not only high-level computation but also low-level computation, and it provides a formal basis for correctness verification of computation in both levels.

Keywords: Query-answering problems, equivalent transformation, SAT solvers, all-solution SAT solvers, set-bounded variables.

1 Introduction

A query-answering problem (QA problem) is a pair $\langle K, q \rangle$, where K is a logical formula, representing background knowledge, and q is an atomic formula (atom), representing a query. The answer to a QA problem $\langle K, q \rangle$ is the set of all ground instances of q that are logical consequences of K. QA problems have attracted wider interest in recent years, owing partly to emerging applications involving integration between rules and ontologies in the Semantic Web's layered architecture. Methods for solving QA problems based on the equivalent transformation (ET) principle have been proposed recently, e.g., [1,2]. Meanwhile efficient problem solving by using satisfiability solvers (SAT solvers) [5] has been witnessed in many practical applications. ET-based computation and SAT-based computation have contrasting characteristics. The former transforms high-level representations of problems, which contain expressive terms involving variables ranging over all simple and compound terms. The latter deals only with propositional clauses, which can be processed at low cost since propositions can be represented as low-level data, i.e., binary variables ranging over 0 and 1.

S. Ramanna et al. (Eds.): MIWAI 2013, LNCS 8271, pp. 1–12, 2013.

The success of these two classes of computation methods naturally brings about the following question: what is the relationship between ET-based computation and SAT-based computation? Possible answers are:

1. They are disjoint; one is purely high-level, whereas the other is purely low-level. If this is the case, another question arises: how to combine them to obtain both rich representative power and efficient computation?
2. ET-based computation includes SAT-based computation, i.e., SAT-based computation can be realized by ET-based computation.

In this paper, we show that in the context of QA problems the second one is indeed the true answer. In particular, we show that there exists an ET-based low-level solution for QA problems, and the main constraint solving steps in this solution correspond to computation by an all-solution SAT solver (All-SAT solver). Accordingly, this solution yields a low-level All-SAT-solver-based solution. We also show that computation by an All-SAT solver can be regarded as an ET step for constraint solving. The obtained All-SAT-solver-based solution can thus still be seen as an ET-based low-level solution. Our findings show that the ET principle supports not only high-level computation but also low-level computation, and the correctness of computation in both levels can be rigorously verified based on the ET principle.

For formalizing and transforming QA problems, Section 2 introduces a formula space with set-bounded variables. Section 3 presents our ET-based low-level solution. Section 4 illustrates its application. Section 5 describes an ET-based solution using an All-SAT solver. Section 6 concludes the paper.

Preliminary Notation. The notation that follows holds thereafter. For any set A, $pow(A)$ denotes the power set of A and if A is finite, then $|A|$ denotes the cardinality of A. For any lists L and L', if L is finite, then $L \cdot L'$ denotes the concatenation of L and L'. For any definite clause C, $head(C)$ and $body(C)$ denote the head of C and the set of all atoms in the body of C, respectively. Given a definite clause C and an atom set A, $C \ominus A$ denotes the definite clause obtained from C by removing from its body all occurrences of atoms in A. Given two atom sets A and A', a definite clause C is said to be *from A to A'* iff $body(C) \subseteq A$ and $head(C) \in A'$. Given a set D of definite clauses, $\mathcal{M}(D)$ denotes the least model of D. Assume that TCON is the set of all true ground constraint atoms, which are predetermined. An atom set A is said to be *closed* iff for any atom $a \in A$ and any substitution θ, $a\theta$ belongs to A.

2 QA Problems with Set-Bounded Variables

2.1 QA Problems

A *query-answering problem* (*QA problem*) on a formula space $\mathcal{F}$ is a pair $\langle K, q \rangle$, where K is a formula in $\mathcal{F}$, representing background knowledge, and q is a usual atom, representing a query. The answer to a QA problem $\langle K, q \rangle$, denoted by

$answer_{QA}(K, q)$, is the set of all ground instances of q that are logical consequences of K. According to the choice of $\mathcal{F}$, QA problems can be classified
into several subclasses, e.g., QA problems on definite clauses, where background
knowledge is a set of definite clauses, QA problems on description logics (DLs),
where background knowledge is a conjunction of axioms and assertions in DLs
[3]. QA problems on definite clauses have been extensively discussed in logic
programming [4]. QA problems on DLs have been discussed in [7]. Answering
queries in Datalog and deductive databases [6] can be regarded as solving QA
problems on a restricted form of definite clauses.

The space $\mathcal{F}$ considered in this paper is obtained by extending definite clauses
with set-bounded variables, which are introduced below.

2.2 Set-Bounded Variables

In this paper, set-bounded variables provide a bridge between high-level logical
problem description and low-level computation such as SAT-solving-like computation. They play an important role for generating input for low-level solvers. A
set-bounded variable is an expression of the form $w : S$, where w is a variable
and S is a set of ground usual terms. When a set-bounded variable $w : S$ occurs
in an expression E, w is said to be *bounded by S in E*. Given a set-bounded
variable $w : S$ and a ground term t, the binding w/t is applicable to $w : S$ iff
$t \in S$, and when it is applied, $w : S$ is replaced with t.

When $S = \{0, 1\}$, a set-bounded variable $w : S$ is called a *binary variable*.
When no confusion is caused, a binary variable $w : S$ is also written as w.

The space obtained by extending the usual formula space with set-bounded
variables is used both for problem representation and for problem transformation
in our proposed framework. It is assumed henceforth that (i) for any set-bounded
variable $w : S$ considered herein, S is finite and (ii) in the scope of a single clause,
a variable can only be bounded by a unique set, i.e., if $w : S_1$ and $w : S_2$ appear
in the same clause, then $S_1 = S_2$.

2.3 An Example

A given QA problem can often be transformed equivalently by changing usual
variables into set-bounded variables. To illustrate, consider a QA problem $\langle D,$
$q(X, Y)\rangle$, where D consists of the following five definite clauses C_1–C_5:

C_1: $q(X, Y) \leftarrow member(X, [1, 2, 3, 6]), member(Y, [4, 5, 6, 7]),$
$member(X, [0, 1, 2, 3, 4]), member(Y, [3, 4, 5]), r(X, Y), s(X)$

C_2: $r(1, 4) \leftarrow$ C_3: $r(2, 1) \leftarrow$
C_4: $r(3, 5) \leftarrow$ C_5: $s(X) \leftarrow r(Y, X)$

By the constraints imposed by the *member*-atoms in its body, C_1 can be transformed by replacing the usual variables X and Y with the set-bounded variables
$X : \{1, 2, 3\}$ and $Y : \{4, 5\}$, respectively, and removing the *member*-atoms. The
resulting clause, C_1', is:

C_1': $q(X : \{1, 2, 3\}, Y : \{4, 5\}) \leftarrow r(X : \{1, 2, 3\}, Y : \{4, 5\}), s(X : \{1, 2, 3\}).$

The initial QA problem can therefore be transformed into the QA problem $\langle D',$ $q(X, Y) \rangle$, where D' is obtained from D by replacing C_1 with C_1'.

3   Solving QA Problems with ET Rules

Built-in atoms of two kinds, called *tran*-atoms and *imply*-atoms, are introduced (Section 3.1); they are used for connecting set-bounded variables with binary variables. ET rules for generating these built-in atoms are given (Sections 3.2 and 3.3), along with ET rules for solving constraints imposed by them (Sections 3.4 and 3.5). A procedure for solving QA problems using ET rules is then presented (Section 3.6). The correctness of this procedure and the completeness of its low-level computation steps are shown (Section 3.7).

3.1   *tran*-Atoms and *imply*-Atoms

To relate set-bounded variables with binary variables, a new binary predicate $tran_w$ is introduced for each set-bounded variable $w : S$ and it is defined as follows: For each $k \in \{1, \ldots, |S|\}$, let $\pi(S, k)$ be the kth element in S with respect to a certain predetermined order. Then $tran_w(t, L)$ is defined to be true iff there exists $k \in \{1, \ldots, |S|\}$ such that (i) $t = \pi(S, k)$ and (ii) L is the list $(0^{k-1} \cdot [1] \cdot 0^{|S|-k})$, i.e., the $|S|$-element list containing 1 at the kth position and containing 0 at every other position. An atom with the predicate $tran_w$ is called a *tran*-atom.

To represent constrains for binary variables, a binary predicate *imply* is defined as follows: $imply(L_1, L_2)$ is true iff (i) each of L_1 and L_2 is a finite list each element of which belongs to $\{0, 1\}$ and (ii) 1 occurs in L_1 or 0 occurs in L_2.

3.2   An ET Rule for Introducing Binary Variables

Let C be a definite clause. If $w : S$ is a set-bounded variable occurring in C, then C can be equivalently transformed by introducing $|S|$ new binary variables, referred to as $\mathbf{v}(w, 1), \ldots, \mathbf{v}(w, |S|)$, and adding atoms to the body of C as follows:

1. Add $tran_w(w : S, [\mathbf{v}(w, 1), \ldots, \mathbf{v}(w, |S|)])$.
2. Add $imply([\mathbf{v}(w, 1), \ldots, \mathbf{v}(w, |S|)], [])$.
3. For any $k_1, k_2 \in \{1, \ldots, |S|\}$ such that $k_1 \neq k_2$, add $imply([], [\mathbf{v}(w, k_1), \mathbf{v}(w, k_2)])$.

3.3   An ET Rule for Converting Constraints into *imply*-Atoms

Let C be a definite clause. Let a be an atom in the body of C such that a is neither a *tran*-atom nor an *imply*-atom. Assume that (i) all the set-bounded variables occurring in a are $u_1 : T_1, \ldots, u_p : T_p$ in their appearance order and (ii) for any terms $t_1, \ldots, t_p$, $a(t_1, \ldots, t_p)$ denotes the atom obtained from a by replacing $u_i : T_i$ with t_i for each $i \in \{1, \ldots, p\}$. Then C can be equivalently transformed as follows: First, remove a from the body of C. Next, for any $j_1 \in \{1, \ldots, |T_1|\}$, $\ldots, j_p \in \{1, \ldots, |T_p|\}$ such that

$$a(\pi(T_1, j_1), \ldots, \pi(T_p, j_p)) \notin \mathcal{M}(D) \cup \mathrm{TCon},$$

add $imply([], [\mathbf{v}(u_1, j_1), \ldots, \mathbf{v}(u_p, j_p)])$ to the body of C.

3.4   ET Rules for Manipulating *imply*-Atoms

Let C be a definite clause whose body contains an *imply*-atom $a = imply(L_1, L_2)$. C can be equivalently transformed using the following rules:

1. If all elements of L_1 are 0 and all those of L_2 are 1, then remove C.
2. If L_1 contains 1, then remove a.
3. If L_2 contains 0, then remove a.
4. If L_1 contains a binary variable v, all other elements of L_1 are 0, and all elements of L_2 are 1, then remove a from C and replace all occurrences of v in the resulting clause with 1.
5. If all elements of L_1 are 0, L_2 contains a binary variable v, and all other elements of L_2 are 1, then remove a from C and replace all occurrences of v in the resulting clause with 0.
6. If L_1 contains 0, then remove it from L_1.
7. If L_2 contains 1, then remove it from L_2.
8. If v is a binary variable occurring in both a and the head of C, then make a copy, say C', of C, replace all occurrences of v in C with 0, and replace all occurrences of v in C' with 1.

3.5   An ET Rule for Eliminating a *tran*-Atom

Let C be a definite clause. Assume that the body of C contains a *tran*-atom $tran_w(w : S, L)$ such that (i) L is ground, (ii) the kth element of L is 1, and (iii) all other elements of L are 0. Then C can be equivalently transformed as follows:

1. Remove $tran_w(w : S, L)$ from the body of C.
2. Apply the substitution $\{w/\pi(S, k)\}$ to the resulting clause.

3.6   A Procedure for Solving QA Problems Using ET Rules

Let $\mathcal{A}_1$ be a closed set of usual atoms. Assume that a QA problem $\langle D, q \rangle$ is given, where D is a set of definite clauses from $\mathcal{A}_1$ to $\mathcal{A}_1$ and $q \in \mathcal{A}_1$. To solve the QA problem $\langle D, q \rangle$ using ET rules, perform the following steps:

1. Determine (i) a closed set $\mathcal{A}_2$ of usual atoms such that $\mathcal{A}_1$ and $\mathcal{A}_2$ are disjoint and (ii) a bijection ϕ from $\mathcal{A}_1$ to $\mathcal{A}_2$ such that for any $a \in \mathcal{A}_1$ and any substitution θ for usual variables, $\phi(a\theta) = \phi(a)\theta$.

2. Transform $\{(\phi(q) \leftarrow q)\}$ using ET rules with respect to D into a set Q of definite clauses from $\mathcal{A}_1$ to $\mathcal{A}_2$ such that for any $C \in Q$, every variable occurring in $body(C)$ is a set-bounded variable. If this transformation is not possible, then stop with failure.
3. For each C in Q, perform the following steps:
    (a) For each set-bounded variable in C, apply the rule in Section 3.2 in order to add a *tran*-atom and *imply*-atoms to the body of C.
    (b) For each atom a in the body of C such that a is neither a *tran*-atom nor an *imply*-atom, apply the rule in Section 3.3 in order to convert the constraints imposed by a into *imply*-atoms.
    (c) Transform *imply*-atoms in the body of C by using the rules in Section 3.4 in order to solve the constraints imposed by these *imply*-atoms.
    (d) For each *tran*-atom t whose second argument is ground, transform t by using the rule in Section 3.5 in order to obtain unit ground clauses.
4. Let Q' be the set of all obtained ground unit clauses. Stop with the output

$$\{\phi^{-1}(head(C')) \mid C' \in Q'\}.$$

3.7   Correctness and Completeness

Correctness. The proposed procedure is correct, i.e., when the procedure stops at Step 4 with the output $\{\phi^{-1}(head(C')) \mid C' \in Q'\}$, this output set is the correct answer to the input QA problem $\langle D, q \rangle$, the reason being as follows: By the conditions imposed on $\mathcal{A}_2$ and the bijection ϕ at Step 1,

$$answer_{\mathrm{QA}}(D, q) = \{\phi^{-1}(g) \mid g \in answer_{\mathrm{QA}}(D \cup \{(\phi(q) \leftarrow q)\}, \phi(q))\}. \quad (1)$$

Since Q' is obtained by transforming $\{(\phi(q) \leftarrow q)\}$ using ET rules solely at Steps 2 and 3,

$$answer_{\mathrm{QA}}(D \cup \{(\phi(q) \leftarrow q)\}, \phi(q)) = answer_{\mathrm{QA}}(D \cup Q', \phi(q)). \quad (2)$$

Since each clause in Q' is a ground unit clause,

$$answer_{\mathrm{QA}}(D \cup Q', \phi(q)) = \{head(C') \mid C' \in Q'\}. \quad (3)$$

It follows from (1), (2), and (3) that

$$answer_{\mathrm{QA}}(D, q) = \{\phi^{-1}(head(C')) \mid C' \in Q'\},$$

i.e., the output set is the correct answer to the QA problem $\langle D, q \rangle$.

Completeness. Assume that C is a clause whose body contains *imply*-atoms and C is being processed at Step 3c. All *imply*-atoms in the body of C can be removed in finite steps, since at least one rule in Section 3.4 is applicable to one of these *imply*-atoms and the application of that rule either (i) decreases the number of variable occurrences or (ii) decreases the number of constant occurrences without increasing the number of variable occurrences. Step 3c is thus completed using a finite number of rule applications. Obviously, each of Steps 3a, 3b, and 3d is also completed using a finite number of rule applications.

4   Examples

Application of the proposed procedure to the QA problem in Section 2.3 is shown first. Its application to a logic-based number-placement puzzle is next illustrated.

4.1   A Simple Example

Consider the QA problem $\langle D, q(X, Y) \rangle$ given in Section 2.3. Assume that q-, r- and s-atoms all belong to $\mathcal{A}_1$, ans is a newly introduced predicate symbol, all ans-atoms belong to $\mathcal{A}_2$, and for any terms t_1, t_2, $\phi(q(t_1, t_2)) = ans(t_1, t_2)$. At Step 2, by unfolding using the clause C_1' in Section 2.3, the set $\{(ans(X, Y) \leftarrow q(X, Y))\}$ is transformed into a singleton set $\{C\}$, where C is the definite clause

$$ans(X : \{1, 2, 3\}, Y : \{4, 5\}) \leftarrow r(X : \{1, 2, 3\}, Y : \{4, 5\}), s(X : \{1, 2, 3\}).$$

From $X : \{1, 2, 3\}$, the predicate $tran_X$ is introduced and only three ground $tran_X$-atoms are true, i.e., $tran_X(1, [1, 0, 0])$, $tran_X(2, [0, 1, 0])$, and $tran_X(3, [0, 0, 1])$. Similarly, from $Y : \{4, 5\}$, the predicate $tran_Y$ is introduced and only two ground $tran_Y$-atoms are true, i.e., $tran_Y(4, [1, 0])$ and $tran_Y(5, [0, 1])$. At Step 3a, the rule in Section 3.2 is applied to C as follows, assuming that (i) $\mathbf{v}(X, i) = x_i$ for any $i \in \{1, 2, 3\}$ and (ii) $\mathbf{v}(Y, j) = y_j$ for any $j \in \{1, 2\}$:

1. From $X : \{1, 2, 3\}$, the following atoms are added to the body of C:
   - $tran_X(X : \{1, 2, 3\}, [x_1, x_2, x_3])$
   - $imply([x_1, x_2, x_3], [])$, $imply([], [x_1, x_2])$, $imply([], [x_1, x_3])$, $imply([], [x_2, x_3])$
2. From $Y : \{4, 5\}$, the following atoms are added to the body of C:
   - $tran_Y(Y : \{4, 5\}, [y_1, y_2])$
   - $imply([y_1, y_2], [])$, $imply([], [y_1, y_2])$

Next, at Step 3b, the rule in Section 3.3 is applied as follows:

1. Let the body atom $s(X : \{1, 2, 3\})$ be taken first. Since $s(2)$ and $s(3)$ are false, this body atom is replaced with $imply([], [x_2])$ and $imply([], [x_3])$.
2. Let the body atom $r(X : \{1, 2, 3\}, Y : \{4, 5\})$ be taken next. Since $r(1, 5)$, $r(2, 4)$, $r(2, 5)$, and $r(3, 4)$ are false, this body atom is replaced with $imply([], [x_1, y_2])$, $imply([], [x_2, y_1])$, $imply([], [x_2, y_2])$, and $imply([], [x_3, y_1])$.

At this stage, all the generated $imply$-atoms are:

1. $imply([x_1, x_2, x_3], [])$ 2. $imply([], [x_1, x_2])$ 3. $imply([], [x_1, x_3])$
4. $imply([], [x_2, x_3])$ 5. $imply([y_1, y_2], [])$ 6. $imply([], [y_1, y_2])$
7. $imply([], [x_1, y_2])$ 8. $imply([], [x_2])$ 9. $imply([], [x_3])$

At Step 3c, by using the rules in Section 3.4, the above 9 $imply$-atoms are transformed as follows:

- Applying the 5th rule to the 8th $imply$-atom yields $x_2 = 0$.
- Applying the 5th rule to the 9th $imply$-atom yields $x_3 = 0$.

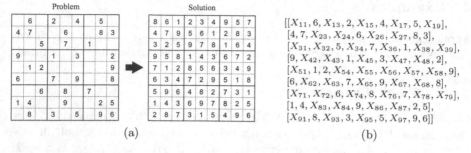

Fig. 1. A Sudoku puzzle and its representation: (a) A puzzle and the solution to it and (b) A matrix representing the puzzle

- Applying the 4th rule to the resulting instantiated version of the 1st *imply*-atom yields $x_1 = 1$.
- Applying the 5th rule to the resulting instantiated version of the 7th *imply*-atom yields $y_2 = 0$.
- Applying the 4th rule to the resulting instantiated version of the 5th *imply*-atom yields $y_1 = 1$.
- By applying the 3rd rule to each of the resulting instantiated versions of the remaining *imply*-atoms, they are all removed.

The body of the resulting clause then consists only of the following two *tran*-atoms: (i) $tran_X(X : \{1, 2, 3\}, [1, 0, 0])$ and (ii) $tran_Y(Y : \{4, 5\}, [1, 0])$. These resulting *tran*-atoms yield $X : \{1, 2, 3\} = 1$ and $Y : \{4, 5\} = 4$. By applying the rule in Section 3.5 at Step 3d, the unit clause $(ans(1, 4) \leftarrow)$ is then obtained. The procedure thus outputs the singleton set $\{q(1, 4)\}$.

4.2   Solving a Sudoku Puzzle

Next, consider the Sudoku puzzle in Fig. 1(a). The goal is to fill a 9×9 grid with digits so that each column, each row, and each of the nine 3×3 sub-grids contain all of the digits from 1 to 9. This problem is represented as a QA problem $\langle D, sudoku(L) \rangle$, where L is the matrix representing the puzzle, which is shown in Fig. 1(b), and D contains the definition of the predicate *sudoku* specifying the constraints on rows, columns, and sub-grids to be satisfied.

At Step 2, a clause $C_1 = (ans(L) \leftarrow sudoku(L))$ is constructed and successively transformed as follows:

- By successively unfolding C_1 using D, a clause C_2 whose body contains only *member*-atoms and *allDiff*-atoms is obtained. An *allDiff*-atom has a single argument; when it is ground, *allDiff*(L) is true iff L is a list whose elements are all different.
- Using the constraints imposed by the *member*-atoms in the body of C_2, all variables occurring in C_2 are changed into set-bounded variables and those *member*-atoms are removed. A clause C_3 whose body consists solely of 27 *allDiff*-atoms is obtained.

- The *allDiff*-atoms in C_3 are transformed into *notMember*-atoms, resulting in a clause C_4 whose body consists of 288 *notMember*-atoms. A *notMember*-atom has two arguments, and for any ground term t and any ground list L, $notMember(t, L)$ is true iff L does not contain t.
- The *notMember*-atoms in C_4 are transformed into *notEqual*-atoms, resulting in a clause C_5 whose body consists of 972 *notEqual*-atoms. A *notEqual*-atom has two arguments, and for any ground terms t and t', $notEqual(t, t')$ is true iff t and t' are different.

Next, when applicable, the rules below are successively applied to C_5. Assuming that $\hat{C}$ is a definite clause whose body contains a *notEqual*-atom a, then:

1. If $a = notEqual(m, n)$, where m and n are numbers, then:
   (a) If $m \neq n$, then remove a from $\hat{C}$.
   (b) If $m = n$, then remove the clause $\hat{C}$.
2. If $a = notEqual(m, w : S)$ or $a = notEqual(w : S, m)$, where m is a number and $w : S$ is a set-bounded variable, then:
   (a) If $m \in S$, then:
       i. Let $S' = S - \{m\}$.
       ii. If S' is a singleton set $\{n\}$, then remove a from $\hat{C}$ and apply the substitution $\{w/n\}$ to the resulting clause.
       iii. If S' contains more than one element, then remove a from $\hat{C}$ and specialize $w : S$ in the resulting clause into $w : S'$.
   (b) If $m \notin S$, then remove a from $\hat{C}$.

Assume that a clause C_6 is obtained from C_5 by repeatedly applying the above rules. Note that the body of C_6 contains only *notEqual*-atoms and the arguments of these atoms are all set-bounded variables. At this stage, Step 2 is completed.

At Step 3a, *imply*-atoms are generated from the set-bounded variables appearing in C_6. At Step 3b, additional *imply*-atoms are generated from *notEqual*-atoms in C_6 as follows: For any atom $notEqual(w : S, w' : S')$ in the body of C_6, any $i \in \{1, \ldots, |S|\}$, and any $j \in \{1, \ldots, |S'|\}$, if $\pi(S, i) = \pi(S', j) \in S \cap S'$, then generate $imply([], [\mathbf{v}(w, i), \mathbf{v}(w', j)])$. Altogether, 3,078 *imply*-atoms are obtained. At Step 3c, the constraints imposed by these *imply*-atoms are solved and all binary variables become ground. At Step 3d, all *tran*-atoms are removed, yielding a ground unit clause from which the solution shown in the right side of Fig. 1(a) is obtained.

5   Using All-SAT Solvers for Constraint Solving

Constraint solving on a set of *imply*-atoms can be realized using a class of SAT solvers, called all-solution SAT solvers (All-SAT solvers). After establishing a one-to-one correspondence between *imply*-atoms and propositional clauses and introducing an All-SAT solver (Section 5.1), we clarify that an All-SAT solver yields an ET step for removing all *imply*-atoms from a clause body (Section 5.2). An ET-based solution using an All-SAT solver is then presented (Section 5.3).

5.1   Transformation to Propositional Clauses and All-SAT Solvers

Propositional Clauses Corresponding to *imply*-Atoms. At Step 3c of the procedure in Section 3.6, all *imply*-atoms are solved and ground instantiations of binary variables are obtained. Given a correspondence between binary variables and propositional symbols, a correspondence between *imply*-atoms and propositional clauses is determined as follows: Assume that ω is a bijection from binary variables to propositional symbols such that for any binary variable v, $\omega(v)$ is the propositional symbol corresponding to v. This bijection determines a bijection $corr_\omega$ from *imply*-atoms to propositional clauses by

$$corr_\omega(imply([u_1, \ldots, u_m], [v_1, \ldots, v_n])) = (\omega(u_1), \ldots, \omega(u_m) \leftarrow \omega(v_1), \ldots, \omega(v_n))$$

for any binary variables $u_1, \ldots, u_m, v_1, \ldots, v_n$. As will be seen in Section 5.2, by the correspondence established by $corr_\omega$, constraint solving on a set of *imply*-atoms can be regarded as computation for finding all models of a set of propositional clauses.

All-SAT Solvers. Let Cs be a finite set of propositional clauses and $prop(Cs)$ denote the set of all propositional symbols occurring in Cs. Taking Cs as input, an *all-solution SAT solver* (*All-SAT solver*) returns the set of all possible truth value assignments that satisfy Cs. A truth value assignment in the output set is called a *model* of Cs and can be represented as a subset $\mathbf{m}$ of $prop(Cs)$ such that for any $p \in prop(Cs)$, $p = true$ iff $p \in \mathbf{m}$.

5.2   ET Steps by All-SAT Solvers

Referring to the bijections ω and $corr_\omega$ in Section 5.1, the next proposition follows directly from the definition of $corr_\omega$ and that of an All-SAT solver. It shows that an All-SAT solver gives an ET step for removing all *imply*-atoms from a clause body.

Proposition 1. Assume that $\mathcal{A}_1$ and $\mathcal{A}_2$ are disjoint closed sets of atoms, C is a definite clause from $\mathcal{A}_1$ to $\mathcal{A}_2$, $ImpSet_C$ is the set of all *imply*-atoms in the body of C, V is the set of all binary variables occurring in $ImpSet_C$, and $Cs = \{corr_\omega(a) \mid a \in ImpSet_C\}$. Suppose that:

1. A set M of models is the output obtained by calling an All-SAT solver with the input Cs.
2. For any model $\mathbf{m} \in M$, $\sigma_\mathbf{m}$ is the substitution

$$\{v/1 \mid (v \in V) \ \& \ (\omega(v) \in \mathbf{m})\} \cup \{v/0 \mid (v \in V) \ \& \ (\omega(v) \notin \mathbf{m})\}.$$

Then for any set D of definite clauses from $\mathcal{A}_1$ to $\mathcal{A}_1$ and any set Q of those from $\mathcal{A}_1$ to $\mathcal{A}_2$, $\mathcal{M}(D \cup Q \cup \{C\}) = \mathcal{M}(D \cup Q \cup \{(C \ominus ImpSet_C)\sigma_\mathbf{m} \mid \mathbf{m} \in M\})$.

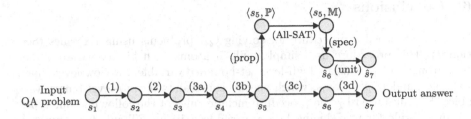

Fig. 2. Computation paths for solving a QA problem

5.3   A Solution Using an All-SAT Solver

Refer to Steps 1, 2, 3a, 3b, 3c, and 3d of the procedure in Section 3.6 and the bijections ω and $corr_\omega$ in Section 5.1. Consider the diagram in Fig. 2, where the diagram components are described as follows: s_1 is an input QA problem; s_2, s_3, s_4, s_5, s_6, and s_7 are the QA problems obtained from s_1 by successive transformations using Steps 1, 2, 3a, 3b, 3c, and 3d, respectively; and (1), (2), (3a), (3b), (3c), and (3d) denote the transformations performed respectively by these steps. Assume that Q is the set of all definite clauses from $\mathcal{A}_1$ to $\mathcal{A}_2$ in s_5 (i.e., the set of all query clauses in s_5), and that PROP denotes the set of all propositional clauses and $\mathcal{G}$ the set of all ground atoms. The components $\mathbb{P}$, (prop), $\mathbb{M}$, (All-SAT), (spec), and (unit) are then given below.

- $\mathbb{P}$ is a mapping from Q to $pow(\text{PROP})$ given by: for each $C \in Q$, $\mathbb{P}(C) = \{ corr_\omega(a) \mid a$ is an $imply$-atom in the body of $C \}$.
- (prop) is a transformation for introducing the mapping $\mathbb{P}$ by applying $corr_\omega$ to $imply$-atoms in each clause in Q.
- $\mathbb{M}$ is a mapping from Q to $pow(pow(\mathcal{G}))$ given by: for each $C \in Q$, $\mathbb{M}(C)$ is a set of models of $\mathbb{P}(C)$ obtained by calling an All-SAT solver.
- (All-SAT) is a transformation for replacing $\mathbb{P}$ with $\mathbb{M}$.
- (spec) is a transformation for specializing each definite clause in Q after removing all $imply$-atoms from it, i.e., (spec) applies the transformation described by Proposition 1 to each clause in Q.
- (unit) is a transformation for producing unit clauses by solving the constraints imposed by $tran$-atoms, which is basically the same as the transformation at Step 3d.

This diagram depicts two possible ET computation paths for solving a QA problem: (i) $[s_1, s_2, s_3, s_4, s_5, s_6, s_7]$ and (ii) $[s_1, s_2, s_3, s_4, s_5, \hat{s}_6, \hat{s}_7]$, where Proposition 1 shows that the transformation of s_5 into $\hat{s}_6$ is an ET step. The first path is directly obtained by applying the procedure in Section 3.6. The second path uses an All-SAT solver as a low-level solver for solving the constraints imposed by $imply$-atoms. Comparing the first path to the second one, when the execution time of (3c) is significantly greater than the sum of the execution time of (prop), (All-SAT), and (spec), All-SAT solvers are useful for improving the efficiency of solving QA problems.

6 Conclusions

We have proposed a procedure for solving QA problems using ET rules (Section 3). This procedure uses simple built-in atoms with binary variables, i.e., *tran*-atoms, which translate high-level set-bounded variables into low-level binary variables, and *imply*-atoms, which specify constraints on low-level binary variables. ET rules used by the procedure include rules of the following four types: (i) an ET rule for introducing binary variables, (ii) an ET rule for converting constraints into *imply*-atoms, (iii) ET rules for manipulating *imply*-atoms, and (iv) an ET rule for eliminating a *tran*-atom. Using rules of the first two types, an initial high-level problem representation with logical variables is transformed into an equivalent problem representation involving low-level binary variables. Using rules of the third and fourth types, *imply*-atoms and *tran*-atoms, respectively, are solved and eliminated. Since all transformation steps are obtained by using ET rules solely, the correctness of the proposed procedure is readily guaranteed. Since the main computation of this procedure is constraint solving using rules of the third type and this computation can be realized by an All-SAT solver, the procedure suggests an All-SAT-based solution (Section 5).

In usual ET computation and logic-based computation, higher-level computation has been preferred; unnecessary variable specializations are avoided since they often generate too many instantiated clauses and instantiated atoms. The theory presented in this paper extends the applicability of ET computation. It reveals that the ET principle provides a basis for computation correctness for high-level computation, low-level computation, and their combinations. Its application to more complicated problem solving will be demonstrated in our future work.

References

1. Akama, K., Nantajeewarawat, E.: Unfolding-Based Simplification of Query-Answering Problems in an Extended Clause Space. International Journal of Innovative Computing, Information and Control 9, 3515–3526 (2013)
2. Akama, K., Nantajeewarawat, E.: Embedding Proof Problems into Query-Answering Problems and Problem Solving by Equivalent Transformation. In: 5th International Conference on Knowledge Engineering and Ontology Development, Vilamoura, Portugal, pp. 253–260 (2013)
3. Baader, F., Calvanese, D., McGuinness, D.L., Nardi, D., Patel-Schneider, P.F. (eds.): The Description Logic Handbook. Cambridge University Press (2007)
4. Lloyd, J.W.: Foundations of Logic Programming, 2nd edn. Springer (1987)
5. Marques-Silva, J.P., Sakallah, K.A.: GRASP: A Search Algorithm for Propositional Satisfiability. IEEE Transactions on Computers 48, 506–521 (1999)
6. Minker, J. (ed.): Foundations of Deductive Databases and Logic Programming. Morgan Kaufmann Publishers (1988)
7. Tessaris, S.: Questions and Answers: Reasoning and Querying in Description Logic. PhD thesis, The University of Manchester, UK (2001)

Incremental Rough Possibilistic K-Modes

Asma Ammar[1], Zied Elouedi[1], and Pawan Lingras[2]

[1] LARODEC, Institut Supérieur de Gestion de Tunis, Université de Tunis
41 Avenue de la Liberté, 2000 Le Bardo, Tunisie
asma.ammar@voila.fr, zied.elouedi@gmx.fr
[2] Department of Mathematics and Computing Science, Saint Mary's University
Halifax, Nova Scotia, B3H 3C3, Canada
pawan@cs.smu.ca

Abstract. In this paper, we propose a novel version of the k-modes method dealing with the incremental clustering under uncertain framework. The proposal is called the incremental rough possibilistic k-modes (I-RPKM). First, possibility theory is used to handle uncertain values of attributes in databases and, to compute the membership values of objects to resulting clusters. After that, rough set theory is applied to detect boundary regions. After getting the final partition, the I-RPKM adapts the incremental clustering strategy to take into account new information and update the cluster number without re-clustering objects. I-RPKM is shown to perform better than other certain and uncertain approaches.

Keywords: Incremental clustering, possibility theory, rough set theory, k-modes method, possibilistic membership, possibility degree.

1 Introduction

Incremental clustering [5], [6], and [7] takes into account new knowledge and updates clustering parameters (e.g., the number of clusters, attributes, and instances) without re-clustering objects. Incremental clustering reduces memory and computing requirements for updating clustering profiles.

As data mining field is pervaded by uncertainty, ignoring these uncertain pieces of knowledge can lead in some cases to inaccurate results. Possibility and rough set theories are well-known uncertainty theories that have been successfully used in clustering [1], [2], [3], [9], and [14]. Possibility theory has been used to deal with uncertainty in databases through the use of possibility distribution and possibiblistic membership degree. Rough set theory was applied to create rough clusters.

In this work, we propose the incremental rough possibilistic k-modes (I-RPKM) which is a soft clustering method that uses the possibility theory to deal with uncertainty in the values of attributes. In addition, an object is not forced to be a member of only one cluster but, it can belong to the k clusters based on its degrees of similarity expressed by possibilistic membership values. I-RPKM also applies the rough set theory to detect boundary regions where

S. Ramanna et al. (Eds.): MIWAI 2013, LNCS 8271, pp. 13–24, 2013.
© Springer-Verlag Berlin Heidelberg 2013

objects are peripheral. It considers new knowledge and updates the partitions by considering the possibility to increase the number of clusters.

The rest of the paper is structured as follows: Section 2 and section 3 give the background concerning the possibility and rough set theories and the k-modes method. Section 4 details our proposal i.e., the incremental rough possibilistic k-modes. Section 5 analyses experimental results. Section 6 concludes the paper.

2   Possibility and Rough Set Theories

2.1   Possibility Theory

Possibility theory is an uncertainty theory devoted to deal with imperfect knowledge. It was proposed by Zadeh in [20] and since then it has been improved by numerous researchers (e.g., Dubois and Prade [10]). This theory of imperfection has been successfully applied in different works [1], [2], and [3].

Possibility Distribution. Based on the universe of discourse $\Omega = \{\omega_1, \omega_2, ..., \omega_n\}$, where ω_i is an event or state [20], the possibility distribution function denoted by π can be defined by:

$$\Pi : \Omega \to L$$
$$\omega \mapsto \Pi(\omega). \tag{1}$$

Note that L is the scale taking values from $[0, 1]$ in the quantitative version of the possibility theory.

In addition, the normalization is defined by $max_i \{\pi(\omega_i)\} = 1$ and the extreme cases of knowledge consist of the complete knowledge ($\exists$ a unique ω_0, $\pi(\omega_0) = 1$ and $\pi(\omega) = 0$ otherwise) and the total ignorance ($\forall \omega \in \Omega, \pi(\omega) = 1$).

Possibilistic Similarity Measure. The information affinity [8] is a possibilistic similarity measure applied on two normalized possibility distributions π_1 and π_2. It measures their similarity by applying Equation (2).

$$InfoAff(\pi_1, \pi_2) = 1 - 0.5 [D(\pi_1, \pi_2) + Inc(\pi_1, \pi_2)]. \tag{2}$$

with $D(\pi_1, \pi_2) = \frac{1}{n} \sum_{i=1}^{n} |\pi_1(\omega_i) - \pi_2(\omega_i)|$, n the number of objects and $Inc(\pi_1, \pi_2) = 1 - \max(\pi_1(\omega) Conj \pi_2(\omega))$ and $\forall \omega \in \Omega$, $\Pi_{Conj}(\omega) = \min(\Pi_1(\omega), \Pi_2(\omega))$ where $Conj$ denotes the conjunctive mode.

2.2   The Rough Set Theory

The rough sets has been proposed by Pawlak in the early 1980s [18], [19] for incomplete knowledge and has been applied for clustering in numerous works such as [3], [14], and [15]. We detail its main concepts as follows.

Information System. The rough set theory uses data sets structured in an information table. It contains objects organized in rows and attributes organized in columns. An information system contains different condition attributes and a decision attribute considered as the class. More generally, we call an information system IS a pair $S = (U, A)$ where U and A are finite and nonempty sets. U is defined as the universe and A is the set of attributes.

Indiscernibility Relation. Given $S = (U, A)$ an IS, the equivalence relation $(IND_S(B))$ for any $B \subseteq A$ is detailed in Equation (3):

$$IND_S(B) = \{(x, y) \in U^2 | \forall a \in B \ a(x) = a(y)\} .\tag{3}$$

where $IND_S(B)$ is B- indiscernibility relation and $a(x)$ and $a(y)$ denote the values of attribute a for the elements x and y.

Approximation of Sets. Given an IS defined by $S = (U, A)$, $B \subseteq A$ and $Y \subseteq U$. The set Y is described using the attribute values from B by defining two sets named the B-upper $\overline{B}(Y)$ and the B-lower $\underline{B}(Y)$ approximations of Y. They are given by Equation (4) and Equation (5).

$$\overline{B}(Y) = \bigcup \{B(y) : B(y) \cap Y \neq \phi\} .\tag{4}$$

$$\underline{B}(Y) = \bigcup \{B(y) : B(y) \subseteq Y\} .\tag{5}$$

The B-boundary region of Y can be deduced using Equation (4) and Equation (5) as follows: $BN_B(Y) = \overline{B}(Y) - \underline{B}(Y)$.

3   The K-Modes Method and Its Extension

3.1   The SKM

The standard k-modes method (SKM) defined in [12] [13] is a clustering approach that handles large categorical data sets. It is based on the the k-means algorithm [17]. Its aim is to cluster objects of the training set into k clusters using a simple matching dissimilarity measure and a frequency-based function.

Given two objects $X_1 = (x_{11}, x_{12}, ..., x_{1m})$ and $X_2 = (x_{21}, x_{22}, ..., x_{2m})$ with m categorical attributes. The simple matching method (SM) is defined by:

$$SM (X_1, X_2) = \sum_{t=1}^{m} \delta (x_{1t}, x_{2t}) .\tag{6}$$

Note that $\delta (x_{1t}, x_{2t}) = 0$ if $x_{1t} = x_{2t}$, and $\delta (x_{1t}, x_{2t}) = 1$ otherwise. Thus, $d \in [0, 1]$ has two extreme values. It is equal to 0 when there is a similarity between all the attributes' values of X_1 and X_2. Otherwise, it is equal to m.

More generally, given a set of n objects $S = \{X_1, X_2, ..., X_n\}$ with its k-modes $Q = \{Q_1, Q_2, ..., Q_k\}$ and k clusters $C = \{C_1, C_2, ..., C_k\}$, we can cluster S into $k \leq n$ clusters. The minimization problem of the clustering is described as follows:

$$\min \ D(W, Q) = \sum_{j=1}^{k} \sum_{i=1}^{n} \omega_{i,j} d(X_i, Q_j). \tag{7}$$

where W is an $n \times k$ partition matrix and $\omega_{i,j} \in \{0, 1\}$ is degree of belonging of X_i in C_j.

SKM has been widely applied and has shown its ability to improve the clustering results. However, it has some drawbacks, especially when handling imperfect knowledge. Real-world databases contain different aspects of uncertainty. We can mention uncertain values of attributes and uncertainty in the belonging of objects to different clusters. SKM is a hard clustering approach where each object is a member of exactly one cluster. However, in practical applications an object can share some similarities with different clusters. To overcome these issues, several researchers have proposed modified versions of SKM such as [1], [2], [3], and [4]. In the next subsections, we briefly describe two soft clustering methods namely RPKM [3] and KM-PF [4] that handle uncertainty.

3.2   The RPKM

The rough possibilistic k-modes method denoted by RPKM [3] is an uncertain soft clustering approach that uses possibility and rough set theories. The possibility theory is applied to deal with uncertainty in the belonging of objects to several clusters by defining possibilistic membership degrees. The rough set theory is used to detect clusters with rough boundary through the set approximations. The RPKM uses a training set where the values are precise and categorical, and the conventional dissimilarity measure.

3.3   The KM-PF

The k-modes under possibilistic framework denoted by KM-PF [4] is a soft clustering approach that clusters objects with uncertain attributes values under possibilistic framework. It uses a modification of SKM in order to cluster instances to k clusters using possibilistic membership degrees. It is an improved version of two possibilistic approaches proposed in [1] and [2]. KM-PF approach is characterized by dealing with uncertain values. For each attribute, it defines a possibility distribution. As a result, each attribute value is assigned a degree of possibility between $[0, 1]$. In addition, each object is assigned a possibilistic degree to denote its degree of belonging to each cluster. To summarize, KM-PM uses possibilistic membership to deal with certain and uncertain attribute values and can assign an instance to multiple clusters.

4 Incremental Possibilistic K-Modes: I-RPKM

The I-RPKM is a generalization of two uncertain approaches namely RPKM [3] and KM-PF [4] described in the previous subsections. Our proposal keeps their advantages and overcomes their limitations especially for dealing with the incremental clustering. It allows the adding of new data without re-clustering the initial objects.

4.1 Parameters

Our proposal uses the following parameters:

1. An uncertain training set: based on categorical and certain real-world databases, our program generates databases containing certain and/or uncertain values of attributes. The new values represent the degree of uncertainty on each real value.

2. The possibilistic similarity measure IA: applied to compute the similarity of an object from the training set to the cluster mode. IA modifies the information affinity measure [8] and it is given by:

$$IA(X_1, X_2) = \frac{\sum_{j=1}^{m} InfoAff(\pi_{1j}, \pi_{2j})}{m}. \tag{8}$$

   where m is the total number of attributes.

3. The possibilistic membership degrees: deduced from the possibilistic similarity measure IA (Equation (8)). Each object has k possibilistic values describing its degrees of belonging to all clusters. ω_{ij} is the possibilistic membership degree of the object i to the cluster j. ω_{ij} is a possibilistic degree defined from $[0, 1]$. When computing ω_{ij} we can get one of the two extreme cases. The first one is when $\omega_{ij} = 1$. This case means that the attributes' values of the object i and the mode of the cluster j are the same. The second one is when $\omega_{ij} = 0$ i.e., the object i is not a member of the cluster j.

4. The update of clusters' modes: takes into consideration the degrees of belonging ω_{ij} and the possibilistic degree defined for each attribute value. The following steps describe how to update the modes.
   - For each cluster j, we compute the number of objects in the training set with the highest membership value (ω_{ij}) using this formula: $NO_j = count_j(\max_i \omega_{ij})$.
   - We define a new parameter Wt_j used as a coefficient of the k' initial modes (and then for the added clusters). It is deduced by Equation (9):

$$Wt_j = \begin{cases} \frac{NO_j}{total\ number\ of\ attributes} & \text{if } NO_j \neq 0, \\ \frac{1}{total\ number\ of\ attributes\ +1} & \text{otherwise.} \end{cases} \tag{9}$$

   - We multiply the computed weight Wt by the initial values of the k' modes (and then by the added modes) in order to get the new modes M'_j.

$$\forall j \in k', M'_j = Wt_j \times Mode_j. \tag{10}$$

5. The detection of boundary region containing peripheral objects: an object is considered as peripheral when it is a member of several clusters. It belongs to boundary region. To detect in which region (upper or lower) these objects belong, we should compute possibilistic membership degrees of all instances. Then, we compute the ratio R_{ij} defined in [9], [16]:

$$R_{ij} = \frac{\max \omega_i}{\omega_{ij}}. \tag{11}$$

First, we set a threshold denoted by $T \geq 1$ then, we compare the value of T to the ratio. In case $R_{ij} \leq T$, the object i belongs to the upper bound of the cluster j. If i is a part of the upper bound of exactly one cluster j, it is necessarily a part of the lower bound of j.

6. The addition of c new clusters: This step comes after computing the final possibilistic membership and detecting rough clusters. This case takes into account new information and updates the cluster number by adding c clusters. We use the final partition and add new clusters without re-clustering initial objects. The added clusters take their c modes from the set of initial instances after removing the k' objects already taken as k' first modes. After that, we have to compute the possibilistic similarity between the objects and the c clusters and deduce the ω_{ij} of these c clusters. Finally, we update our method and re-compute the rough clusters.

4.2 Algorithm

Begin

1. *Randomly select ($k' = k - c$) initial modes, one mode for each cluster.*
2. *Use Equation (8) to calculate the possibilistic similarity measure IA applied between instances and modes. Deduce the membership degree ω_{ij} of each object to the k' clusters.*
3. *Allocate an object to the k' clusters using ω_{ij}.*
4. *Use Equation (9) to compute the weight Wt_j for each cluster j then, use Equation (10) to update the cluster mode.*
5. *Retest the similarity between objects and modes. Reallocate objects to clusters using possibilistic membership degrees then update the modes.*
6. *Repeat (4) until all clusters are stable.*
7. *Derive the rough clustering by computing first the possibilistic membership degrees then, the ratio of each object R_{ij} using Equation (11). Assign each object to the upper or the lower region of the rough cluster.*
8. *Add c new clusters and compute the possibilistic similarity measure IA between the objects and new clusters. Determine the ω_{ij} of the objects relative to the added clusters.*
9. *Re-compute the weight Wt_j for each cluster j using Equation (9) then, update the cluster mode using Equation (10).*
10. *Repeat (4) to (7) until all clusters are stable.*

End.

---

5 Experiments

5.1 The Framework

For the experiments, we choose six databases from UCI machine learning repository [11]. The databases are *Balloons (Bal), Soybean (S), Post-Operative Patient (POP), Balance Scale (BS), Solar-Flare (SF)* and *Car Evaluation (CE)*. Based on these certain databases, we have artificially created uncertain data sets. This pre-treatment procedure consists of replacing each categorical value of attribute by a possibility degree from [0, 1] with respect to the real values. Table 1, Table 2 and Table 3 illustrate the pre-treatment process with the help of examples.

Table 1. Example of four instances of Balloons data set

Attribute information			Color	Size	Act	Age
Color	yellow,	purple	X_1	yellow	small	stretch	adult
size	large,	small	X_2	yellow	small	stretch	child
act	stretch,	dip	X_3	yellow	small	dip	adult
age	adult,	child	X_4	yellow	small	dip	child
Classes inflated			True,	False		

1. Certain case (certain attributes' values): Each true value takes the possibility degree 1. All remaining values of the data sets are replaced by the possibility degree 0. The certain case translates to the case of complete knowledge.

Table 2. Pre-treatment of the attributes' values of objects in certain case

	yellow	purple	small	large	stretch	dip	adult	child
X_1	1	0	1	0	1	0	1	0
X_2	1	0	1	0	1	0	0	1
X_3	1	0	1	0	0	1	1	0
X_4	1	0	1	0	0	1	0	1

2. Uncertain case (uncertain attributes' values): Values take different possibility degrees from]0, 1[except true values which are replaced by 1.

Table 3. Pre-treatment of the attributes' values under uncertainty

	yellow	purple	small	large	stretch	dip	adult	child
X_1	1	0.248	1	0.381	1	0.192	1	0.013
X_2	1	0.143	1	0.027	1	0.348	0.094	1
X_3	1	0.316	1	0.248	0.045	1	1	0.483
X_4	1	0.346	1	0.251	0.051	1	0.276	1

5.2   Evaluation Criteria

To test and evaluate our proposal, we use four evaluation criteria as follows:

1. The accuracy [12]: $AC = \frac{\sum_{j=1}^{k} a_j}{T}$ where a_j is the correctly classified objects from the total number of object n. The error rate is deduced from AC by $ER = 1 - AC$, low value of ER implies better clustering results.
2. The iteration number (IN): denotes the number of iterations needed to classify the objects into $(k' + c)$ clusters.
3. The execution time (ET): is the time taken to get the final partition after adding c new clusters.

5.3   Experimental Results

We use the evaluation criteria cited in the previous subsection and data sets created artificially from [11]. After running our program ten times, the average of the accuracy is calculated to get unbiased results. Furthermore, we cross validate our data sets by dividing them into two sets mainly the training and test sets.

In the following subsections, we study and analyze the results of our proposal i.e., I-RPKM compared to SKM and two modified versions of SKM namely RPKM and KM-PF.

Certain Case. This case corresponding to the complete knowledge is obtained by defining a possibility distribution for each attribute. The values known with certainty are replaced by the degree 1. The remaining values take the degree 0.

Table 4 details the results based on different evaluation criteria. For an initial number of clusters k', we notice that I-RPKM provides better results (i.e., less error rate, IN and ET) than SKM and the uncertain approaches KM-PF and RPKM.

When increasing k', I-RPKM leads to lower error rate and lower execution time compared to other methods. The primary reason behind this improvement is the fact that I-RPKM does not re-cluster the instances. The other three methods, on the other hand, need to re-cluster the instances.

Uncertain Case. The databases are modified using possibilistic values between $]0, 1]$. Only the true values take the degree 1. Table 5 shows the error rate criterion of I-RPKM compared to KM-PF, where A represents percentage of uncertain values and d is possibilistic degree. Note that, the uncertainty in the databases proportionally increases with the value of A.

Looking at Table 5, it is obvious that our proposal i.e., I-RPKM has considerably improved the results by decreasing the error rate. For example, for the Solar-Flare data set, the ER of I-RPKM is 0.09 which is the smallest in the table for k' clusters. After adding a new cluster, I-PRKM is still the best approach with the lowest ER (0.07 for the Solar-Flare data set) compared to KM-PF (0.09 for the Solar-Flare data set).

Table 4. The incremental possibilistic k-modes I-RPKM vs. SKM, KM-PF, RPKM

		Bal	S	POP	BS	SF	CE
k'		1	3	2	2	2	3
SKM	ER	0.57	0.54	0.42	0.36	0.27	0.32
	IN	5	7	8	7	10	9
	ET/s	10.35	11.9	13.73	29.41	1794.35	2580.03
KM-PF	ER	0.37	0.38	0.3	0.29	0.16	0.28
	IN	2	4	4	2	3	2
	ET/s	0.27	1.07	1.16	7.3	42.27	76.53
RPKM	ER	0.36	0.34	0.3	0.29	0.16	0.28
	IN	2	4	4	2	8	8
	ET/s	10.54	12.57	12.81	29.43	76.47	167.183
I-RPKM	ER	0.27	0.27	0.25	0.17	0.11	0.16
	IN	2	2	4	2	4	2
	ET/s	1.11	1.15	1.52	5.42	32.17	63.41
k		2	4	3	3	3	4
SKM	ER	0.48	0.4	0.32	0.22	0.13	0.2
	IN	9	10	11	13	14	11
	ET/s	14.55	16.08	17.23	37.81	2661.634	3248.613
KM-PF	ER	0.26	0.22	0.25	0.17	0.07	0.11
	IN	3	6	6	2	6	4
	ET/s	0.9	1.34	1.4	8.51	55.39	89.63
RPKM	ER	0.32	0.27	0.23	0.12	0.06	0.08
	IN	4	6	8	2	12	12
	ET/s	13.14	15.26	16.73	35.32	95.57	209.68
k= k'+c				c=1			
I-RPKM	ER	0.22	0.2	0.21	0.1	0.08	0.09
	IN	3	4	6	2	4	4
	ET/s	0.86	1	1.1	3.75	26.11	42.81

Table 6 presents the IN and ET of our proposal and KM-PF. As shown in Table 6, the IN and the ET of the new incremental approach are the lowest. By using the k' final partitions produced in the previous step instead re-clustering, I-RPKM improves the results. On the other hand, KM-PF is unable to update its final partitions. It re-clusters instances which makes its clustering process longer.

Table 5. Average of error rates of I-RPKM vs. KM-PF

		Bal	S	PO	BS	SF	CE
k'		1	3	2	2	2	3
A < 50% and 0<d<0.5	KM-PF	0.43	0.31	0.35	0.3	0.28	0.19
	I-RPKM	0.4	0.33	0.35	0.43	0.2	0.14
A < 50% 1 and 0.5≤d≤ 1	KM-PF	0.42	0.35	0.39	0.29	0.17	0.23
	I-RPKM	0.43	0.3	0.37	0.27	0.1	0.15
A ≥ 50% and 0<d<0.5	KM-PF	0.27	0.29	0.31	0.19	0.11	0.18
	I-RPKM	0.23	0.27	0.29	0.14	0.09	0.14
A ≥ 50% and 0.5≤d≤ 1	KM-PF	0.34	0.3	0.38	0.3	0.2	0.19
	I-RPKM	0.36	0.34	0.3	0.28	0.2	0.17
c=1 k=k'+c		2	4	3	3	3	4
A < 50% and 0<d<0.5	KM-PF	0.36	0.23	0.27	0.21	0.14	0.13
	I-RPKM	0.22	0.13	0.19	0.15	0.09	0.1
A < 50% 1 and 0.5≤d≤	KM-PF	0.35	0.27	0.29	0.2	0.11	0.17
	I-RPKM	0.26	0.21	0.19	0.13	0.08	0.12
A ≥ 50% and 0<d<0.5	KM-PF	0.19	0.2	0.22	0.13	0.09	0.1
	I-RPKM	0.12	0.13	0.17	0.09	0.07	0.06
A ≥ 50% and 0.5≤d≤ 1	KM-PF	0.27	0.21	0.28	0.2	0.13	0.12
	I-RPKM	0.25	0.22	0.18	0.11	0.1	0.13

Table 6. The IN and ET of I-RPKM vs. KM-PF

		Bal	S	PO	BS	SF	CE
k'		1	3	2	2	2	3
The IN of the main program	KM-PF	3	4	8.	2	8	4
	I-RPKM	2	4	6	2	6	2
The elapsed time in seconds	KM-PF	0.67	0.76	0.95	9.65	56.78	90.3
	I-RPKM	0.45	0.51	0.63	5.81	39.87	71.42
c=1 k=k'+c		2	4	3	3	3	4
The IN of the main program	KM-PF	3	4	8	2	8	4
	I-RPKM	2	2	4	2	4	2
The elapsed time in seconds	KM-PF	0.68	0.79	0.98	9.67	56.79	90.43
	I-RPKM	0.32	0.47	0.54	5.21	31.42	68.94

The I-PRKM makes it possible to use existing results and saves computing time by adding new clusters. It also improves the performance of SKM by providing lower error rate and taking into account new knowledge. In addition, lower execution time is another advantage of I-RPKM.

6 Conclusion

In this work, we have highlighted three main problems when clustering categorical data sets. The first one is dealing with uncertainty in the values of attributes and in the belonging of objects to several clusters using possibility theory. The second one is relative to detecting peripheral objects that belong to boundary regions through rough sets. The last one consists of the update of the cluster results by considering new knowledge. As a result, the cluster number can be increased without re-clustering instances. Based on accuracy and execution time, our proposal outperformed SKM and its two uncertain versions, i.e., KM-PF and RPKM.

References

1. Ammar, A., Elouedi, Z.: A New Possibilistic Clustering Method: The Possibilistic K-Modes. In: Pirrone, R., Sorbello, F. (eds.) AI*IA 2011. LNCS, vol. 6934, pp. 413–419. Springer, Heidelberg (2011)
2. Ammar, A., Elouedi, Z., Lingras, P.: K-Modes Clustering Using Possibilistic Membership. In: Greco, S., Bouchon-Meunier, B., Coletti, G., Fedrizzi, M., Matarazzo, B., Yager, R.R. (eds.) IPMU 2012, Part III. CCIS, vol. 299, pp. 596–605. Springer, Heidelberg (2012)
3. Ammar, A., Elouedi, Z., Lingras, P.: RPKM: The rough possibilistic K-modes. In: Chen, L., Felfernig, A., Liu, J., Raś, Z.W. (eds.) ISMIS 2012. LNCS, vol. 7661, pp. 81–86. Springer, Heidelberg (2012)
4. Ammar, A., Elouedi, Z., Lingras, P.: The K-Modes Method under Possibilistic Framework. In: Zaïane, O.R., Zilles, S. (eds.) Canadian AI 2013. LNCS, vol. 7884, pp. 211–217. Springer, Heidelberg (2013)
5. Lin, J., Vlachos, M., Keogh, E.J., Gunopulos, D.: Iterative Incremental clustering of time series. In: Bertino, E., Christodoulakis, S., Plexousakis, D., Christophides, V., Koubarakis, M., Böhm, K. (eds.) EDBT 2004. LNCS, vol. 2992, pp. 106–122. Springer, Heidelberg (2004)
6. Charikar, M., Chekuri, C., Feder, T., Motwani, R.: Incremental clustering and dynamic information retrieval. In: Proceedings of the 29th Annual ACM Symposium on Theory of Computing, pp. 626–635 (1997)
7. Ester, M., Kriegel, H.P., Sander, J., Wimmer, M., Xu, X.: Incremental clustering for mining in a data warehousing environment. In: Proceedings of the 24th International Conference on Very Large Data Bases, pp. 323–333. Morgan Kaufmann (1998)
8. Jenhani, I., Ben Amor, N., Elouedi, Z., Benferhat, S., Mellouli, K.: Information Affinity: a new similarity measure for possibilistic uncertain information. In: Proceedings of the 9th European Conference on Symbolic and Quantitative Approaches to Reasoning with Uncertainty, pp. 840–852 (2007)
9. Joshi, M., Lingras, P., Rao, C.R.: Correlating Fuzzy and Rough Clustering. Fundamenta Informaticae 115(2-3), 233–246 (2011)
10. Dubois, D., Prade, H.: Possibility theory: An approach to computerized processing of uncertainty. Plenum Press (1988)
11. Murphy, M.P., Aha, D.W.: UCI repository databases (1996),
http://www.ics.uci.edu/mlearn

12. Huang, Z.: Extensions to the k-means algorithm for clustering large data sets with categorical values. Data Mining and Knowledge Discovery 2, 283–304 (1998)
13. Huang, Z., Ng, M.K.: A note on k-modes clustering. Journal of Classification 20, 257–261 (2003)
14. Lingras, P., Hogo, H., Snorek, S., Leonard, B.: Clustering Supermarket Customers Using Rough Set Based Kohonen Networks. In: International Syposium on Methodologies for Intelligent Systems, pp. 169–173 (2003)
15. Lingras, P., West, C.: Interval Set Clustering of Web Users with Rough K-means. Journal of Intelligent Information Systems 23, 5–16 (2004)
16. Lingras, P., Nimse, S., Darkunde, N., Muley, A.: Soft clustering from crisp clustering using granulation for mobile call mining. In: Proceedings of the International Conference on Granular Computing, pp. 410–416 (2011)
17. MacQueen, J.B.: Some methods for classification and analysis of multivariate observations. In: Proceedings of the Fifth Berkeley Symposium on Mathematical Statistics and Probability, vol. 1, Statistics, pp. 281–297 (1967)
18. Pawlak, Z.: Rough Sets. International Journal of Information and Computer Science 11, 341–356 (1982)
19. Pawlak, Z.: Rough Sets: Theoretical Aspects of Reasoning about Data. Kluwer Academic Publishers (1992)
20. Zadeh, L.A.: Fuzzy sets as a basis for a theory of possibility. Fuzzy Sets and Systems 1, 3–28 (1978)

Probabilistic Neural Network for the Automated Identification of the Harlequin Ladybird (*Harmonia Axyridis*)

Mohd Zaki Ayob[1] and E.D. Chesmore[2]

[1] Universiti Kuala Lumpur British Malaysian Institute
Gombak, Selangor Malaysia
mohdzaki@bmi.unikl.edu.my
[2] Department of Electronics
University of York YO10 5DD York, UK
david.chesmore@york.ac.uk

Abstract. This paper describes recent work in the UK to automate the identification of Harlequin ladybird species (*Harmonia axyridis*) using color images. The automation process involves image processing and the use of probabilistic neural network (PNN) as classifier, with an aim to reduce the number of color images to be examined by entomologists through pre-sorting the images into correct, questionable and incorrect species. Two major sets of features have been extracted: color and geometrical measurements. Experimental results revealed more than 75% class match for the identification of taxa with similar-colored spots.

Keywords: automated identification, feature extraction, image processing, probabilistic neural network.

1   Introduction

The use of automated identification of ladybirds using color images has considerable potential for biodiversity monitoring and has not been previously explored. In spite of the many advantages, automating the identification process is not trivial. Traditionally, an expert will manually obtain details from the elytra and other features for species identification. Dichotomous keys have been most commonly used, which are based on morphological criteria such as color and size. These physical traits are too small to examine without magnification, considering the size of typical ladybirds themselves which are around 2-8 mm, and harlequin ladybirds vary around 6-8 mm [1]. Furthermore, most ladybird species have a variety of color forms and spot patterns. Both intra-species and inter-species variations can be very large, making species identification a difficult skill to master. Part of the objectives of this research is to capture specific features that would make pre-sorting of ladybird species easier. These factors contribute to the development of an automated system for ladybird identification. In this paper the focus is the identification of harlequin ladybirds (*Harmonia axyridis*), an alien ladybird species found in the UK.

S. Ramanna et al. (Eds.): MIWAI 2013, LNCS 8271, pp. 25–35, 2013.

2   Background

The Harlequin ladybird is a voracious ladybird species that has been invading the UK since 2004. It feeds on aphids and other native ladybird species [2-3]. The beetle has now spread to most parts of England and Wales. It is receiving considerable attention due to its potential impact on ecological and biological balance [4]. To make things worse, there has been a general decline in the taxonomic workforce which in effect has been part of the taxonomic impediment to biodiversity studies [5]. In response, The UK Ladybird survey is a web-based recording system for the general public to send in their sightings and photographs. These color images enable the physical details of the ladybird forewings (called 'elytra') to be visually examined through image processing steps, which will be explained further in subsequent sections.

There are up to 15 described color forms of *H. axyridis*; however, only three different forms are studied here. Fig. 1 shows three color forms of *H. axyridis*. It shows the variation in colors; hence the problem to perform automated identification is based on color images only. Table 1 contains the physical description, which serves as an identification guide for harlequin ladybirds in UK. In terms of size and shape, the Harlequin is generally large and the length is between 6 mm to 8 mm. The pronotum pattern can be white or cream in color. It can contain up to 5 spots or fused lateral spots forming 2 curved lines, M-shaped mark or solid trapezoid. The wing cases have wide keel at the back, and legs are almost always brown.

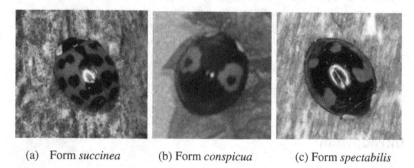

(a)   Form *succinea*   (b) Form *conspicua*   (c) Form *spectabilis*

Fig. 1. Top-view of the three forms of *H. axyridis* (Source: CEH Wallingford, UK)

Table 1. Identification guide for *H. axyridis* f. *succinea*, f. *conspicua* and f. *spectabilis*

Character	Form succinea	Form conspicua	Form spectabilis
Color	Orange with 0 to 21 black spots	Black with 2 red/orange spots, and inner black spots	Black with 4 red/orange spots
Shape	Round and domed	Round and domed	Round and domed
Size	6-8 mm	6-8 mm	6-8 mm

There are two major development works in this investigation, which are image processing and intelligent systems. The work in image processing involves two major processes: greyscale operations and color image processing. Section 3 covers the concepts and experimental work that has been carried out in image processing. These operations are implemented in MATLAB 2012 and GIMP2 [6]. Subsequently, the intelligent system consisting of PNN been implemented in WEKA [7] is elaborated. Section 4 discusses the results of experiments on PNN, and in Section 5 the paper concludes with future recommendation.

3   Methodology

3.1   System Block Diagram

A prototype automated species identification system has been developed to distinguish UK ladybird species using image processing and probabilistic neural networks (PNN). The block diagram is shown in Fig. 2.

Fig. 2. Block diagram of the prototype of an automated ladybird identification system

3.2   Image Processing

Based on color variation, the author has initiated investigating the use of color as the leading feature for identifying ladybird species. Based on visual observation, color represents the natural characteristic of ladybirds. Ladybirds commonly possess large variation in body color. Some are quite obvious, for instance, there is a species commonly called 'orange ladybird' which has orange-colored elytra and sixteen white spots. There are also 'striped ladybirds' with brownish elytra and cream-white stripes. The sightings can be confirmed based on publications by the UK ladybird identification guide, Leicester & Rutland color key and Southampton Natural History Society [1, 8-9].

In this work, CIELAB color space is used for representing pixel colors obtained from the elytra and body markings. CIELAB is an approximate uniform color scale to represent visual difference in the form of color plane, and able to represent chroma separately from lightness [10]. It is also able to represent the chroma values in the form of color plane. The maximum values for a* and b* are +120, while minimum values are -120. The range for L-axis is 0-100. Unlike RGB, CIELAB is also device independent.

CIELAB was selected as the desired color space to reduce illumination problems since the majority of ladybird images obtained for the study were photographed in their natural habitat. These images are prone to illumination issue, which is a variable that is quite difficult to control. By separating lightness and chroma using CIELAB as the color plane, subsequent work has been made simpler to solve [11-12]. CIELAB is useful as it distinguishes one specific color to be distinct from another visually similar color, and the difference between chroma values can be calculated [13]. Hence, errors between visual perception and actual values due to bad illumination are significantly reduced.

Another issue with the images is the background clutter. This is shown in Fig. 3, where work on the image of a scarce 7-spot ladybird (*Coccinella magnifica*) without background clutter revealed elytra markings better than the same image with background. Based on testing, it has been decided that the use of CIELAB is limited to capturing pixel values on the elytra and spots. Segmentation and elimination of background clutter are done in RGB color space. Body marking measurements are performed in greyscale and the image then converted into binary format. Only the colors of the body markings are captured in CIELAB values.

For each image, CIELAB values are obtained by reading the average L*, a* and b* values from a user-interactive pixel capture box. The size of the capture box is not fixed. It varies between 25x25 to 100x100 square pixels depending on the image resolution, hence user and image-dependent. Higher resolution images require a smaller capture box, and vice versa. If the size of the capture box were fixed, and the image is of low-resolution then the border pixels become indistinct and edges are difficult to locate. Once the average values were obtained, each value was normalized to [-1, 1].

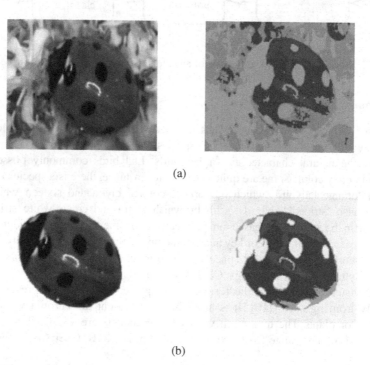

(a)

(b)

Fig. 3. (a) Image of scarce 7-spot (with background) after segmentation showing background clutter, and (b) same image with clean background

The following formulae were applied for normalization:

$$Normalized\ L^* = \left(\frac{L^*}{100}\right) \tag{1}$$

$$Normalized\ a^* = \left(\frac{a^*+120}{240}\right) \tag{2}$$

$$Normalized\ b^* = \left(\frac{b^*+120}{240}\right) \tag{3}$$

Fig. 4 shows the representation of the spot color and elytra/BG color (in CIELAB values) on normalized color planes. Refer Table 2 for species names and acronyms.

3.3 Geometrical Characters Extraction

In addition to color, geometrical characters have been utilized for identification. In order to get geometrical measurements, greyscale and binary image processing were performed rather than using color image processing as it involves minimal complications to perform binary processing in one channel. Initially images have been converted to greyscale via the MATLAB function 'rgb2gray' and resized to 640x480 pixels. The image is then converted to CIELAB for extracting pixel values. Geometrical measurements on spots and elytra are then performed.

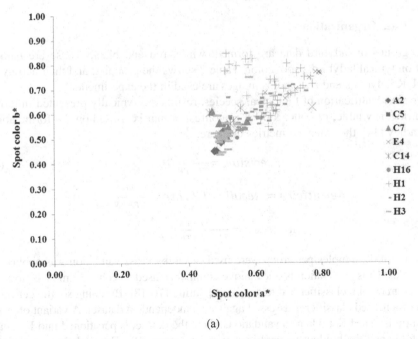

(a)

Fig. 4. CIELAB color distributions among local ladybird species and *H. axyridis*

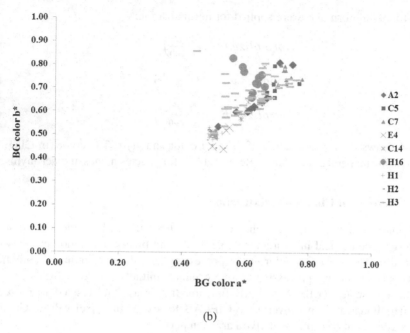

(b)

Fig. 4. (*continued*)

3.4   Data Organization

Three groups of ladybird data are formed: white, red and black. These are named based on typical ladybird's spot color. Table 2 shows the grouping and their acronym. Both UK ladybirds and Harlequins images are used in the experiments.

For the identification of biological species, results are typically presented in terms of contingency table, or better known as confusion matrix. Based on the formulation by Bradley [14], the extended metrics used are:

$$Precision = \frac{TP}{TP+FP} \qquad (4)$$

$$Sensitivity = Recall = TP\ rate = \frac{TP}{TP+FN} \qquad (5)$$

$$Accuracy = \frac{TP+TN}{TP+FP+TN+FN} \qquad (6)$$

A total of 40 samples per group were used, whereby cross-validation was applied to all samples. Cross validation is a technique commonly used to compare models, to estimate accuracy of a classifier and to avoid over fitting [16-18]. By doing so, the generalisation of a trained classifier is assessed against an independent dataset. A variant of cross validation is called K-fold cross validation, where the dataset is partitioned into K equalsized folds and the holdout method is repeated K times [19]. For each run, one of the folds is used as the test set and the (K-1) remaining folds used for training. Each of the K

Table 2. Ladybird groups, species and acronym (in brackets)

	Groups		
	White	**Red**	**Black**
Taxa	*Calvia quattuordec-imguttata* Linnaeus (C14) *Halyzia sedecimgut-tata* Linnaeus (H16)	*Exochomus quadri-pustulatus* Linnaeus (E4) *Harmonia axyridis* f. *spectabilis* Pallas (H1) *Harmonia axyridis* f. *conspicua* Pallas (H2)	*Adalia bipunctata* Linnaeus (A2) *Coccinella quinque-punctata* Linnaeus (C5) *Coccinella sep-tempunctata* Linnaeus (C7) *Harmonia axyridis* f. *succinea* Pallas (H3)

folds will be used once as validation data. After K runs, the average cross validation error across all runs is computed. The error is an estimate of how the classifier would perform if the data collected is an accurate representation of the real world.

3.5   Probabilistic Neural Networks (PNN)

Probabilistic neural networks (PNN) are used as the classifier. A classifier functions to map unlabeled instances to a class label using internal data structures [19]. PNN is chosen as classifier over other techniques because it implements kernel discrimination analysis, meaning the operations are organized into a multilayer feed forward neural network consisting of input layer, radial basis layer and competitive layer [20-21]. For a supervised classifier such as PNN to work, it contains exemplars and targets. The network structure is shown in Fig. 5.

The input layer consists of nodes that receive the data. The radial basis layer contains a probability density function (pdf), using a given set of data points as centers. The PNN uses normalized Gaussian radial basis functions as a network [22]. The output layer selects the highest value and determines the class label. In general, a PNN for M classes is defined as the following [23]:

$$y_j(x) = \frac{1}{n_j}\sum_{i=1}^{n_j} e^{\left(-\frac{(\|x_{j,i}-x\|)^2}{2\sigma^2}\right)} \tag{7}$$

where j = 1,..., M and n_j is the number of data points in class j.

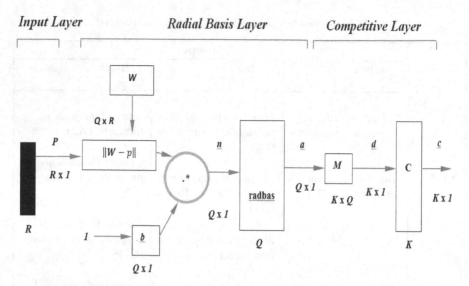

Fig. 5. PNN network structure (Source: MathWorks)

A decision boundary is found by finding the numerical solution to the above for each class. For instance, for a two-class problem this is done by equating $y_1(x)$ to $y_2(x)$ and finding solution using grid search [24].

4   Results and Discussion

In this work, 10-fold cross validation was applied and the following results were obtained:

Table 3. Confusion matrix for C14H16 (white group)

	C14	H16
C14	40	0
H16	0	40

Table 4. Detailed metrics for C14H16 (white group)

Class	TP rate	FP rate	Precision	Recall	AUC
C14	1	0	1	1	1
H16	1	0	1	1	1
Weighted average	1	0	1	1	1

Note: *Results obtained at MinStdDev = 0.1, no. of clusters = 2, AUC = area under curve.*

Table 5. Confusion matrix for E4H1H2 (red group)

	E4	H1	H2
E4	40	1	0
H1	0	28	17
H2	0	11	23

Table 6. Detailed metrics for E4H1H2 (red group)

Class	TP rate	FP rate	Precision	Recall	AUC
E4	1	0.013	0.976	1	0.993
H1	0.7	0.213	0.622	0.7	0.825
H2	0.575	0.138	0.676	0.575	0.847
Weighted average	0.758	0.121	0.758	0.758	0.888

Note: *Results obtained at MinStdDev = 0.1, no. of clusters = 3, AUC = area under curve.*

Table 7. Confusion matrix for A2C5C7H3 (black group)

	A2	C5	C7	H3
A2	39	0	0	0
C5	0	34	0	10
C7	0	0	39	0
H3	1	6	1	30

Table 8. Detailed metrics for A2C5C7H3 (black group)

Class	TP rate	FP rate	Precision	Recall	AUC
A2	0.975	0	1	0.975	1
C5	0.85	0.083	0.773	0.85	0.954
C7	0.975	0	1	0.975	1
H3	0.75	0.067	0.789	0.75	0.935
Weighted average	0.888	0.038	0.891	0.888	0.972

Note: *Results obtained at MinStdDev = 0.1, no. of clusters = 4, AUC = area under curve.*

From the results in Table 3, it has been shown that PNN is able to identify ladybirds in the white-spotted ladybirds group to 100% accuracy. The fact that they have been grouped based on their spot colors means PNN is able to completely discriminate the two species. Next, test results on the red-spotted ladybirds group reveals 75.83% accuracy. For H1, one instance of misidentification as E4 and eleven misidentifications as H2. For H2, the 0.575 TP rate means seventeen instances of H1 have been confused as H2.

Table 7 shows 88.75% accuracy for the identification of ladybirds in the black-spotted group. The TP rate of 0.975 for A2 shows only one misidentification as H3. This shows that between the two species A2 and H3, there are inherent similarities which contribute to the low numbers of misidentification. Similar observation can be said for C7 and H3. Interesting enough for H3, 10 out of 40 instances have been misidentified as C5 giving only 0.75 TP rate.

Use of CIELAB values to represent spot colors and applying PNN in ladybird identification are some contributions to knowledge. The observations in this work will be further investigated in the near future to see any correlations, for instance, to understand which character contributes more towards the identification process. Finding this 'contribution factor' will involve deeper investigations into network model, structure and the PNN algorithm itself. It will also involve shuffling between taxa to investigate intra-species and inter-species variations, as experimented by Ayob and Chesmore using multilayer perceptron (MLP) [25]. At present, it can be said that PNN works well to identify up to 4 taxa with a minimum of 75.83% accuracy.

5   Conclusion

In short, the use of PNN can be regarded as highly useful in the automated identification of ladybirds, including alien species. This is supported by experimental results on three groups of ladybirds involving UK species and *Harmonia axyridis*, which have been pre-grouped by their spot colours. Two-species identification of white-spotted taxa reveals 100% accuracy, while the identification of red-spotted taxa and black-spotted taxa reveals 75.83% and 88.75% accuracies respectively. Future investigations will look into the role of characters towards species identification.

Acknowledgment. The authors would like to extend thanks to Universiti Kuala Lumpur, University of York and Majlis Amanah Rakyat (MARA) for sponsoring the research work. Special thanks to Centre for Ecology & Hydrology (CEH) Wallingford, UK for supplying stock ladybird images.

References

1. UK Ladybird Survey, http://www.coleoptera.org.uk
2. Ware, R.L., Majerus, M.E.: Intraguild Predation of Immature Stages of British and Japanese Coccinellids by the Invasive Ladybird Harmonia Axyridis. BioControl 53, 169–188 (2008)
3. Majerus, M.E.N., Strawson, V., Roy, H.: The Potential Impacts of the Arrival of the Harlequin Ladybird, Harmonia Axyridis (Pallas) (Coleoptera:Coccinellidae). Ecological Entomology 31, 207–215 (2006) (in Britain)
4. UK Harlequin Survey,
   http://www.harlequin-survey.org/
   recognition_and_distinction.htm
5. Hopkins, G.W., Freckleton, R.P.: Decline In the Numbers of Amateur and Professional Taxonomists: Implications for Conservation. Animal Conservation 5, 245–249 (2002)
6. GIMP2, http://www.gimp.org/
7. WEKA, http://www.cs.waikato.ac.nz/ml/weka/

8. Mabbott, P. (ed.): Community Heritage Initiative. A colour key for identifying ladybirds in Leicester & Rutland, http://www.leics.gov.uk/celebrating_wildlife

9. Southampton Natural History Society, Ladybirds of Southampton, http://sotonnhs.org/docs/LadybirdAll.pdf

10. CIELAB colour models – Technical Guides, http://www.dba.med.sc.edu/price/irf/Adobe_tg/models/cielab.html

11. Torres, L., Reutter, J.Y., Lorente, L.: The Importance of the Color Information in Face Recognition. In: International Conference on Image Procesing (ICIP 19999), vol. 3, pp. 627–631 (October 1999)

12. Yip, A., Sinha, P.: Role of Color in Face Recognition. In: MIT tech report (ai.mit.com) AI Memo 2001-035, pp. 2001–2035. Massachusetts Institute of Technology, Cambridge, USA (2001)

13. Vízhányó, T., Felföldí, J.: Enhancing Colour Differences in Images of Diseased Mushrooms. Computers and Electronics in Agriculture 26, 187–198 (2000)

14. Bradley, A.: The Use of the Area Under the ROC Curve in the Evaluation of Machine Learning Algorithms. Pattern Recognition 30, 1145–1159 (1997)

15. Fawcett, T.: An Introduction to ROC Analysis. Pattern Recognition Letters 27(8), 861–874 (2006)

16. Omid, M.: Design of an Expert System for Sorting Pistachio Nuts Through Decision Tree and Fuzzy Logic Classifier. Expert Systems with Applications 38, 4339–4347 (2011)

17. Wolpert, D.H.: Stacked Generalization. Neural Networks 5, 241–259 (1992)

18. Prechelt, L.: Automatic Early Stopping Using Cross Validation: Quantifying the Criteria. Neural Networks 11(4), 761–767 (1998)

19. Kohavi, R.: A Study of Cross-Validation and Bootstrap for Accuracy Estimation and Model Selection. In: Proceedings of the International Joint Conference on Artificial Intelligence (IJCAI), pp. 1137–1143. Morgan Kaufmann, San Francisco (1995)

20. Wu, S.G., Bao, F.S., Xu, E.Y., Wang, Y.X., Chang, Y.F., Xiang, Q.L.: A Leaf Recognition Algorithm for Plant Classification Using Probabilistic Neural Network. In: 2007 IEEE International Symposium on Signal Processing and Information Technology, Giza, December 15-18, pp. 11–16 (2007)

21. MathWorks documentation, http://www.mathworks.com/help/nnet/ug/probabilistic-neural-networks.html

22. Hagan, M.T., Demuth, H.B., Beale, M.: Neural Network Design. PWS Publishing Company, Beijing (2002) ISBN 7-111-10841-8

23. Foody, G.M.: Thematic Mapping from Remotely Sensed Data with Neural Networks: MLP, RBF and PNN based Approaches. Journal of Geographical Systems 3, 217–232 (2001)

24. Hong, X.: Probabilistic neural network (PNN), http://www.personal.reading.ac.uk/~sis01xh/teaching/CY2D2/Pattern3.pdf

25. Ayob, M.Z., Chesmore, E.D.: Hybrid Feature Extractor for Harlequin Ladybird Identification Using Color Images. In: 2012 IEEE Symposium on Computational Intelligence in Bioinformatics and Computational Biology (CIBCB), San Diego, pp. 214–221 (2012)

Unifying Multi-level Business Process Discovered by Heuristic Miner Algorithm

Yu-Cheng Chuang[1], PingYu Hsu[2], and Hung-Hao Chen[3]

[1,2] Department of Business Administration, National Central University
[3] Institute of Industrial Management, National Central University
tocasper@hotmail.com, pyhsu@mgt.ncu.edu.tw, hhchen0424@gmail.com

Abstract. Process mining techniques are commonly used in business management while conformance checking becomes an important issue in business process management. Researchers derive the actual business process from event logs to draw a comparison with the business process model. When the two are inconsistent, a lack of internal controls happens.

This research proposes the consistence checking in the event logs. Because of the different granularity in the same event logs, the process can be demonstrated as grain 1 to grain n, in which the smaller the grain means the finer the granularity of the process. While using a process log to retrace the business process, different business processes might be shown in the processes with different granularities in the same event logs. The dependency threshold of Heuristic miner algorithm is used to deal with the differences of consistency automatically in this research.

This research uses the event logs from a marble processing industry for the case conformance. Focusing on the fine and coarse two granularities of the business process matrix, the conformance checking is applied for a consistent business process via the setting of the dependency threshold and the consistent ratio. The result show the valuable information to audit or re-design the business process model.

Keywords: Process Mining, Conformance checking, Heuristic miner algorithm, Petri Net.

1   Introduction

Business process management is the key factor for survival of a company. While a company develops and the internal risk cannot be found promptly, then any contingency measures become unworkable. Therefore, if there are no systematic process planning, good management team and clear business process, the departments are hard to share their responsibility, and finally will seriously impact on the development of the enterprises.

Because of the developed information system, integrating information in the database is essential to improve the readability; and many companies apply process mining techniques in management. The business process model is derived from the

S. Ramanna et al. (Eds.): MIWAI 2013, LNCS 8271, pp. 36–49, 2013.
© Springer-Verlag Berlin Heidelberg 2013

event logs in the information system while the actual business process is derived for a complete inspection of the company business processes. Meanwhile, the gap analysis is used for the comparison of the actual business process and the business process model to summarize the best rules of the business process for decision-makers as an important reference [20].

When the business of an enterprise increasingly expands, it is more difficult for business executives to manage the huge and non-integrated information [19]. Corporate governance and internal audit trails are getting attention in many companies. Besides, business process consistency in the business process management becomes particularly important. Business processes can be derived from the event logs; meanwhile, business processes can also be obtained by the flow chart. However, both should be consistent. If the actual business process model is inconsistent with the expected, there might be fraud, a risk of error, also a lack of internal control embedded [1, 6].

When process records of the event logs in the enterprise are checked, the event logs retracing can be used in the business processes. Because of the different granularities in the event logs, the granularities of the process in the event logs are not the same. The process can be demonstrated as grain 1 to grain n, in which the smaller the grain means the finer the granularity of the process. For example, in Table 1, the process of grain 1 is smaller than grain 2's. While using a process log to retrace the business process, different business processes might be shown in the processes with different granularities in the same event logs. However, because of the different granularities, different business processes are produced in retracing.

Table 1. Event logs

no.	Case ID	Grain 1	Grain 2	no.	Case ID	Grain 1	Grain 2
1	case 1	A-1	A	10	case 3	C-1	C
2	case 1	B-2	B	11	case 3	D-2	D
3	case 1	C-1	C	12	case 4	A-1	A
4	case 1	D-3	D	13	case 4	C-2	C
5	case 2	A-1	A	14	case 4	B-1	B
6	case 2	B-2	B	15	case 4	D-2	D
7	case 2	C-1	C	16	case 5	A-1	A
8	case 3	A-1	A	17	case 5	B-3	B
9	case 3	B-2	B	18	case 5	D-1	D

Business processes in Figure 1, 2 are the recorded processes with two different granularities in Table 1 derived by Heuristic miner algorithm. While the dependency threshold is set to be 0.6, two different business processes are produced though both processes are derived from the same event logs in the database.

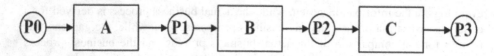

Fig. 1. Process Model of grain 1 (dependency threshold=0.6)

It is unreasonable that in the case of the same dependency threshold, a business process derived from the same event logs generates different business processes. The conformance checking is to solve the inconsistent business processes from the event logs in the same database because of the different granularities. By adjusting the dependency threshold to solve problems of the discrepancy of business processes because of the different granularities of the event logs in same database, the system can automatically generate the best business process from different business processes according to users' requirements.

When the conformance checking result show both dependency threshold and consistent ratio is low, it provide us with valuable objective information that the business process model should be audit, analysis or re-design.

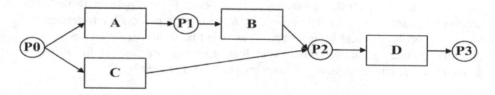

Fig. 2. Process Model of grain 2 (dependency threshold=o.6)

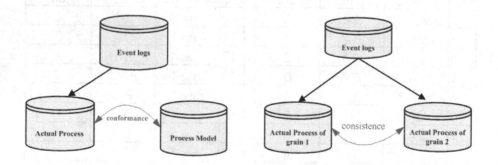

Fig. 3a. Conformance checking   **Fig. 3b.** Consistence checking

In other scholars' studies, the consistency of business processes proposed are based on whether the event logs and the business processes models are consistent. Aalst et al. (2010) have proposed business process mining and used conformance checking to examine the differences of consistency between the business processes models and the actual process from the event logs [2], Jans et al. suggested that process mining techniques can reduce the risk of fraud in the company and check if there is any discrepancy in the reconstruction of process mining between the business process model and the actual process [7].

In addition, Christian et al. proposed in 2007 a fuzzy mining to simplify views of reality business process. Many scholars also used methods of fitness and appropriateness by Aalst et al. proposed in 2006 to detect the consistency of the event logs and the process models [9]. The consistency of business processes in our study is different from such studies, is not about the discrepancy of the business model and the actual process. This study aims to solve the discrepancy amongst actual business processes derived from the same event logs because of the process with different granularities. Further, the processes with different granularities from the same event logs can be automatically identify a particular consistent business process.

This study mainly uses Heuristic miner algorithm for process mining. The steps of Heuristic miner algorithm for deriving business processes are shown as follows: firstly, finding out the correlation amongst processes in the records of the event logs for identifying the right activity links by the formula in Heuristic miner algorithm. Secondly, setting a standard dependency threshold, that is for the system to show business processes with a higher dependency threshold while users input it. That means the higher the dependency threshold is, the correlation between two activities is higher. Finally, the dependency graph obtained earlier is converted into a cause and effect matrix, and then to show the final process model by a Petri Net.

2   Literature Review

2.1   Process Mining

To identify process-related information from the event logs is the main skill of the process mining, Song and Aalst (2008) found out the business process model from such information [11]. Cook and Wolf (1998) have proposed three methods of process mining: the first method is to use neural network; the second method is to use Purely algorithm; the last one is to use markovian algorithm [3,4]. Their largest part for process mining is based on the latter two methods. Purely algorithm is to build a finite state and links possible future cases together to form the final process model; Markovian algorithm is to use hybrid Purely algorithms and statistical methods, which have the ability to deal with the error data, extending to the processes that use graphical approach to research in 1999[18].

Mining process is currently and mainly used in three perspectives: (1) a process perspective, (2) an organizational perspective and (3) the case perspective [7]. There are many mining techniques to meet process mining with different requirements, just as Alpha miner, Alpha + + miner, Heuristic miner and so on. Alpha miner, is a relatively intuitive and simple method, in which all data is applied in the model for mining all possible processes in the database [15, 16]. And Alpha + + miner covers some disadvantages of Alpha miner, such as loops [13]. Heuristic miner algorithm is a method to minimize the impact caused by errors if there is any error in the record file [18]. The main concept is to build links amongst all the activities, and then to analyze the error information. This study compares the consistency, which is adopted by the adjustment of Heuristic miner algorithm. Therefore, this study finally selects Heuristic miner algorithm to derive business processes.

2.2   Heuristic Miner Algorithm

Heuristic miner algorithm proposed by Weijters in 2006 is a process mining algorithm [18]. Such an algorithm divides the correlation amongst event records into three types, namely, direct dependency, concurrent and non-direct connectivity. The direct dependency sees the start of an activity as the judgment so while the information shows activity B just follows after activity A's process, there is direct dependency between them. The concurrent means that when activity *a* is over, there are more than two activities to be selected as the next activity. The non-direct connectivity means that the correlation between the two activities does not meet the direct dependency.

The procedures of Heuristic miner algorithm are as follows: first, constructing a dependency graph, then using heuristic search method to identify the correlation amongst each process, finally finding out the right activity link. This algorithm calculates the dependency coefficient between two activities according to the correspondence between two activities, and the value is between -1 to 1. Besides, this algorithm sets dependency threshold to handle noise data and low frequency behavior. When the dependency coefficient between two activities is calculated, the result is changed into a cause and effect matrix and a Petri Net will show the final process.

Heuristic miner algorithm can also find dependencies of short loop and non-local. The dependency of short loop means after activity A's process, there is another activity A follows. The dependency of non-local means the repetition of the activities in the same order. For example, the sequence of *abcbcbcde* activities, in which the repetition of *bc* in the process is called non-local dependency. Heuristic miner algorithm can handle abnormal information magnificently and derive process models with precise results.

2.3   Conformance and Consistence

In order to learn whether the process models by mining is the correct interpretation of the event logs, the consistency between the event logs and the process models becomes an important research topic [2, 5, 17]. At present, while the consistency of business process is discussed in the field of process mining algorithms, many scholars use the methods of fitness and appropriateness by Aalst et al. proposed in 2006. Fitness, is the main standard for the conformance checking to measure whether the process model can simulate the processes of activities in the event logs, further to assess a possible order of each process model in the event logs. Appropriateness assesses the quantity of the allowed models and the quantity of models that never actually used in the event logs. If the business models contain redundant and unworkable activities, it possibly is unnecessary [9].

In related studies, Rafael Accorsi and Thomas Stoker used the conformance checking to audit whether the bank's loan application process is consistent with the model; also for security audits [1, 14]. Dirk Fahland also used the conformance checking to audit the process of customers, CD retail business and supplier orders in which if there is any repetition in orders [10]. Many related researches of business processes are extending based on the fitness and appropriateness proposed by Aalst et al.. However, the extensions almost use the process information in the event logs to audit if the business models are consistent and if there is any fraud because of discrepancy. Our research explores the consistency of business processes by Heuristics miner algorithm for comparison.

3   Methodology

3.1   Conceptual Framework

The framework is divided into three steps: first, Heuristic miner algorithm is used to derive actual business processes from the event logs with different granularity. After that, the dependency coefficient is adjusted to resolve the consistency between actual business processes with different granularity. Finally, a Petri Net is generated by the adjusted business processes. Details of the process are as shown in Figure 4.

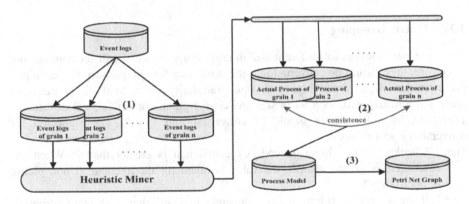

Fig. 4. Research framework

3.2   Finding Activity Relationship

First, the correlation amongst activities from the event logs is calculated. This study, referenced Heuristic miner algorithm proposed by Weijters in 2006, respectively calculates the quantity of activities and further calculates the dependency coefficient. For example, suppose activity b operates right after activity a, that means there is no other activity interrupted, then activities a and b are directly dependent $a >_W b$. Besides, $a \Rightarrow_W b$ represents the dependency coefficient with direct dependency coefficient between two activities, the value is between -1 to 1.

Suppose there are n activities in the event logs

Step1. Calculating the quantity of activity's occurrence with the direct dependency in each group $a >_w b$.

Step2. Using the formula of Heuristic miner algorithm to calculate the dependency coefficient amongst activities.

$$a \Rightarrow_w b = \left(\frac{|a > w\, b| - |b > w\, a|}{|a > w\, b| + |b > w\, a| + 1} \right)$$

Step3. Using a matrix to show the dependency coefficient between two activities.

The short loops in Heuristics miner algorithm, $a \Rightarrow w\, a$, presents the dependency of activities in the same kind of activity. The dependency coefficient value is between 0 to 1.

Suppose there are n activities in the event logs

Step1. Calculating the quantity of activity's occurrence with the direct dependency in each group a $>_w$ a.

Step2. Using the formula of Heuristic miner algorithm, $a \Rightarrow_w a = \left(\frac{|a > w\, a|}{|a > w\, a| + 1} \right)$, to calculate the dependency coefficient of the short loops.

Step3. Using a matrix to show the dependency coefficient between two activities.

In this step, the matrix derived contains the direct dependency coefficient amongst each group. Its main purpose is to adjust the differences of consistency in actual business processes in the next step.

3.3   Matrix Grouping

There are two matrixes for different size in granularity. It is difficult for comparison. Therefore, they should be converted into the same size for comparison. For example, the activities of the event logs with fine granularity are a total of 94 items so presented a matrix with 94 x 94 in size and coarse granularity are a total of 8 items. Therefore, 94 x 94 matrix should be converted into 8 x 8. The algorithm for conversion is as follows:

Step1. Checking if the direct dependency coefficient is greater than 0. When the coefficient is greater than 0, the value is taken for the average; if not, please refers to step 2.

Step2. If the projects of 0 have a direct dependency coefficient with other activities, the value is not taken for the average, and vice versa.

Step3. The cases with a direct dependency coefficient are taken into the average.

Step4. Generating the integrated direct dependency coefficient

3.4   Consistence Checking

The derived matrixes are used for the adjustment of consistency. Besides, this study applies the dependency threshold in the matrixes while the two thresholds for the matrixes are set to be m, n. If the dependency coefficient is greater than a given dependency threshold, the value becomes 1, otherwise 0. Then, the given dependency threshold is used to adjust and compare the consistency between two matrixes.

Table 2. Two Matrices with the dependency threshold *m* and *n*

	A	B	C	D			A	B	C	D
A	0	1	0	1		A	0	1	0	1
B	0	0	1	0		B	0	0	0	1
C	0	0	0	1		C	0	0	0	1
D	0	0	0	0		D	0	0	0	0

Assume the two corresponding items be x and y in two matrixes (Table 2) and the consistent ratio of two matrixes be z.

Calculation of consistent ratio step:

Step1. Setting the dependency threshold of two matrixes be *m* and *n*

Step2. When the coefficient of matrix 1 is greater than *m*, the value becomes 1, otherwise 0

Step3. When the coefficient of matrix 2 is greater than *n*, the value becomes 1, otherwise 0

Step4. When $x = y$, x and y both are 0 or 1, z plus 1, otherwise z plus 0

Step5. The sum of all values of z divides the total number of matrixes is the consistent ratio.

While the dependency threshold is *m* in matrix 1, and the dependency threshold is *n* in matrix 2, the consistent ratio of two matrixes come out. Our research is based on the consistent ratio to compare the consistency between two actual business processes.

3.5   Create Petri Net

The Petri Net uses the matrix derived in the previous step to construct actual business process. When the consistent ratio is adjusted, the matrix is converted into a Petri Net to denote business processes. The method of derivation is as follows: dependent matrixes are derived from the event logs. First, the consistent ratio threshold is set. While the consistent ratio is greater than the threshold, it is the derived business process. Table 3 is a dependent matrix. Assume the consistent ratio threshold be 0.7. While the consistent ratio is greater than 0.7, the business process is the derived one (red sash part is shown in Table 3). The final business processes are activity *B* follows activity *A*, and activity *C* follows activity *B*. Figure 7 is the scheme to show the Petri Net for Table 3.

Table 3. A dependent matrix with consistent ratio

	A	B	C	D
A	0	0.8	0.5	0
B	-0.8	0	0.7	-0.3
C	-0.5	-0.7	0.4	0
D	0	0.3	0	0

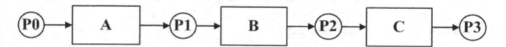

Fig. 5. Petri Net Example of business process

The Petri Net is used to indicate processes as it allows users to have an easier and more direct understanding towards business processes.

4   Experiments and the Evaluation

4.1   Real Case

In this research, the manufacturing process of the marble processing industry is the real case for analysis. The data of the event logs comes from the enterprise resource planning system (ERP) of a small and medium enterprise (SME) from 2011/12 to 2012/07. The activities mainly are CUT, QL, QW, AL, AW, B, IPQC and PACK, each main process can be subdivided into several activities.

The data of marble processing is divided into two different granularities. The activity with small granularity is named as grain 1, which is the original activity's name in the event logs. The activity with eight main processing above is named as grain 2. Two different granularities in the event logs are used in the research to derive particular business processes. In the followings, grain 1 is represented by fine, and grain 2 is represented by coarse.

4.2   Finding a Dependency Relation between Both Activities

The first step of the algorithm is to calculate the quantity of activity's occurrence between the two activities. The activities of the event logs with fine granularity are a total of 94 items so the matrix presented is a matrix with 94 x 94 in size. Table 4a is the derived matrix with 8 x 8 in size from the activities with coarse granularity of the event logs.

Table 4a. A frequency matrix with coarse granularity   **Table 4b.** A dependency matrix with coarse granularity

	CUT	QL	QW	AL	AW	B	IPQC	PACK
CUT	0	122	17	45	8	0	110	0
QL	0	145	73	0	0	0	23	0
QW	0	0	104	0	0	0	35	0
AL	0	0	0	59	31	0	4	0
AW	1	0	0	0	38	0	12	0
B	222	4	0	4	0	50	0	0
IPQC	0	0	0	0	0	0	0	184
PACK	0	0	0	0	0	1	0	0

	CUT	QL	QW	AL	AW	B	IPQC	PACK
CUT	0	0.992	0.944	0.978	0.7	-0.996	0.991	0
QL	-0.992	0.993	0.986	0	0	-0.833	0.958	0
QW	-0.944	-0.986	0.99	0	0	0	0.972	0
AL	-0.978	0	0	0.983	0.969	-0.8	0.8	0
AW	-0.7	0	0	0	0.974	0	0.923	0
B	0.996	0.833	0	0.8	0	0.98	0	0
IPQC	-0.991	-0.958	-0.972	-0.8	-0.923	0	0	0.995
PACK	0	0	0	0	0	0	-0.995	0

4.3 Caculating the Dependency Coefficient between Both Activities

From the calculated matrix in 4-2, the quantity of occurrence of the direct dependency between two activities is learnt. The second step is to calculate the direct dependency coefficient by Heuristic miner algorithm.

After calculation, a matrix is used to make a record. The value come out is the dependency coefficient between two activities while the value calculated by the formula is between -1 and 1. In this step, the matrix with the direct dependency coefficient derived from the matrix with the direct dependency coefficient in the fine granularity is a matrix with 94 x 94 in size. Table 4b is the derived matrix from the activities with coarse granularity of the event logs by Heuristic miner algorithm.

4.4 Checking the Consistence between Both Matrices

The fine business processes and the coarse business processes are adjusted by the direct dependency coefficient of activities that derived in the previous two parts. While business processes need to derived, the direct dependency coefficient, between 0 and 1, is used to find out a closer correlation between activities. Since the fine event logs have 94 activities, and the coarse event logs only have 8 major activities, it is needed to multiply the direct dependency coefficient of the fine event logs to a course 8 x 8 matrix before the comparison of consistency. Then, after the dependency threshold and the consistent ratio threshold are set, the business processes are derived out.

When the direct dependency coefficient of the fine activities is converted into a 8 x 8 matrix, sum of the direct dependency coefficient between activities is taken to make an average. However, if the projects of 0 have a direct dependency coefficient with other activities, the value is not taken for the average.

Table 5a. A dependency coefficient matrix with fine

Table 5b. A dependency coefficient matrix with coarse

	CUT	QL	QW	AL	AW	B	IPQC	PACK
CUT	0	0.754	0.681	0.67	0.65	0	0.991	0
QL	0	0.653	0.6	0	0	0	0.746	0
QW	0	0	0.591	0	0	0	0.833	0
AL	0	0	0	0.63	0.61	0	0.5	0
AW	0.5	0	0	0	0.62	0	0.653	0
B	0.876	0.5	0	0.5	0	0.702	0	0
IPQC	0	0	0	0	0	0	0	0.995
PACK	0	0	0	0	0	0	0	0

	CUT	QL	QW	AL	AW	B	IPQC	PACK
CUT	0	0.992	0.944	0.978	0.7	0	0.991	0
QL	0	0.993	0.986	0	0	0	0.958	0
QW	0	0	0.99	0	0	0	0.972	0
AL	0	0	0	0.983	0.969	0	0.8	0
AW	0	0	0	0	0.974	0	0.923	0
B	0.996	0.833	0	0.8	0	0.98	0	0
IPQC	0	0	0	0	0	0	0	0.995
PACK	0	0	0	0	0	0	0	0

Since this research is grouping a 94 x 94 matrix into a 8 x 8 matrix, the calculation does not count the activities in the corresponding category with the value of 0 but only counts the activities with a positive value. Then, the matrix with the fine original dependency coefficient integrated into a 8 x 8 matrix (Table 5a) is adjusted with a coarse 8 x 8 matrix (Table 5b) for consistency.

4.5   Adjusting the Dependency Threshold between Both Matrices

Next, a pair of matrixes is compared. Figure 8, 9 explain the process of the derivation from the process models with fine granularity to the process models with course granularity. Figure 8 is an interface for users to enter information at the beginning. First, the consistent ratio is filled; that is the expected consistency between two actual business processes. In the second blank, the dependency threshold of the fine process model between activities is set. Finally, in the third blank, if the consistent ratio threshold of the fine and course matrixes cannot be match, the consistent ratio of this blank is needed to be adjusted, and the default value is 5%.

Find the Business Model of Coarse	
Consistent ratio	
	%
The dependency threshold of fine	
adjustment interval	
[enter]	

Fig. 6. Set up consistence ratio and dependency threshold

When the user sets 85% as the consistent ratio, 0.7 as the dependency threshold of the fine process model, and 0.5% as the adjustment range, the consistent ratio of the fine and coarse process models is 85%. If there is not exact 85% consistent ratio, the system will automatically adjust upwards, 0.5% as an interval in every adjustment. When there is a conforming business process, it is presented in the form of a matrix. Otherwise, the system asks the user to enter a new consistent ratio threshold as well as another dependency threshold. Figure 9 is an example. In the process, the user set 85% as the consistent ratio and 0.75 as the dependency threshold. However, there is no matching matrix in the coarse process model, and then the system automatically adjusts and finds a matrix with 86% consistent ratio, in which the dependency threshold of the fine

Consistent ratio:86%
The dependency threshold of coarse:0.95

Process Model of fine(Dependency Threshold:0.75)

	CUT	QL	QW	AL	AW	B	IPQC	PACK
CUT	0	1	0	0	0	0	1	0
QL	0	0	0	0	0	0	0	0
QW	0	0	0	0	0	0	1	0
AL	0	0	0	0	0	0	0	0
AW	0	0	0	0	0	0	0	0
B	1	0	0	0	0	0	0	0
IPQC	0	0	0	0	0	0	0	1
PACK	0	0	0	0	0	0	0	0

Process Model of coarse(Dependency Threshold:0.95)

	CUT	QL	QW	AL	AW	B	IPQC	PACK
CUT			1				1	
QL		1	1				1	
QW			1					
AL				1	1			
AW					1			
B					1			
IPQC								
PACK								

Fig. 7. System UI when consistent ratio is 86% and coarse dependency threshold is 0.95

process model is 0.95. When the user enters a different percentage, the system will also automatically generate the corresponding business model.

4.6 Creating Petri Net

The information in Figure 7 can be converted into a Petri Net in Figure 8. The black line between two Petri Nets indicates the consistent part between two business processes while the red dotted line indicates the differences between two business processes.

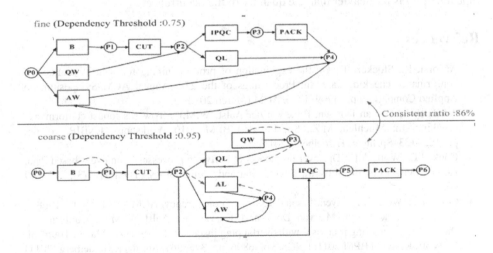

Fig. 8. Consistent ratio: 86% (coarse dependency: 0.95, fine dependency: 0.75)

Figure 8 illustrate the automatic derivation of business processes with different granularities. In the same event log, if there is any difference of consistency in business processes because of the difference granularities, the algorithms in this research will automatically solve the problem by the setting of the dependency threshold and the consistent ratio for deriving a business process with another granularity.

5　Conclusion and Future Research

In business process mining, companies can derive actual business process from event logs, in which the actual process should be consistent with the expected model. In the past researches about the conformance checking of the business processes, scholars mainly explored whether the actual processes are consistent with the business process models. This approach helps companies to solve the problem of consistency amongst actual business processes with different granularities, and allows users to set the dependency threshold and the consistent ratio of the process to derive the corresponding business processes with different granularities automatically. The manufacturing processes of the event logs in the marble processing industry have been verified the method we suggested.

Heuristic miner algorithm only considers the id, name and the time of start of the activity. In the algorithm, the order of processes is used to derive business processes. For future studies, the quantity of each activity in the process can be accessed. For example, there are 10 direct occurrences between activity a and b however the input between them is one ton; and there are 50 occurrences between activity a and c however the input between them is 50 kilograms. Then, the occurrence between activity a and b is more than that between activity a and c however the proportion of input quantity is far heavier than the quantity of the occurrence.

References

1. Accorsi, R., Stocker, T.: On the exploitation of process mining for security audits: the conformance checking case. In: Proceedings of the 27th Annual ACM Symposium on Applied Computing, pp. 1709–1716. ACM (March 2012)
2. Adriansyah, A., van Dongen, B.F., van der Aalst, W.M.P.: Towards robust conformance checking. In: Muehlen, M.Z., Su, J. (eds.) BPM 2010 Workshops. LNBIP, vol. 66, pp. 122–133. Springer, Heidelberg (2011)
3. Cook, J.E., Wolf, A.L.: Discovering models of software processes from event-based data. ACM Transactions on Software Engineering and Methodology (TOSEM) 7(3), 215–249 (1998)
4. Cook, J.E., Wolf, A.L.: Event-based detection of concurrency. ACM 23(6), 35–45 (1998)
5. Fahland, D., de Leoni, M., van Dongen, B.F., van der Aalst, W.M.P.: Conformance checking of interacting processes with overlapping instances. In: Rinderle-Ma, S., Toumani, F., Wolf, K. (eds.) BPM 2011. LNCS, vol. 6896, pp. 345–361. Springer, Heidelberg (2011)
6. Jans, M., Lybaert, N., Vanhoof, K.: A framework for internal fraud risk reduction at it integrating business processes: the IFR[2] framework. The International Journal of Digital Accounting Research 9(15), 7 (2010)
7. Mans, R.S., Schonenberg, M.H., Song, M., Van der Aalst, W.M.P., Bakker, P.J.M.: Application of process mining in healthcare–a case study in a dutch hospital. In: Fred, A., Filipe, J., Gamboa, H. (eds.) BIOSTEC 2008. CCIS, vol. 25, pp. 425–438. Springer, Heidelberg (2009)

8. Murata, T.: Petri nets: Properties, analysis and applications. Proceedings of the IEEE 77(4), 541–580 (1989)
9. Rozinat, A., van der Aalst, W.M.P.: Conformance testing: Measuring the fit and appropriateness of event logs and process models. In: Bussler, C.J., Haller, A. (eds.) BPM 2005. LNCS, vol. 3812, pp. 163–176. Springer, Heidelberg (2006)
10. Rozinat, A., van der Aalst, W.M.: Conformance checking of processes based on monitoring real behavior. Information Systems 33(1), 64–95 (2008)
11. Song, M., van der Aalst, W.M.: Towards comprehensive support for organizational mining. Decision Support Systems 46(1), 300–317 (2008)
12. Van der Aalst, W.M.P.: Petri-net-based workflow management software. In: Proceedings of the NFS Workshop on Workflow and Process Automation in Information Systems, pp. 114–118. IEEE Computer Society (May 1996)
13. van der Aalst, W.M.P., van Dongen, B.F.: Discovering workflow performance models from timed logs. In: Han, Y., Tai, S., Wikarski, D. (eds.) EDCIS 2002. LNCS, vol. 2480, pp. 45–63. Springer, Heidelberg (2002)
14. Van Aalst, W.M., van Hee, K.M., van Werf, J.M., Verdonk, M.: Auditing 2.0: Using process mining to support tomorrow's auditor. Computer 43(3), 90–93 (2010)
15. Van der Aalst, W.M., Weijters, A.J.M.M.: Process mining: a research agenda. Computers in Industry 53(3), 231–244 (2004)
16. Van der Aalst, W., Weijters, T., Maruster, L.: Workflow mining: Discovering process models from event logs. IEEE Transactions on Knowledge and Data Engineering 16(9), 1128–1142 (2004)
17. Weijters, A.J.M.M., Van der Aalst, W.M.P.: Rediscovering workflow models from event-based data. In: Proceedings of the 11th Dutch-Belgian Conference on Machine Learning (Benelearn 2001), pp. 93–100 (2001)
18. Weijters, A.J.M.M., van der Aalst, W.M., De Medeiros, A.A.: Process mining with the heuristics miner-algorithm. Technische Universiteit Eindhoven, Tech. Rep. WP, 166 (2006)
19. Günther, C.W., van der Aalst, W.M.P.: Fuzzy Mining - Adaptive Process Simplification Based on Multi-perspective Metrics. In: Alonso, G., Dadam, P., Rosemann, M. (eds.) BPM 2007. LNCS, vol. 4714, pp. 328–343. Springer, Heidelberg (2007)

User Distributions in N-Tier Platform with Effective Memory Reusability

Yu-Cheng Chuang[1,*], Pingyu Hsu[1], Mintzu Wang[1], Ming-Te Lin[1], and Ming Shien Cheng[2,*]

[1,2] Department of Business Administration, National Central University, Taiwan
tocasper@hotmail.com
[2] Department of Industry Technology and Management,
Ming Chi University of Technology, Taiwan
mscheng@mail.mcut.edu.tw

Abstract. Due to the vigorous development of the network, a variety of application software has been widely used. In order to serve more users, it is necessary to build middleware servers (Application Servers). If the middleware server was overloaded, it will result in poor performance. Besides, it is simply waste of resources if loading was idle. So the server with or without load balancing becomes an important issue. The proposed LAPO algorithm in this paper can dynamically allocate each middleware server for each user. And firstly, it improves the POCA algorithm that spends a lot of computation time to determine the optimal combination of solutions. Secondly, it can uniformly distribute each user on the servers; and finally, propose the model for best combination of load balancing solution. By using the SAP ERP ECC 6.0 for implementation, this study can verify that the LAPO is not only more efficient in computation time than POCA, but also more in line with the actual situation of the enterprises use. Moreover, we comment the results of experiments and some limitations.

Keywords: Load Balancing, Transaction, User Distribution, Performance Evaluation.

1 Introduction

The n-Tier architecture has been widely used in a variety of information systems, such as Enterprise Resource Planning (ERP), Supply Chain Management (SCM). In Figure 1, application servers (APs) are designed to share the database access requirements from many users. So the APs' load balance will determine whether the system is running at its best condition.

Fig. 1. N-Tier architecture[1]

---

* Corresponding authors.

S. Ramanna et al. (Eds.): MIWAI 2013, LNCS 8271, pp. 50–58, 2013.
© Springer-Verlag Berlin Heidelberg 2013

Regarding load balancing, many studies have proposed algorithms to do matching respectively in a clustering manner in accordance with the characteristics of the user or the capacity of the server [3]. The POCA (Profile Oriented Clustering Algorithm) has proposed the load-balancing method with higher memory sharing for different users performing the same program, but this algorithm consumes a lot of computation time to yield the best combination of solutions [8].

Therefore, this study proposes LAPO (Load Adjusted Profile Oriented User Allocation Algorithm), under the conditions that the number of APs is known, groups all users, and assigns them to various APs in order to shorten computation time of the algorithm, according to the method with highest memory sharing.

The data in this study was taken from the SAP ERP ECC 6.0 system. Based on the system log, the execution of Transaction Code (Tcode) can deduce the assigned status of all users for different number of APs, and finds the best combination.

2   Literature Review

Popular information systems with large number of users such as ERP, will normally configure APs to achieve Load balancing. Take the SAP ERP ECC 6.0 as an example, after a user logs on, the Web Dispatcher, as Figure 2, will assign the user to specific APs, while the APs' main job is handling client request [2, 9, 10, 11, 12] and retains the operation records of the individual user [3] that this study will use for implementation usage.

The APs may execute different applications concurrently such that will cause overload in peak hour, while at other periods in idle state. According to the study by P. Krueger, et al. [4]: The average utilization rate of the system is 9%, and even under the most peak load periods, in average still more than 87% of computing capacity is wasted. Also, the study of M.L. [5, 8] pointed out that in a medium load, the system will have better performance. Appropriate redistribution of the work load between APs is what we call Load Balancing. If the purpose of redistribution is to make certain performance measures tend to be equalized, for example, the ratio of the server idle time becoming equal is known as Load Balancing [6, 13]. Also, if the purpose of redistribution is to improve some of the performance measures such as average job response time, then we call it Load Sharing [7, 14].

Under the framework of Load Balancing, a Client-based architecture will distribute user requests at the client site, and need to use the browser software to determine which APs to handle the request. But the problem is this calculation will not only increase the work load of the browser, but also may cause the phenomenon of network traffic. A Dispatcher-based architecture works through a transmission channel by a central dispatcher, all requests must go through the dispatcher to be assigned to each server, but this also incurs bottleneck situation when large amount of data being processed at the same time. So in this paper, the concept of Cluster-based web server distributed system [15,16] has been used to avoid situation that work heavily loaded, work lightly loaded or idle state may occur.

POCA [8] proposed to turn the operating characteristics of the user into a distribution index. Because most ERP have a variety of modules, for example, SAP has FI, CO, PP, SD, etc., and users have their own common Tcode, POCA, based on this feature, proposed the Regular Transaction code (Regular Tcode) and the AR (Application Reusability) to measure the degree of similarity of the regular transactions between users, and establish the AR threshold, find all clustering methods that meet the threshold. However, this algorithm, when in the search for all clustering combinations, requires longer computation time and does not propose how to pick out the best clustering combination. The proposed LAPO algorithm improves the POCA algorithm that spends a lot of computation time to determine the optimal combination of solutions. Secondly, it can uniformly distribute each user on the servers; and finally, propose the model for best combination of load balancing solution.

3   Methodology

3.1   Finding Users' Regular Transactions

Most enterprise systems can save some recordable data and application during a user's operation session. For example, ST01 (system trace) command in SAP can be traced the historical record of a user's operating system usage. This paper will treat Usage Profile as a storage unit, which contains user information: {user-id, transaction-set}, as shown in Table 1.

Table 1. User Profile

Seq	User-ID	T-set	Seq	User-ID	T-set
1	1	{A,B,E,F,H }	8	5	{B,I,J,K }
2	1	{A,B,E,F}	9	6	{B,I,J,L}
3	2	{A,B,E,F,G}	10	6	{B,I,J,K}
4	2	{A,B,E,H}	11	7	{O,P,Q,R}
5	3	{A,B,E}	12	8	{O,P,Q,R}
6	3	{B,E,F,H}	13	9	{P,Q,R,K}
7	4	{I,J,K,L}	14	10	{W,X,Z}

For example, user1 was online twice that day, and the Tcode he executed were {A, B, E, F, H} and {A, B, E, F} respectively. The executed Tcode will be temporarily stored into to a buffer. If a rarely used Transaction has been kept stored in buffer, it may occupy buffer space and lead to reduce the utilization rate of the buffer. So we want to store and only store the most frequently executed Tcode into APs Memory, we call it a Regular Transaction.

Definition 1: Give a user s, a user profile U, and a transaction t, t is a regular transaction in U, if

$$\frac{|t \in s.transaction_{set}, s \in U|}{|U|} \geq profile\ support\ threold \tag{1}$$

Profile support threshold is defined by the system administrator, and it aims to filter regular transactions out of the usage profile. It qualifies as "regular" only when it is higher than the threshold value.

Assume the profile support threshold is set to 20%, take A and W as an example, Support is 5/14 = 35.7%, 1/14 = 7.14% respectively, thus A is a regular Tcode. On the contrary, w is not, so the Regular transaction set in Table 1 is {A, B, E, F, H, I, J, K, P, Q, R}.

Definition 2: Give a user s, a user profile U, and a transaction t, t is one of s's regular transactions in U and s is regular user in U,

$$\frac{|\{s|s \in U, s.user - id = u \wedge t \in s.transaction - set\}|}{\{s|s \in U, s.user - id = u\}}$$
$$\geq user\ support\ threshold. \tag{2}$$

The purpose of setting User support threshold is to identify whether the user executing regular Tcode is also a Regular User. Let User support threshold set to 40%, <userid, Tcode> is set to <1, A> and <1, H> respectively, we get 2/2 = 100%, 1/2 = 0.5% if any group passes 40%, the user 1 belongs the regular users. In addition, the X in <10, X> is not a regular transaction, so user 10 is not a regular user. As shown in Table 2, there are nine regular users totally except user 10.

Table 2. Regular Transaction

User-id	Regular Transaction	User-id	Regular Transaction
1	{A,B,E,F,H }	6	{B,I,J,K}
2	{A,B,E,F,H}	7	{P,Q,R}
3	{A,B,E,F,}	8	{P,Q,R}
4	{I,J,K}	9	{K,P,Q,R}
5	{B,I,J,K}	10	φ

3.2   Profile Oriented Clustering Algorithm (POCA)

The concept of the POCA algorithm is to put the users with similar Tcode usage onto the same APs, for the purpose of sharing memory. If there are users executing similar regular Tcode, the users belong to the same Cluster. And the POCA algorithm uses AR to measure similarity.

Definition 3: The AR value of each cluster is P / (u * q), where P calculates the repeat times of a regular Tcode in a cluster being executed, u counts the number of users in the cluster, q denotes how many different kinds of Tcode there are in the cluster. The Cluster AR is the average of the Transaction AR, and therefore its value also falls between 0 and 1.

For example, if a cluster contains two users {4, 6}, and jointly execute {I, J, K, L} and {B, I, J, L}, because L is not a regular transaction and should be excluded, the total execution times for {I, J, K} and {B, I, J, L} is seven, and there are four transaction codes, so AR = 7/(2 * 4) = 87.5%.

Moreover, if there are three AP servers to service 9 individuals, whom were orderly assigned as {1,2,3,5} {7,8,9} {4,6}, and the AR values were {19 / (4 * 8) = 59.4%}, {9 / (3 * 3) = 100%} and {7 / (2 * 4) = 87.5%} respectively, the average AR is equal to (59.4% +100% +87.5%) / 3 = 82.3%.

Definition 4: When the AR value exceeds AR Threshold in a cluster, this cluster is called qualified cluster. The Tcode sets in this cluster are mutually exclusive and disjointed.

3.3 Clustering and Distributing by Load Adjusted Profile Oriented User Allocation Algorithm (LAPO)

The LAPO algorithm is derived to improve the POCA problem of too much calculation time. To meet the business situation, LAPO must determine the number of APs firstly.

Definition 5: Under the premise of a known number of APs, it then can determine the number of users that each of the APs can service. Table 3 shows a distribution case when there are 9 users and APs=2, 3 or 4

$$\text{Users(Min)} = \lfloor (\text{User} / \text{Apserver}) \rfloor \text{ -}\varepsilon; \ 0 \leq \varepsilon \leq U_{\text{average}}, \ \varepsilon \text{ is an integer}$$

$$\text{Users(Max)} = \lceil (\text{User} / \text{Apserver}) \rceil \text{ +}\varepsilon; \ 0 \leq \varepsilon \leq U_{\text{average}}, \varepsilon \text{ is an integer}$$

Table 3. Users=9, APs=2,3,4, Distribution Table of APs and Users

case	AP s	Uaverage	ε	Umax<=Users	Umin>=1
1	2	4.5	1	5+1=6	4-1=3
2	3	3.0	1	3+1=4	3-1=2
3	4	2.3	1	3+1=4	2-1=1

Definition 6: Given C_1 is 1-users qualified clusters and a user distribution UD is $UD = \{\langle C_{Users(Min)}, ..., C_{Users(Max)} \rangle\}$, if

$(|C_{Users(Min)}|,...,|C_{Users(Max)}|) = |C_1|$ and $\cup (C_{Users(Min)}, ..., C_{Users(Max)}) = \cup C_1$

Definition 7: Given known APs, to identify the best cluster sets. Let U represent the number of users, ε denotes deviation value, APs are the number of known servers.

$$\max \left(\frac{\sum c^i (AR)}{APs} \right)$$

$$subject \ to \begin{cases} c^i(AR) > AR_{Threshold} \ for \ 1 \leq i \leq APs \\ \frac{|U|}{APs} - \varepsilon \leq |c^i| \leq \frac{|U|}{APs} + \varepsilon \end{cases} \tag{3}$$

Table 4. Distribution status for different user count, when APs= 3, Umax=4, Umin=2

Combination #	Aps 1	Aps 2	Aps 3	Total Users
Combination 1(C1)	4	3	2	9
Combination 2(C2)	3	3	3	9

The POCA algorithm does not constrain the number of APs. For the case of 9 users, its number of APs can be up to nine units, and the minimum is one unit, and all possible combinations is

$$\sum_{1}^{i} C_i^n = C_1^9 + C_2^9 + C_3^9 + C_4^9 + C_5^9 + C_6^9 + C_7^9 + C_8^9 + C_9^9$$

$$= 9 + 36 + 84 + 126 + 126 + 84 + 36 + 9 + 1.$$

Equivalent to $2^n - 1 = 512 - 1$, therefore, as we can see LOPA can save a lot of computing time.

In Table 5, we can see the best combination solution is when AR = 91.7

Table 5. Distribution of the best combination for each user count, APs = 3

	Cluster 1 , AR	Cluster 2, AR	Cluster 3, AR	$AR_{AVERAGE}$
Top AR of C1	{1,2,3} , 100	{4,5,6} , 91.6	{7,8,9}, 83.3	91.7
Top AR of C2	{1,2,3,5}, 59.4	{7,8,9} , 83.3	{4,6} ,87.5	76.7

The POCA algorithm has to spend a lot of time to calculate all qualified clusters that meet the AR Threshold, and by determining how to configure AR value to be the highest it can determine how the system should combine the cluster set into the server, but POCA does not limit the number of APs nor the accommodated user count. For this, given known number of AP server, LAPO chooses to add the deviation value evaluation of the user count into each server, so given a fixed number of APs, it can limit the maximum and minimum user counts that each APs can cover. As a result, taking average results from AR excludes the need to calculate all combinations of qualified cluster, and therefore enhance the efficiency and also more in line with the actual situation of the corporate use.

4   Experiments and the Evaluation

The purpose of this study is to provide the system the capability to decide every day how to assign users to the servers, and to expect that the resources are taken full advantage of and gain a balance with processing efficiency. Through finding the users' regular transactions and AR threshold conditions setting, this study can find the user clustering regular transaction set, and recommends the best solution combination in the system while still meets the condition of highest AR value.

In this study, The data were collected between 2012/03 and 2012/04 with 1001 records of user profile from the SAP ERP ECC 6.0 system, including 39 users and 336 kinds of the Transaction code. The data have detail description of which transaction has been executed after a user's login to the system. Also, the data have been set the regular transaction support threshold to 20% and the regular user support threshold to 35%. After calculation by definition 1 & 2, it obtains 25 Regular Transactions and 22 Regular Users.

Table 6a. Umax and U_{min} when APs=2-7   **Table 7b.**   User allocation when APs=6

APs	users	Uaverage	ε	U_{max}	U_{min}
2	22	11	1	12=11+1	10=11-1
3	22	7.3	1	9 = 8+1	6=7-1
4	22	5.5	1	7=6+1	4=5-1
5	22	4.4	1	6=5+1	3=4-1
6	22	3.7	1	5=4+1	2=3-1
7	22	3.1	1	5=4+1	2=3-1

case 1	APs1	APs2	APs3	APs 4	APs 5	APs 6
C1	5	5	5	3	2	2
C2	5	5	4	4	2	2
C3	5	5	4	3	3	2
C4	5	5	3	3	3	3
C5	5	4	4	4	3	2
C6	5	4	4	3	3	3
C7	4	4	4	4	4	2
C8	4	4	4	4	3	3

Next, follow the definition 3 & 4; the qualified Cluster for AR is 50, so we get the qualified Cluster of the combination of different user count. And then, by definition 5, we must determine the APs value. For the case of twenty two users, if the APs value is 1, there is no need to perform load balancing analysis, and if the APs value is 8, there will be one server being assigned to only one user such that not enough cost-effective, so the above cases are not taken into consideration. In Table 7a, we analyze the status of the upper and lower limits of the users when the APs value is 2, 3, 4, 5, 6, or 7. Then, by definition 6, if the APs value is 6, all permutations and combinations for the Cluster Set and User_ids are listed as shown in Table 7b.

The POCA algorithm calculates all solution combinations for n users, r server, its number of calculations is

$$\sum_{1}^{22} C_i^{22} = 4194303$$

When Users = 22, there are 4,194,303 combinations in total. For LAPO, in the case of the same six hosts, simply eight calculation phases is enough to be able to figure out all combination solutions that meet AR Threshold. Table 9 shows the first six records with highest AR average among the thirty three kinds of qualified Cluster sets of combinations.

The best solutions combination is AR = 61.7 Cluster-sets, and the user cluster of the six APs are as follows, {03,04,06,09,11}, {05,12,19,20,22}, {01,10,14,15,17}, {02,07,08}, {13,18}, {16,21}

Table 8, Figure 2 and 3 shows that the LAPO method outperformed the other two methods in both application reusability and hit rate.

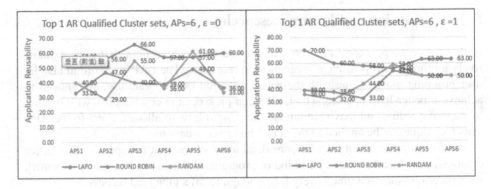

Fig. 2. The Evaluation of Application Reusability

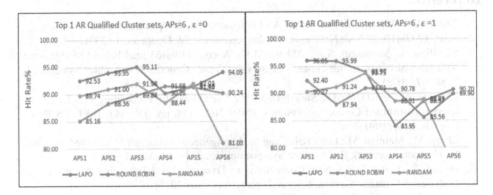

Fig. 3. The Evaluation of hit rate

Table 7. Evaluation Metrics

	The methods		
	LAPO	*Round Robin*	*Random*
	When APs=6 and ε = 0		
Application Reusability (AR)	59.00%	40.67%	42.33%
Hit rate (HR)	**92.89%**	89.42%	89.04%
	When APs=6 and ε = 1		
Application Reusability (AR)	61.67%	44.00%	45.17%
Hit rate (HR)	**91.54%**	89.60%	88.19%

5   Conclusion and Future Research

In this paper we propose LAPO to improve the POCA algorithm that must exhaustively go through all possible combinations of all users. The LAPO algorithm also let user grouping more in line with the actual situation of corporate use, and indeed to achieve a better load balance. The LAPO can pick out the combination with highest average AR value, and may recommend the system how to allocate users to the APs. This technique can be applied to the actual business situations.

For follow-up study, it may be considerable to add weight for specific Tcode, for example, month end process. This kind of program, although not frequently executed, the execution time each time is very long enough to affect the load balance.

References

1. SAP, TADM10_1 : SAP NetWeaver AS Implementation & Operation 1 (2008)
2. SAP, TADM10_2 : SAP NetWeaver AS Implementation & Operation 1 (2008)
3. Sharifian, S., Motamedi, S.A., Akbari, M.K.: A content-based load balancing algorithm with admission control for cluster web servers. Future Generation Computer Systems 24, 775–787 (2008)
4. Krueger, P., Chawla, R.: The Stealth Distributed Schedular. In: Procedding 11th Int'l Conference Distributed Computing Systems, Order No. 2144, pp. 336–343. IEEE CS Press, Los Alamitos (1991)
5. Livny, M., Melman, M.: Load Balancing in Homogeneous Broadcast Distributed Systems. In: Computer Network Performance Symposium, pp. 336–343 (May 1991)
6. Huang, Z., Liang, B.: A New Content-Aware Dynamic Load Balancing Algorithm for Web Server Clusters. Sciverse Science Direct (December 2011)
7. Sriram Iyengar, M., Singhal, M.: Effect of network latency on load sharing in distributed systems. J. Parallel Distrib. Comput. Sciverse Science Direct 66, 839–853 (2006)
8. Ting, P.-H.: A Research of User Distributions in Enterprise Systems. Universal Computer Science 12(2), 160–186 (2006)
9. Song, J., Iyengar, A., Levy-Abegnoli, E., Dias, D.: Architecture of a Web server accelerator. Computer Networks 38, 75–97 (2002)
10. Challenger, J., Dantzig, P., Iyengar, A.: A scalable and highly available system for serving dynamic data at frequently accessed websites. In: Proceedings of ACM/IEEE SC 1998 (November 1998)
11. Challenger, J., Iyengar, A., Dantzig, P.: A scalable system for consistently caching dynamic web data. In: Proceedings of IEEE INFOCOM 1999 (March 1999)
12. Iyengar, A., Challenger, J.: Improving web server performance by caching dynamic data. In: Proceedings of the USENIX Symposiumon Internet Technologies and Systems (December 1997)
13. Maddah, B., El-Taha, M., Tayeh, R.A.: Optimal allocation of servers and processing time in a load balancing system. Computers & Operations Research 37, 2173–2181 (2010)
14. Tari, Z., Broberg, J., Zomayab, A.Y., Baldoni, R.: A least flow-time first load sharing approach for distributed server farm. Science Direct 65, 832–842 (2005)
15. Sharifian, S., Motamedi, S.A., Akbari, M.K.: A predictive and probabilistic load-balancing algorithm for cluster-based web servers. Science Direct, Applied Soft Computing 11, 970–981 (2011)
16. Casalicchio, E., Cardellini, V., Colajanni, M.: Content-aware dispatching algorithms for cluster-based web servers. Cluster Computing 5, 65–74 (2002)

Bat Algorithm, Genetic Algorithm and Shuffled Frog Leaping Algorithm for Designing Machine Layout

Kittipong Dapa, Pornpat Loreungthup, Srisatja Vitayasak,
and Pupong Pongcharoen[*]

Centre of Operations Research and Industrial Applications,
Department of Industrial Engineering,
Faculty of Engineering, Naresuan University, Phitsanulok, Thailand 65000
{kittipongben,mint_lo_ruangsup}@hotmail.com,
{srisatjav,pupongp*}@nu.ac.th

Abstract. Arranging non-identical machines into a limited area of manufacturing shop floor is an essential part of plant design. Material handling distance is one of the key performance indexes of internal logistic activities within manufacturing companies. It leads to the efficient productivity and related costs. Machine layout design is known as facility layout problem and classified into non-deterministic polynomial-time hard problem. The objective of this paper was to compare the performance of Bat Algorithm (BA), Genetic Algorithm (GA) and Shuffled Frog Leaping Algorithm (SFLA) for designing machine layouts in a multiple-row environment with the aim to minimise the total material handling distance. An automated machine layout design tool has been coded in modular style using a general purpose programming language called Tcl/Tk. The computational experiment was designed and conducted using four MLD benchmark datasets adopted from literature. It was found that the proposed algorithms performed well in different aspects.

Keywords: Machine layout, Bat Algorithm, Genetic Algorithm, Shuffled Frog Leaping, Multiple-row, Parameter setting.

1 Introduction

In the context of operations research and computational intelligence, a computational method optimises a problem by iteratively trying to improve a candidate solution considering a given measure of quality. Stochastic search algorithms have been successfully applied to solve various problems especially for which are very large in size but they do not guarantee the optimum solution [1]. The inspirations for the computational intelligence can be categorised into three main groups [2]: physically-based inspiration e.g. as Simulated Annealing [3], Harmony Search [4]; socially-based inspiration such as Tabu Search [5]; and biologically-based inspiration including Genetic Algorithm [6, 7], Ant Colony Optimization [8], Particle Swarm Optimization [9], Shuffled Frog Leaping Algorithm (SFLA) [9], Firefly Algorithm [10], and Bat Algorithm [11].

---

[*] Corresponding author.

S. Ramanna et al. (Eds.): MIWAI 2013, LNCS 8271, pp. 59–68, 2013.

Machine layout design (MLD) is the process of assigning a number of machines into working area in such a way that the efficiency of the layout is maximised. The effective facility layout can help to reduce the production cost by 10-30% [12]. Shorter handling distance of material flow between machines required for manufacturing products leads to quicker transfer time within the shop floor area. MLD problems are classified as a NP-hard problem [13], which means that the amount of computation required to find the solutions increases exponentially with problem size. Solving this kind of problem by full numerical methods especially for a large sized problem can be computationally expensive. If there are n machines, there are n! possible solutions. For example, for a layout design of ten machines, the number of possible solutions is 3,628,800 (10!).

Bat Algorithm (BA), introduced by Yang in 2010, is a metaheuristic search algorithm, inspired by the echolocation behavior of bats with varying pulse rates of emission and loudness. The primary purpose for a bat's echolocation is to act as a signal system to sense distance and hunt for food/prey. BA was claimed to be another powerful algorithm in exploitation using local search [14]. This algorithm has been applied to solve several problems, e.g. scheduling [15], constrained optimisation tasks [16], multi-objective optimisation [17] and nonlinear engineering optimisation problems [18]. Yang (2010) reported that the BA is potentially more powerful than Particle Swarm Optimisation, Genetic Algorithm and Harmony Search. The primary reason is that BA uses combination of the major advantages of the other algorithms in some way. There has also been no report in the international academic databases on the application of the BA for designing machine layout [19].

The objective of this paper is to compare the performance of Bat Algorithm (BA), Genetic Algorithm (GA) and Shuffled Frog Leaping Algorithm (SFLA) for machine layout design in a multiple-row environment with the aim to minimise the total handling distance of material flow between machines. The remaining sections are organised as follows. Section 2 describes non-identical rectangular machine layout design followed by the metaheuristics for solving MLD problem and its pseudo-code in section 3. The experimental results are presented in section 4. Finally, a conclusion is drawn in section 5.

2   Non-identical Rectangular Machine Layout Design

Machines are usually designed in rectangular shape of different sizes and different models. Arranging non-identical rectangular machines in multiple-row environment means that machines are placed row by row within restricted area such as that shown in Fig. 1. Machines are placed in parallel row by row. Flow path means movement of material handling equipment, e.g. automated guided vehicles, which can move to left or right side of the row and then move up or down to the destination row. The distance of material flow was evaluated looking for the shortest distance.

In this work, the following assumptions were made in order to simplify and formulate the problem: i) the material handling distance between machines was determined from the machine's centroid; ii) there were enough sizes of shop floor area for machine arrangement; iii) the movement of AGV was in a straight line; iv) the gap between machines was similar; and v) the quantity of products, processing time and moving time were not taken into consideration.

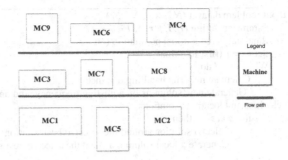

Fig. 1. Example of multi-row MLD modified from Drira et al. [20]

The objective function for this study is to minimise the material handling distance as illustrated in equation (1).

$$Z = \sum_{j=1}^{M} \sum_{i=1}^{M} f_{ij} d_{ij} \tag{1}$$

M is a number of machines, i and j are machine sequences (i and $j = 1, 2, 3,..., M$), f_{ij} is frequency of material flow between machines i and j, d_{ij} is the distance between machines i and j ($i \neq j$).

3 Metaheuristics for Machine Layout Design Problem

A metaheuristic is formally defined as an iterative generation process which guides a subordinate heuristic by combining intelligently concepts for exploring and exploiting the search space, learning strategies are used to structure information in order to find efficiently near-optimal solutions [21]. In this work, three metaheuristics approaches were applied to design the machine layout problem as described in the following subsections.

3.1 Bat Algorithm (BA)

The pseudo-code of the proposed BA for MLD shown in Fig. 2 can be described as follows: i) encode the problem to produce a list of machines using a numeric string (see Fig. 3). Each bat contains a number of machines to be arranged, so that a single bat represents a candidate solution; ii) prepare input data (number of machines: N_m and dimension of machines (width: w x length: l), number of parts: N_p and its machine sequences: S_i), and identify parameters (bat population size: P, number of iterations: I, frequency: f, pulse rate: r, loudness: A, floor length: F_L, floor width: F_W and gap between machines: G); iii) randomly create an initial population based on population size; iv) arrange machines row by row based on F_L and F_W and evaluate the fitness function value; v) adjust frequency and update velocities and locations/solutions; vi) a new solution is generated. If any solution is found with has a better solution, the pulse rate and loudness are updated; vii) select the best bat having the shortest material handling distance; and viii) stop the BA process according to the I. When BA process is terminated, the best-so-far solution is presented.

```
Input problem dataset (Nm , w, l, Np , Si )
      Parameter setting (P, I, f, r, A, FL, Fw, G)
      Randomly create initial population (P)
      Set i  = 1 (first iteration)
      While i  ≤ I do
            Calculate material handling distance
              Generate new solutions by adjusting frequency, and updating
velocities and locations/solutions
            if (rand > ri) then
                  Select a solution around the selected best solution
                  Generate a local solution around the selected best solution
            end if
            Generate a new solution by flying randomly
            If (rand < Ai & f(xi) < f(x∗)) then
                  Accept the new solutions
                  Increase ri and Ai
            end if
            Rank the bats and find the current best x∗
            i = i + 1
      End loop while
Output the best solution
```

Fig. 2. Pseudo-code of BA for MLD problem modified from Yang (2010) [11]

4	8	2	9	5	7	6	3	1	10

Fig. 3. Representation of a candidate solution

3.2   Genetic Algorithm

GA is classified as an evolutionary algorithm and starts with an initial population of random solutions called chromosomes, in which chromosomes are reproduced and mutated. The pseudo-code of the proposed GA for MLD is shown in Fig. 4 and can be described as follows: i) encode the problem to produce a list of genes using a numeric string (see Fig. 3). Each chromosome contains a number of genes, each representing machine number, so that the length of chromosome is equal to the total number of machines that are to be arranged; ii) prepare input data and identify parameters (population size: P, number of generations (iterations): I, probability of crossover: P_c, probability of mutation: P_m, floor length: F_L, floor width: F_W and gap between machines: G); iii) randomly generate an initial population based on population size; iv) apply crossover and mutation operators to generate new offspring respecting P_c and P_m; v) arrange machines row by row based on F_L and F_W; vi) evaluate the fitness function value; vii) select the best chromosome which has the shortest material handling distance using the elitist selection; viii) choose chromosomes for next generation by using the roulette wheel selection [22]; and ix) stop the GA process according to the I. When GA process is terminated, the best-so-far solution is reported.

```
Input problem dataset (N_m, w, l, N_p, S_i)
    Parameter setting (P, I, P_c, P_m, F_L, F_W, G)
    Randomly create initial population (Pop)
    Set i = 1 (first generation)
    While i ≤ I do
        For j = 1 to cross do (cross = round ((P_c x Pop)/2)))
            Crossover operation
        End loop for
        For k = 1 to mute do (mute = round(P_m x Pop))
            Mutation operation
        End loop for
        Arrange machines row by row based on F_L, F_W and G
        Calculate material handling distance
        Elitist selection
        Chromosome selection using roulette wheel method
        i = i + 1
    End loop while
Output the best solution
```

Fig. 4. Pseudo-code of GA for MLD problem modified from Yang (2008) [10]

3.3   Shuffled Frog Leaping Algorithm (SFLA)

Shuffled Frog Leaping Algorithm (SFLA) developed by Eusuff et al. (2003) [9] is a population-based cooperative search metaphor combining the benefits of the genetic-based memetic algorithm and the social behaviour based particle swarm optimisation. The pseudo-code of the proposed GA for MLD shown in Fig. 5 can be described as follows: i) encode the problem to produce a list of genes using a numeric string (see Fig. 3). Each solution refers to a frog consisting of the number of machines that are to be arranged; ii) prepare input data and identify parameters (population of frogs: P, number of generations (iterations): I, number of memplexes: M, iterations for frog improving: F, floor length: F_L, floor width: F_W and gap between machines: G); iii) randomly generate an initial population based on population of frogs; iv) evaluate the fitness function value; v) all frogs are partitioned into a number of parallel subsets referred to as memeplexes. For each memeplex, determine the best and worst frogs and improve the worst frog position; vi) if the worst frog is not improved, the worst frog is replaced with a new frog generated randomly; vii) stop the improving process according to iterations for frog improving and number of memeplex; viii) combine the whole frogs together; and ix) stop the SFLA process according to the I. When SFLA process is terminated, the best-so-far solution is then reported.

All three metaheuristics based machine layout design programs have been developed and coded in modular style using the Tool Command Language and Tool Kit (Tcl/Tk) programming language [24]. This program is user-friendly and designed to redefine the parameters for each computational run. The computational experiments were designed and conducted on personal computer with Intel Core i7 3.4 GHz and 16 GB DDR3 RAM.

```
Input problem dataset (Nm, w, l, Np, Si)
    Parameter setting (P, I, M, I, FL, Fw, G)
    Randomly create initial population of frog (P)
    Set i  = 1 (first generation)
    While i ≤ I do
        Arrange machines row by row based on FL, Fw and G
        Calculate material handling distance
        Divide the population into several memeplexes
        For j = 1 to M
            Determine the best and worst frogs
            Improve the worst frog position x(w)
            If no improvement becomes possible in this case,
                then x(w) is replaced by a randomly generated solution within
                the entire feasible space
        End for
        Combine the evolved memeplexes
        i = i + 1
    End loop while
Output the best solution
```

Fig. 5. Pseudo-code of SFLA for MLD problem modified from Pan et al. 2010 [23]

4   Experimental Results

In this work, the computational experiments were conducted using four MLD benchmark datasets (As shown in Table 1) adopted from literature [25]. These had different sizes according to the number of machines and products. Dataset M10N3 means that there are three products to be processed on ten non-identical rectangular machines.

Table 1. Benchmark datasets

Dataset	Number of machines (M)	Number of products (N)
M10N3	10	3
M20N5	20	5
M15N9	15	9
M30N10	30	10

The number of searches conducted within metaheuristics is one of the key issues on its performance. In order to conduct a fair comparison on the metaheuristics' performance, the total number of searches assigned for the proposed algorithms must be the same. Generally, the number of searches is determined by combination of population size (P) and number of iterations (I). In this work, the total number of searches was set to 2,500. Refer to the previous work [26], the appropriate setting of GA parameters for machine layout design has been reported at $P = 100$, $I = 25$, $P_c = 0.9$ and $P_m = 0.5$. The genetic operators including the Two-point Centre Crossover (2PCX) and Two Operation Random Swap (2ORS) were based on the recommendation by Vitayasak and Pongcharoen (2011) [27]. For SFLA, the best setting of SFLA parameters $(P, I, F$ and $M)$ for machine layout design were suggested to be set at 25, 100, 90 and 6, respectively [28]. The experimental results on investigation of the appropriate setting of BA parameters are described in the subsection below.

4.1   BA Parameter Setting

The aim of this experiment was to investigate the appropriate setting of BA parameters including a combination of population size and the number of iterations (P/I), pulse rate (r) and loudness (A). The rate of pulse can simply be in the range of [0, 1] and the initial loudness is typically around [1, 2] [11]. All BA parameters were investigated at three levels. The experimental design and the range of values considered for each parameter are shown in Table 2. With four datasets, each of which used five replications, and three values of three parameters, a total of 540 computational runs were carried out. The results obtained from the computational experiment were analysed using the analysis of variance (ANOVA) as shown in Table 3, in which the degree of freedom (DF), F value and P value for adjusting the values are given.

Table 2. Experimental factors and its levels

Parameters	Number of levels	Values
P/I	3	25/100, 50/50, 100/25
r	3	0.1, 0.5, 0.9
A	3	1.1, 1.5, 1.9

Table 3. Analysis of variance (ANOVA)

Source	DF	M10N3		M20N5		M15N9		M30N10	
		F	P	F	P	F	P	F	P
P/I	2	11.53	0.000	0.42	0.657	1.18	0.312	12.23	0.000
r	2	0.03	0.966	0.21	0.814	0.02	0.984	0.08	0.926
A	2	16.38	0.000	1.92	0.151	75.02	0.000	44.50	0.000
$P/I*r$	4	0.03	0.998	0.06	0.994	0.03	0.998	0.04	0.996
$P/I*A$	4	5.96	0.000	0.23	0.920	2.40	0.055	0.56	0.689
$A*r$	4	0.03	0.998	0.11	0.978	0.03	0.998	0.07	0.991
$P/I*r*A$	8	0.03	1.000	0.23	0.984	0.02	1.000	0.04	1.000
Error	108								
Total	134								

From the ANOVA table, the loudness (A) was a significant factor for all problems except M20P5 with 95% confident interval by having the P values at less than or equal to 0.05, and the combination of the amount of bats and number of iterations (P/I) has a significant effect on the total material handling distances in M10N3 and M30N10. An interaction between main factors $(P/I*A)$ was statistically significant in M10P3. The appropriate parameter setting for each dataset was shown in Table 4.

Table 4. Appropriate parameter setting

Dataset	P/I	r	A
M10N3	100/25	0.5	1.1
M20N5	25/100	0.9	1.5
M15N9	50/50	0.9	1.5
M30N10	25/100	0.9	1.9

4.2 Performance Comparison

For each dataset, each algorithm was computationally repeated thirty times by adopting its optimised parameter setting. The computational results obtained from the proposed algorithms were analysed in terms of the minimum, maximum, mean and standard deviation (SD) as shown in Table 5. It can be seen that the GA produced the lowest on the average of material handling distance. While the solutions obtained from the SFLA generated the lowest values of SD in almost all datasets. The proposed BA produced the highest mean and standard deviation. The mean computational time required by BA was quicker than SFLA in all datasets. Since, in SFLA, the candidate solutions were divided into several groups (Memeplex) and then the solution's improving process was conducted in each memeplex. This was the reason of the longer execution time but the solution quality was better than those obtained by BA.

Table 5. Comparison of solutions produced by BA, GA [29] and SFLA [28, 30]

Dataset	Algorithm	Value				
		Mean	SD	Min	Max	Time (second)
M10N3	BA	193.84	5.91	186.97	209.97	69
	GA	**187.36**	0.29	186.98	187.58	6
	SFLA	187.92	1.81	186.97	194.07	107
M20N5	BA	1,387.96	57.57	1,250.10	1,473.50	173
	GA	**1,361.16**	49.23	1,231.65	1,448.55	20
	SFLA	1,375.37	38.43	1,312.50	1,434.80	352
M15N9	BA	1,427.78	32.58	1,345.05	1,487.95	108
	GA	**1,382.01**	22.12	1,347.75	1,417.35	30
	SFLA	1,412.38	21.87	1,355.25	1,452.55	365
M30N10	BA	5,013.51	154.78	4,602.37	5,286.27	418
	GA	**4,770.54**	141.75	4,524.43	5,041.18	73
	SFLA	4,884.42	96.28	4,680.48	5,045.23	837

Using the Student's t-test, there were statistically significant differences in mean of total cost of BA and the other algorithms as shown in Table 6 except M20N5. However, the minimum value obtained from all datasets was lower than SFLA except M10N3, and the SFLA took more execution time than the BA.

Table 6. P-value of Student's t-test

Dataset	BA&GA	BA&SFLA	GA&SFLA
M10N3	0.000	0.000	0.104
M20N5	0.058	0.324	0.218
M15N9	0.000	0.036	0.000
M30N10	0.000	0.000	0.001

5   Conclusions

This paper presents the performance comparison of Bat Algorithm (BA), Genetic Algorithm (GA) and Shuffled Frog Leaping Algorithm (SFLA) for machine layout design in a multiple-row environment. The algorithms were aimed to minimise the total material handling distance. The computational experiment was based on four benchmark datasets adopted from the literature. The experimental results indicated that the BA parameters setting had an influence on its performance. The GA performance is the best in terms of mean material handling distance and computational time. While the SFLA produced solutions with the lowest values of SD in almost all datasets. The BA can solve MLD problems with quicker average computational times than SFLA in all datasets. Future research should be focused on investigating the mechanism of generating the new solutions and adjusting BA parameters during the searching process to improve the solution quality.

Acknowledgements. This work was part of the research project supported by the Naresuan University Research Fund under the grant number R2556C091.

References

1. Pongcharoen, P., Hicks, C., Braiden, P.M., Stewardson, D.J.: Determining optimum Genetic Algorithm parameters for scheduling the manufacturing and assembly of complex products. International Journal of Production Economics 78, 311–322 (2002)
2. Pongcharoen, P., Chainate, W., Pongcharoen, S.: Improving Artificial Immune System Performance: Inductive Bias and Alternative Mutations. In: Bentley, P.J., Lee, D., Jung, S. (eds.) ICARIS 2008. LNCS, vol. 5132, pp. 220–231. Springer, Heidelberg (2008)
3. Kirkpatrick, S., Gelatt, C.D., Cevvhi, M.P.: Optimization by Simulated Annealing. Science New Series 220, 671–680 (1983)
4. Geem, Z.W., Kim, J.H., Loganathan, G.V.: A new heuristic optimization algorithm: Harmony search. Simulation 76, 60–68 (2001)
5. Glover, F.: Tabu search: Part I. ORCA Journal on Computing 1, 190–206 (1989)
6. Goldberg, D.E.: Genetic algorithms in search, optimization and machine learning. Addison-Wesley Publishing Company, Inc. (1989)
7. Holland, J.H.: Outline for a logical theory of adaptive systems. Journal of ACM 3, 297–314 (1962)
8. Dorigo, M.: Optimization, Learning and Nature Algorithms. PhD, Politecnico di Milano, Italy (1992)

9. Eusuff, M.M., Lansey, K.E.: Optimization of water distribution network design using the Shuffled Frog Leaping Algorithm. Journal of Water Resources Planning and Management-Asce 129, 210–225 (2003)
10. Yang, X.-S.: Nature-Inspired Metaheuristic Algorithm. Luniver Press (2008)
11. Yang, X.-S.: Nature-Inspired Metaheuristic Algorithm, 2nd edn. Luniver Press (2010)
12. Tompkins, J.A., White, J.A., Bozer, Y.A., Tanchoco, J.M.A.: Facilities Planning, 4th edn. John Wiley & Sons, Inc. (2010)
13. Loiola, E.M., de Abreu, N.M.M., Boaventura-Netto, P.O., Hahn, P., Querdo, T.: A survey for the quadratic assignment problem. European Journal of Operational Research 176, 657–690 (2007)
14. Wang, G., Guo, L.: A novel hybrid bat algorithm with harmony search for global numerical optimization. Journal of Applied Mathematics 2013 (2013)
15. Musikapun, P., Pongcharoen, P.: Solving multi-stage multi-machine multi-product scheduling problem using Bat Algorithm. In: 2nd International Conference on Management and Artificial Intelligence, Thailand (2012)
16. Gandomi, A.H., Yang, X.S., Alavi, A.H., Talatahari, S.: Bat algorithm for constrained optimization tasks. Neural Computing and Applications, 1–17 (2012)
17. Yang, X.S.: Bat algorithm for multi-objective optimisation. International Journal of Bio-Inspired Computation 3, 267–274 (2011)
18. Yang, X.S., Gandomi, A.H.: Bat algorithm: A novel approach for global engineering optimization. Engineering Computations (Swansea, Wales) 29, 464–483 (2012)
19. Vitayasak, S.: Facility layout problem: a 10-year review and reserach perspectives. Naresuan Unversity Engineering Journal 5, 46–62 (2010) (in Thai)
20. Drira, A., Pierreval, H., Hajri-Gabouj, S.: Facility layout problems: A survey. Annual Reviews in Control 31, 255–267 (2007)
21. Osman, I.H., Laporte, G.: Metaheuristics: a bibliography. Annual of Operations Research 63 (1996)
22. Gen, M., Cheng, R.: Genetic Algorithms and Engineering Design. John Wiley & Sons, Inc. (1997)
23. Pan, Q.-K., Wang, L., Gao, L., Li, J.: An effective shuffled frog-leaping algorithm for lot-streaming flow shop scheduling problem. International Journal of Advanced Manufacturing Technology (2010)
24. Ousterhout, J.K.: Tcl and Tk tookit, 2nd edn. Addison Wesley (2010)
25. Nearchou, A.C.: Meta-heuristics from nature for the loop layout design problem. International Journal of Production Economics 101, 312–328 (2006)
26. Vitayasak, S.: Multiple-row rotatable machine layout using Genetic Algorithm. Research report, Naresuan Univeristy, Phitsanulok, Thailand (2011) (in Thai)
27. Vitayasak, S., Pongcharoen, P.: Interaction of crossover and mutation operations for designing non-rotatable machine layout. In: Operations Research Network Conference, Bangkok, Thailand (2011)
28. Leechai, N., Iamtan, T., Pongcharoen, P.: Comparison on Rank-based Ant System and Shuffled Frog Leaping for design multiple row machine layout. SWU Engineering Journal 4, 102–115 (2009)
29. Singpraya, S., Dedklen, S., Vitayasak, S., Pongcharoen, P.: Adaptive Genetic Algorithm for designing non-identical rectangular machine layout. In: Operations Research Network Conference, Thailand (2012)
30. Iamtan, T., Pongcharoen, P.: Swap and adjustment techniques in Shuffled Frog Leaping algorithm for solving machine layout. Thai VCML Journal 3, 25–36 (2010)

Hue Modeling for Object Tracking in Multiple Non-overlapping Cameras

Minho Han* and Ikkyun Kim

Cyber Security Research Department, ETRI,
138 Gajeongno Yuseong-gu, Daejeon, 305-700, Korea
{mhhan,ikkim21}@etri.re.kr

Abstract. Images collected through CCTV can be expressed using an RGB color channel. As a result, studies on object tracking using a color histogram consisting of an RGB color channel are currently underway. However, the color histogram of the same object may be displayed differently in different cameras when installed under variable lighting conditions. Under such an environment, object tracking using a color histogram cannot be applied. To resolve this problem, we propose the use of hue modeling to find the same object in multiple non-overlapping cameras located at different physical sites.

Keywords: video surveillance, object tracking, multiple non-overlapping cameras.

1 Introduction

Tracking objects across multiple cameras is a challenging area of research in visual computing, especially when the cameras have non-overlapping fields of view. For economic reasons, most cameras do not have an overlapping field of view, which makes tracking objects of interest a difficult task as they move around. In the following section, we discuss the research related to this issue. In section 3, we present a method for finding the same object in multiple non-overlapping cameras located at different physical sites. Finally, in section 4, we discuss our experiments, which validate the proposed approach.

2 Related Work

In multiple non-overlapping cameras $(C_1, C_2, ..., C_n)$ located at different physical sites, to re-establish a match of the same object O, the color histogram H_i of camera C_i, and the color histogram H_j of camera C_j, must be similar. However, the different lighting conditions between camera views (a problem of changes in illumination between camera views) generate an unexpected situation,

---

* This work was supported by the IT R&D program of the Ministry of Knowledge Economy, Korea. [KI002140, "Development of Video Surveillance Security Technologies for Preserving Personal Security"].

S. Ramanna et al. (Eds.): MIWAI 2013, LNCS 8271, pp. 69–78, 2013.

such as $H_i{\neq}H_j$, for the same object. Figure 1 shows example images collected from a public residential area CCTV network used for our experiments. Simple methods such as a model function [1], mean bright transfer function (BTF) [2], and cumulative BTF [3] have been proposed to handle changes in lighting condition between cameras. In other words, $H_i = f(H_j)$ has been established as a way to deal with such changes. However, these methods have the following disadvantages.

- A training phase to create a color mapping function (f) is required prior to operation. In other words, a preparation (calibration) process using training data is needed.

- If $H_1 = fa(H_2)$ and $H_2 = f_b(H_3)$, $H_1 = f_a (f_b(H_3))$ is established, and in the case of N cameras, N-1 mapping functions (f) are needed.

- After the mapping function (f_i) for camera C_i is computed, if the lighting conditions of camera C_i change, the mapping function (f_i) for camera C_i must be recomputed. In other words, it is impossible to apply (f) for a camera with frequent changes in brightness.

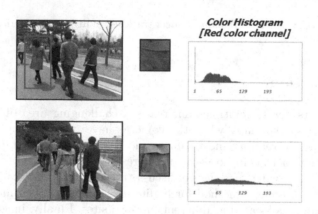

Fig. 1. Images and color histograms of the same person in two non-overlapping cameras located at different physical sites

3   Hue Modeling

Hue is an angle of 0 to 360 degrees: red is typically 0 degrees; yellow, 60 degrees; green, 120 degrees; cyan, 180 degrees; blue, 240 degrees; and magenta, 300 degrees. In other words, hue is represented by a value between 0 to 359 degrees. Figure 2 shows a color range based on the hue value [4].

Fig. 2. Colors based on hue values

Red, green, and blue (RGB) colors can be converted into hue, saturation, and intensity (HSI) through the following formula [5]:

$$H = \cos^{-1}\left[\frac{\frac{1}{2}([(R-G)+(R-B)]}{\sqrt{(R-G)^2+(R-B)(G-B)}}\right] \tag{1}$$

$$S = 1 - \frac{3}{R+G+B}[\min(R,G,B)] \tag{2}$$

$$I = \frac{1}{3}(R+G+B) \tag{3}$$

If we compare the same object viewed from cameras installed at different locations, we can see that several changes in the RGB values have occurred. However, the hue values calculated using Eq. (1) show almost no changes. Figure 3 shows an example of the changes in both RGB and hue values based on the changes in illumination. This chapter describes how to detect the same person in multiple cameras using robust hue characteristics under varying lighting conditions.

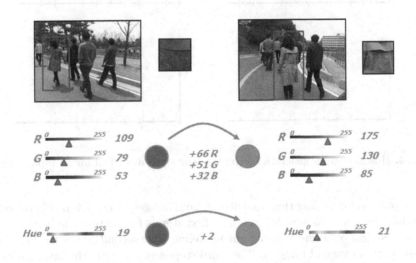

Fig. 3. Changes in RGB and hue values

Four steps are used in the hue modeling of an image. The first step is to partition the hue space into a small number of hues. The second step is to assign one of these hues to each pixel in the image. Prior to implementing the second step, the RGB values must be converted into HSV values using Eqs. (1) through

(3). The third step is to select partitions for the average value. The fourth step is to group adjacent partitions and compute the center value of each partition. Figure 4 shows a hue histogram made using these four steps.

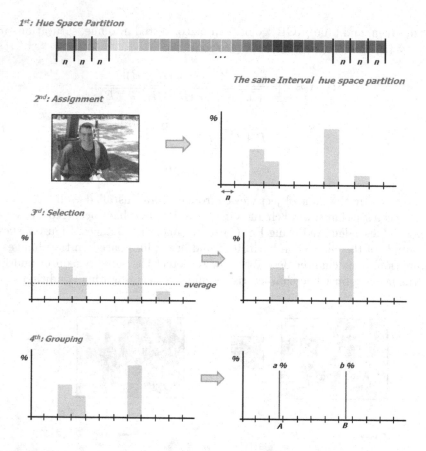

Fig. 4. Hue histogram generated using the four steps described for an input image

The process for finding the same object from images (images 1 and 2) collected from different cameras is as follows. The first step is to create a hue histogram of each image (images 1 and 2) using hue modeling. The second step is to extend the hue histogram range of image 1. The third step is to compare the hue histogram of image 2 using the extended hue histogram of image 1. If the hue histogram of image 2 is in the extended hue histogram of image 1, the two images are similar. Figure 5 shows the process used to find the same object from two images using these three steps.

The hue modeling and process presented above for finding the same object are more efficient in cameras installed under variable lighting conditions. An example of this can be seen in Fig. 6, which shows a sample taken from two

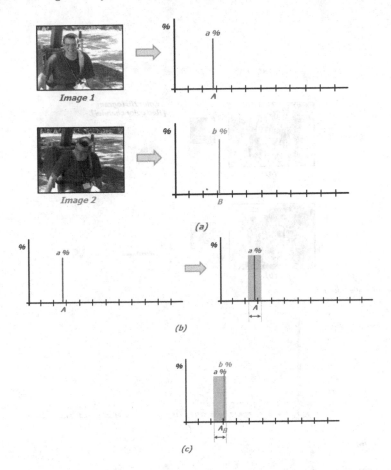

Fig. 5. Process used for finding the same object from two images: (a) step 1, creation of a hue histogram of each image using hue modeling; (b) step 2, extension of the hue histogram range of image 1; and (c) step 3, a comparison of the two histogram ranges

non-overlapping cameras located at different physical sites. Objects O_a and O_b shown in the red lines of Figs. 6(a) and 6(b), respectively, have different colors under variable lighting conditions, but are the same object.

However, the use of hue alone without the use of saturation and intensity can sometimes cause a false positive to occur (i.e., the hue histogram generates similar results for different objects). To prevent a false positive, the saturation and intensity characteristics can be applied. Even if two objects have a similar hue histogram, if the saturation and intensity difference between the two objects is above a certain level, the two objects are different. If so, the method used to set the range of saturation and intensity difference between the two objects should be determined. In this paper, when the saturation and intensity differences are 16 and 32 or more, respectively, we determine that the two objects are different. At this point, the range of saturation and intensity value are from 0 to 255.

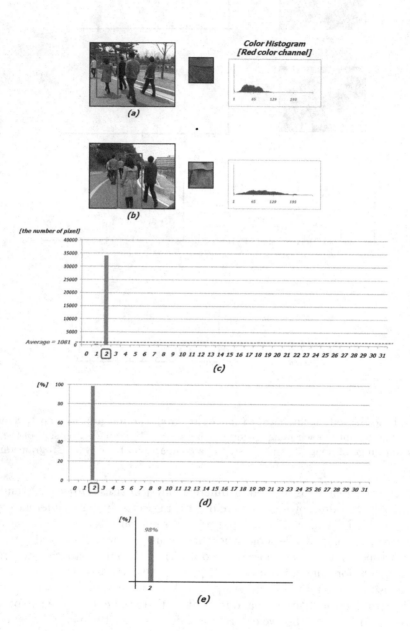

(a)

Color Histogram
[Red color channel]

(b)

[the number of pixel]

Average = 1081

(c)

[%]

(d)

[%]

98%

2

(e)

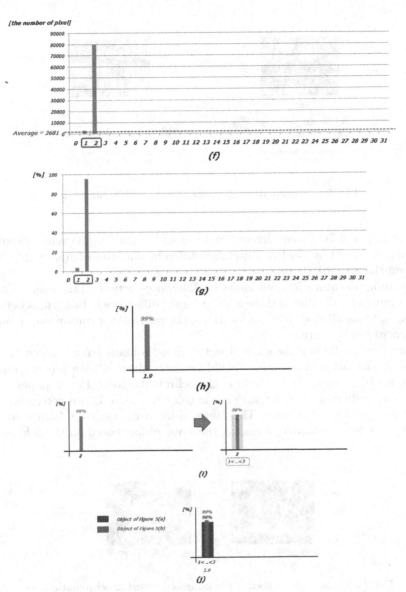

Fig. 6. (a), (b) Images and color histograms for the same person in two non-overlapping cameras located at different physical sites. (c) Each pixel of object O_a in (a) is assigned to one hue partition. In this example, the number of partitions is 32. In other words, n (the size of the partition) is 8. (d) and (e) are the steps used to select partitions over an average value, to group adjacent partitions and to compute the center value of the partition. (f) Each pixel of object O_b in (b) is assigned to a hue partition. (g) and (h) are the steps used to select partitions over an average value, to group adjacent partitions, and to compute the center value of partition. (i) is a step used to extend the hue histogram range of object O_a in Fig. 6(a). In this example, the range of extension is -1 to +1. (j) Because the hue histogram of object O_b is in the extended hue histogram of object O_a, the two objects have a similarity of 98% or more.

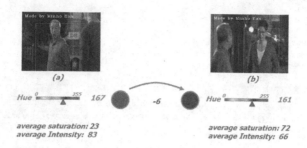

Fig. 7. An example of different objects with similar hues

Figures 7(a) and 7(b) show different objects with a similar hue value. However, we can determine that the two objects are different since the saturation difference between the two objects is 49.

The range of saturation and intensity differences between two objects do not always have a fixed value. If there is a significant difference in brightness between the cameras installed at different locations, the range of saturation and intensity differences can be increased.

A method for finding the same object with achromatic color without the use of hue or saturation information should be determined [6]. We propose finding the same object using the intensity of the achromatic color. In this paper, when the intensity difference of the achromatic color is within 32, we determine that the two objects are the same. The range of intensity values is from 0 to 255. Figure 8 shows an example of finding the same object based on the achromatic color.

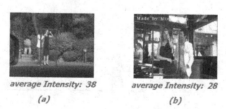

Fig. 8. An example of finding the same object using achromatic color

4   Experiments

In this section, we present the results of the proposed method using actual non-overlapping cameras. The first experiment was conducted using five cameras in an outdoor environment. The camera topology used is shown in Fig. 9. The purpose of the experiment was to find the same object correctly when it moves across cameras (Cam 1 through Cam 5). The first camera (Cam 1) created a hue histogram for the object using hue modeling. The four other cameras (Cam 2

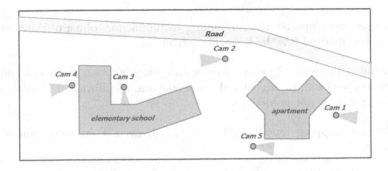

Fig. 9. Camera configuration used in the experiments

(a) Scenario 1: Cam 1   *(b) Scenario 1: Cam 2*   *(c) Scenario 1: Cam 3*

(d) Scenario 1: Cam 4   *(e) Scenario 1: Cam 5*

Fig. 10. (a) through (e) show the same object. The red box in (b) through (e) indicates the same found object. The results show an accuracy of 98% or more.

(a) Scenario 2: Cam 1   *(b) Scenario 2: Cam 2*

Fig. 11. (a) and (b) show different objects. The results of this experiment are close to 0% in terms of accuracy.

through Cam 5) then found the same object using the hue histogram created by Cam 1. Figure 10 shows an example of matching an object using the proposed method.

A second experiment was conducted using two cameras (Cam 1 and Cam 2), as shown in Fig. 9. The purpose of the experiment was to distinguish between different objects of a similar color. Figure 11 shows an example of an unmatched object using the proposed method.

5   Conclusions

In this paper, we proposed a hue modeling technique for robust object tracking. The proposed method has the following advantages.

– A training phase for creating a color mapping function (f) is not required prior to operation. In other words, a preparation (calibration) process using training data is not needed.

– Only one mapping function (f) is required, even if multiple cameras are present.

– Even when the lighting conditions of camera C_i change, the mapping function (f) for camera C_i does not need to be recomputed. In other words, it is possible to apply (f) for a camera with frequent changes in brightness.

Our method for finding the same object in multiple non-overlapping cameras under variable lighting condition has been confirmed to work well. However, if objects with the same color appear in multiple cameras simultaneously, a method for selecting and tracking one of the objects is required. To resolve this problem, we plan to add the spatio-temporal information of the object.

References

1. Porikli, F.: Inter-Camera Color Calibration by Correlation Model Function. In: IEEE Int. Conf. on Image Processing (2003)
2. Javed, O., Shafique, K., Shah, M.: Appearance Modeling for Tracking in Multiple Non-overlapping Cameras. In: CVPR, vol. 2, pp. 26–33 (2005)
3. Prosser, B., Gong, S., Xiang, T.: Multi-camera Matching using Bi-Directional Cumulative Brightness Transfer Function. In: British Machine Vision Conference (BMVC) (2008)
4. Adventures in HSV Space, http://www.buena.com/article.html
5. Foley, van Dam, Fiener, Hughes: Computer Graphics. Addison-Wesley (1997)
6. Fairchild, M.D.: Color Appearance Models. Addison-Wesley (1997)
7. Stone, M.C.: A survey of color for computer graphics, in Proc. Ann. In: Conf. Special Interest Group on Graphics and Interactive Techniques (2001)

Using HMMs and Depth Information
for Signer-Independent Sign Language Recognition

Yeh-Kuang Wu, Hui-Chun Wang, Liung-Chun Chang, and Ke-Chun Li

Institute for Information Industry, Tanshui, Taipei, Taiwan

Abstract. In this paper, we add the depth information to effectively locate the 3D position of the hands in the sign language recognition system. But, the information will be changed by the different testers and we can't do the recognition well. So, we use the incremental changes of the three-dimensional coordinates on a unit time as the feature parameter to fix the above problem. We record the changes of the three-dimensional coordinates in time, then using the hidden Markov models to recognize the variety of sign language movement changing on the time domain. Experiment verifies the proposed method is superior to traditional ones.

1 Introduction

Gesture is a nonverbal communication way. It can generally be classified to several types as follows: joining some auxiliary gesture in a dialogue, manipulation gesture and communication gesture. In communication gesture, the sign language is considered that is the most variable and has the integral structure [1]. Stokoe et al. [2], define four parts of sign language in combination: the shape of hands, position, the moving direction and trajectory. The four parts are classified to two categories: hand gestures and space gestures [1]. Hand gestures include the shape of hands and the moving direction; Space gestures include the position and trajectory. Most researches obtain this information by tracking the position of hands with skin-color detection, or wearing the specific color gloves [3] [4]. For example, Koki Ariga et al. [3] use the HMMs as recognizer and skin-color as feature to detect face and hands. Then they use the centroid of face with K-means as the reference point to obtain 2D dimensional coordinates of hands' centroid. Also, they include the 1st-order and 2nd-order differential coefficients of the hands' coordinates, which called Dynamic Feature; M.Mohandes et al. [4] apply Gaussian Skin Color Model and the Region-growing Technique to track the face position of signers. Moreover, the tester's hands respectively wear the yellow and orange glove to obtain higher recognition rate. However, background noise will easily affect the accuracy of feature extraction. Since, using colors as features will be influenced by similar color in background and the light changing in environment.

In 2011, Microsoft released Kinect. This equipment has a RGB camera, a depth sensor and a multi-array microphone. Thus, it can track the action of the players, who can interact with the game station by the motion of the body and the voice. Sign language is also according to the difference of the facial expressions with the variety

S. Ramanna et al. (Eds.): MIWAI 2013, LNCS 8271, pp. 79–86, 2013.

of body gesture to expresses the vocabulary and grammar. Simon Lang et al. [5], use Kinect to obtain body skeleton and depth information and combine the HMMs to recognize the Deutsche Gebärdensprache Sign Language (DGS). They try to raise the recognition rate by including the depth information, but the features they choose is not satisfied. That they can't achieve the higher recognition rate in Signer Independent experiments.

Also, W.Gao et al. [6][7][8] use a CyberGlove and three-dimensional position tracker released by Pohelmus to obtain the hands' features. Then integrate the Simple Recurrent Network (SRN) and Hidden Markov Models (HMMs) to establish the recognition system for Chinese Sign Language (CSL). In [8], W.Gao et al. apply the Self-Organizing Feature Maps (SOFMs) and HMMs, to raise the recognition rate with Self-adjusting Recognition Algorithm. The three-dimensional position can be correctly detected with the equipment, but the price is too high and it is hard to set up. So, it doesn't suit for the general users.

In sign language recognition, feature extraction and recognition are the most important. In this paper, we use Kinect to capture the features, to avoid lighting effect of environment. We can easily turn the feature information from dimensional coordinates to three-dimensional coordinates with infrared depth sensor. Owing to the underflow effect of HMMs, we apply HMMs which add Scaling factor [9][10]. Therefore, we can get much higher recognition rate.

2   Proposed Method

The organization of the proposed method is as follows. We separate the system into two parts. First, we extract the features by Kinect. We use the middle software NITE developed by PrimeSense to obtain the human skeleton. Then we normalize the coordinates to get the observation sequence. Second, we apply the Hidden Markov Models as the recognizer to identify the sign language. The Figure 1 summarizes the method in a system flow chart.

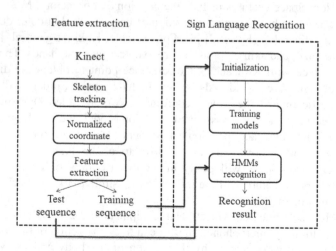

Fig. 1. The system flow chart

2.1   Feature Extraction

In this section, we first apply Kinect to track the signer's skeleton, then we will normalize the coordinates. We use the Homogeneous coordinates to replace the skeleton coordinate system. In the practical application, the user would be at different position. So, it would make the feature we captured unstable. In this system, we do the geometric conversion to the Cartesian coordinate system for setting up the torso center as the coordinate origin. The Figure 2 shows the geometry conversion in 3D Cartesian coordinates; The Figure 3 shows the torso as the origin of the coordinate system [11]; The Figure 4 shows the normalization of Kinect skeleton coordinates.

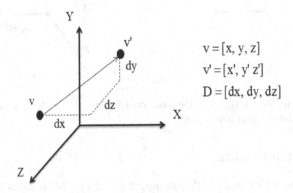

$$v = [x, y, z]$$
$$v' = [x', y' \, z']$$
$$D = [dx, dy, dz]$$

Fig. 2. Geometry conversion in the 3D Cartesian coordinates

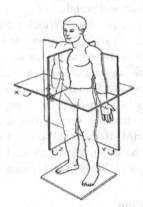

Fig. 3. The torso as the origin of the coordinate system [11]

After that, we use the characteristic of Kinect which can build the three-dimensional space to capture the three-dimensional coordinates of the human joint points. In this paper, we consider the Signer-Independent experiments. Therefore, choosing the feature is important to us. At first, we choose the absolute distance between the three-dimensional coordinates of the joint points and each of the joint points to be our features. But in the signer-independent experiment, this parameter will decrease the ability of recognition. And we found out that the value of this features will be affected by the size of the signers. The above features are very close with the position in the space. So, we use the variation of the movement as our feature. It means that we use the X-axis, Y-axis and Z-axis coordinates offset to be our feature. Then we can solve the recognition problem caused by the different signers.

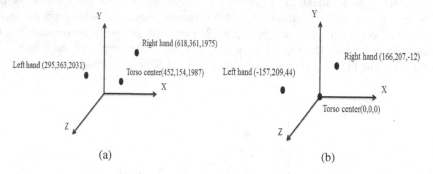

Fig. 4. This figure shows that (a) Original coordinates of the Kinect skeleton and (b) The skeleton coordinates after normalizing

2.2   HMMs Model Training

In this section, we will introduce the training method of HMMs model. In this paper, we use the continuous HMMs model. Therefore, we first initialize the parameters including mean vector, covariance and weight coefficient. After that, we can start to train the HMMs models. We will apply Baum-Welch method for training to adjust the parameters to get the maximum. We set up 5 signers and 20 words for training. Each signer can produce 10 sets of observation sequence to 20 words. And 20 words can train 20 HMMs models. The training data of every model respectively is the 6 sets of observation sequence. The others are for testing. In Singer-Independent experiment, we follow the above mentioned way to train the 5 signers in turn. Otherwise, the initialized data we use is the training data. After all, we can obtain a set of the initialized parameter from HMMs model. We can input this parameters and the training data, then we through the Baum-Welch method to produce a new parameter for recognizing. The Figure 6 is the HMMs model training flow chart.

2.3   Sign Language Recognition

The most difficult thing in sign language recognition is to recognize the same movement. Even we ask the same signers to do the same sign language; they may not do exactly the same. So, for solving this problem to get the higher recognition rate is how to calculate the statistical variations. In this system, we change the mean to the mean vector and the covariance to the covariance matrix. This solution is Continuous Hidden Markov Model. We have to initialize the mean vector, the covariance matrix and the weight coefficient. Also we need to re-estimate. After that, we use Forward-backward Algorithm to calculate the probability of the testing sequence in the HMMs

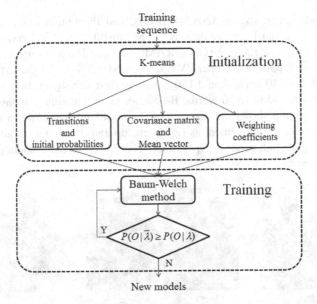

Fig. 5. HMMs model training flow chart

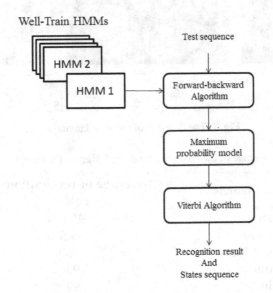

Fig. 6. The recognition flow chart

models. Then we apply Viterbi Algorithm to search the state sequence corresponded to the testing sequence in the model. Therefore, the model and the state sequence are the recognition results. The Figure 7 is the recognition flow chart.

3   Experiment Result

The input depth image size is 320x240 pixels, and the output color image size is 640x480 pixels. The hardware we used is the Microsoft Kinect and computer CPU of Intel ® Core (TM) i5-2400M 3.1GHz, RAM 3.49GB. The software we used is the Microsoft Visual Studio 2010, openCV2.3, OpenNI 1.5.4.0. In the experiment, we set up the Kinect at 140 cm high and 150 cm away from the signer. In our experiment, we invite 5 signers and have 20 words. Below we show our sign language movement in chart. The Figure 8 shows the diagram of the sign language. And in the Figure 8, the red line presents the movement as right, left, up and down. The table 1 shows the results of the Signer-Independent experiment. Therefore, we can calculate the average of the recognition rate is 97 %.

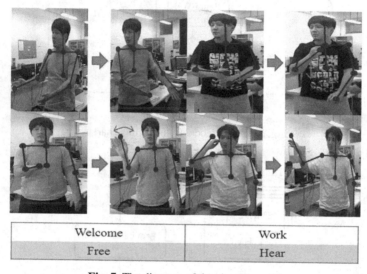

Welcome	Work
Free	Hear

Fig. 7. The diagram of the sign language

Table 1. The average of the recognition rate in 5 signers for Singer-Independent

Words	Average of recognition rate
Work	95.5 %
No good	94.5 %
Do not know	98 %
Center	93 %
Hello	97 %
Free	98 %
Patience	98.5 %
Photograph	94%

Table 1. (*continued*)

Service	98.5 %
Know	99.5%
Consult	98 %
Visit	97 %
Enter	97.5 %
Open	98.5%
Sorry	99.5 %
Excuse me	100 %
Thank you	94 %
Welcome	99 %
Hear	95%
Toilets	97 .5%
Mean	97 %

4   Conclusions

In this paper, we use the depth information obtained from Kinect and combine the skeleton tracking system by OpenNI to capture the human skeleton coordinates. And we get the feature parameters through the simple algorithm for the system. We apply the HMMs to conduct the independent sign language recognition experiment. Since we use the feature parameters have low correlation with the space position to reduce the problem about the size of the signers. Also we solve the problem of the different postures. Therefore, we can increase the recognition rate.

We hope that we can join the feature information of the speed and the direction, etc. And we cannot be affected by the signer's habit or the speed of the sign language movement. Moreover, we hope that we can recognize the much fine movement to increase the number of words. In addition to using the HMMs as the recognizer, we can combine another algorithm. So that, we can make our system more complete and apply more widely.

Acknowledgment. This study is conducted under the "III Innovative and Prospective Technologies Project" of the Institute for Information Industry which is subsidized by the Ministry of Economy Affairs of the Republic of China.

References

1. Kelly, D., McDonald, J., Markham, C.: Recognizing Spatiotemporal Gestures and Movement Epenthesis in Sign Language. In: IMVIP 2009. IEEE Computer Society, Washington, DC (2009) ISBN: 978-0-76953-769-2

2. Stokoe Jr., W.C.: Sign Language Structure:An Outline of the Visual Communication Systems of the American Deaf. Journal of Deaf Studies and Deaf Education 10(1), 3–37 (2005)
3. Ariga, K., Sako, S., Kitamura, T.: HMM-based Sign Recognition in Consideration of Motion Diversity. In: IUCS 2010 (2010)
4. Mohandes, M., Deriche, M., Johar, U., Ilyas, S.: A signer-independent Arabic Sign Language recognition system using face detection, geometric features, and a Hidden Markov Model. Computers and Electrical Engineering 38, 422–433 (2012)
5. Lang, S., Block, M., Rojas, R.: Sign Language Recognition Using Kinect. In: Rutkowski, L., Korytkowski, M., Scherer, R., Tadeusiewicz, R., Zadeh, L.A., Zurada, J.M. (eds.) ICAISC 2012, Part I. LNCS, vol. 7267, pp. 394–402. Springer, Heidelberg (2012)
6. Fang, G., Gao, W.: A SRN/HMM System for Signer-independent Continuous Sign Language Recognition. In: Proceedings of the Fifth IEEE International Conference on Automatic Face and Gesture Recognition (FGR 2002) (2002)
7. Wang, C., Gao, W., Shan, S.: An Approach Based on Phonemes to Large Vocabulary Chinese Sign Language Recognition. In: Proceedings of the Fifth IEEE International Conference on Automatic Face and Gesture Recognition (FGR 2002) (2002)
8. Fang, G., Gao, W., Ma, J.: Signer-Independent Sign Language Recognition Based on SOFM/HMM. In: Proceedings of the IEEE ICCV Workshop on Recognition, Analysis, and Tracking of Faces and Gestures in Real-Time Systems, p. 90 (2001)
9. Rabiner, L.R.: A Tutorial on Hidden Markov Models and Selected Applications in Speech Recognition. Proceedings of the IEEE 77(2), 257–286 (1989)
10. Rahimi, A.: "An Erratum for 'A Tutorial on Hidden Markov Modelsand Selected Applications in Speech Recognition", website of Ali Rahimiat MIT. MediaLaboratory (2000), http://xenia.media.mit.edu/~rahimi/rabiner/rabinererrata/rabiner-errata.html
11. Hsieh, C.-T., Chung, R.-C.: Physical rehabilitation assistant system based on Kinect. In: Proceedings of 2012 National Symposium on System Science and Engineering National Taiwan Ocean University, Keelung (2012)

Unscented Kalman Filter for Noisy Multivariate Financial Time-Series Data

Said Jadid Abdulkadir and Suet-Peng Yong

Department of Computer and Information Sciences,
Universiti Teknologi PETRONAS,
31750 Tronoh, Perak, Malaysia
jadid86@gmail.com, yongsuetpeng@petronas.com.my

Abstract. Kalman filter is one of the novel techniques useful for statistical estimation theory and now widely used in many practical applications. In this paper, we consider the process of applying Unscented Kalman Filtering algorithm to multivariate financial time series data to determine if the algorithm could be used to smooth the direction of KLCI stock price movements using five different measurement variance values. Financial data are characterized by non-linearity, noise, chaotic in nature, volatile and the biggest impediment is due to the colossal nature of the capacity of transmitted data from the trading market. Unscented Kalman filter employs the use of unscented transformation commonly referred to as sigma points from which estimates are recovered from. The filtered output precisely internments the covariance of noisy input data producing smoothed and less noisy estimates.

Keywords: multivariate, noise, sigma points.

1 Introduction

Financial planning is the manner in which companies envision and strategize for the future, by making marketing decisions based on current and past stock prices. Planning is a dynamic process and perplexing task in the financial division for numerous reasons. First, it helps financial market analysts to evade stock trading losses and obtain huge profits by coming up with promising business policies. Hence, financial companies can make precise predictions by being able to "see what interventions are required to meet their business performance targets" [1].

Secondly, the shareholders who are commonly referred to as stockholders have expectations which are the main cause for concern among trading companies because there are usually scrutinized by short and long term shareholders. Stockholders analyse their investments which are in the form of shares by comparing their analysis from the forecaster's analysis. This scrutiny has an effect on any financial institution because it leads to the lack of trust on any failed financial model presented by financial institutions.

An example of financial time-series is stock market indices, nevertheless the process of forecasting stock market indices is met with numerous difficulties

S. Ramanna et al. (Eds.): MIWAI 2013, LNCS 8271, pp. 87–96, 2013.

which are obtained by the continuous fluctuations in the daily trading market. Financial data are characterized by non-linearity, noise, chaotic in nature and volatile thus making the process of planning cumbersome. The biggest impediment is due to the colossal nature of the capacity of transmitted data from the trading market [2].

Precise timing of buying and selling stock shares is very important because of the continuous variations of stock prices within the stock exchange. Market volatility is a regular occurrence in any stock market hence in order to avoid the regular instability, accurate timing is important [3]. The main difficulty of stock data is the size of data obtained from daily intervals which are massive hence having a direct impact on the ability of performing predictions through planning.

The intricate nature of stock market planning and prediction has led to the need for further improvements in the use of intelligent prediction techniques that would drastically decrease the dangers of inaccurate decision making [4]. Predictors goal is to develop and change numerous techniques that can effectively forecast stock indices with one main objective of following legal trade strategies and avoiding losses. "The central idea to successful stock market prediction is achieving best results using minimum required input data and the least complex stock market model" [5].

2   Time-Series Data

Time series is a data order spaced in defined time intervals [1]. Examples include KLCI sectoral indices, trading returns, consumer returns etc. The data orders are in time intervals which are graphically presented using line charts. Time-series data are commonly used in measurements [6], signal [7], recognition [8], economics [9], agricultural forecasting [10], electroencephalography [11], control engineering, communications engineering [12] etc.

Time series data have a natural temporal ordering. This makes time series analysis distinct from other common data analysis problems, in which there is no natural ordering of the observations. Time series analysis is also distinct from spatial data analysis where the observations typically relate to geographical locations. A stochastic model for a time series will generally reflect the fact that observations close together in time will be more closely related than observations further apart [13]. In addition, time series models will often make use of the natural one-way ordering of time so that values for a given period will be expressed as deriving in some way from past values, rather than from future values. Additionally methods of time series analysis may be divided into linear and non-linear, univariate and multivariate [14].

2.1   Multivariate Time-Series Data

In multivariate time-series data, several variables are involved in the input, and the output is influenced by these input variables which themselves are time variant. Multivariate time series analysis are used to establish cause and effect relationships between dependent and independent variables [15].

The practical implementation of multivariate statistics to a particular problem may involve several types of univariate and multivariate analysis in order to understand the relationships between variables and their relevance to the actual problem being studied. Example of multivariate time-series data include opening and closing stock prices and sectoral indices of stock markets. Financial time-series data are characterized by non-linearity, noise, chaotic in nature and volatile thus making the process of planning cumbersome. Hence, the need for smoothing and noise reduction through the use of Unscented Kalman Filter (UKF). There are three distinct types of noise found in time series data; Process noise which represents the shocks that drive the dynamics of the stochastic process, Measurement noise which is encountered when observing and measuring the time series and Arrival noise which reflects uncertainty concerning whether an observation will occur at the next time step.

3 Kalman Filter

The Kalman filter is a recursive predictive filter that is based on the use of state space techniques and recursive algorithms. It estimates the state of a dynamic system which can be disturbed by some noise, mostly assumed as white noise. To improve the estimated state, Kalman filter uses measurements that are related to the state but disturbed as well. The Kalman filter consists of two steps: prediction and correction steps [16].

In the first step the state is predicted with the dynamic model. In the second step it is corrected with the observation model, so that the error covariance of the estimator is minimized. In this sense it is an optimal estimator. This procedure is repeated for each time step, with the state of the previous time step as initial value. Therefore the Kalman filter is called a recursive filter. The basic components of the Kalman filter are the state vector, the dynamic model and the observation model.

3.1 Kalman Filtration Process

The Kalman filter has been applied extensively to the field of non-linear estimation for both state-estimation and parameter-estimation. The basic framework for the Kalman filter involves estimation of the state of a discrete-time non-linear dynamic system,

$$x_{k+1} = F(x_k, u_k) + v_k \tag{1}$$

$$y_k = F(x_k) + n_k, \tag{2}$$

where x_k represent the unobserved state of the system, u_k is a known exogenous input, and y_k is the observed measurement signal. The process noise v_k drives the dynamic system, and the observation noise is given by n_k. The Kalman filter involves the recursive estimation of the mean and covariance of the state under a Gaussian assumption [17].

In contrast, parameter-estimation, sometimes referred to as system identification, involves determining a non-linear mapping $y_k = G(x_k, w)$ where x_k is the input, y_k is the output, and the non-linear map, $G(.)$, is parametrized by the vector w. Typically, a training set is provided with sample pairs consisting of known input and desired outputs, (x_k, d_k). The error of the machine is defined as $e_k = d_k - G(x_k, w)$, and the goal of learning involves solving for the parameters w in order to minimize the expectation of some given function of the error. While a number of optimization approaches exist (e.g., gradient descent and Quasi-Newton methods), parameters can be efficiently estimated on-line by writing a new state-space representation

$$w_{k+1} = w_k + r_k \tag{3}$$

$$d_k = G(x_k, w_k) + e_k, \tag{4}$$

where the parameters w_k correspond to a stationary process with identity state transition matrix, driven by process noise r_k (the choice of variance determines convergence and tracking performance). The output d_k corresponds to a non-linear observation on w_k.

4   Unscented Kalman Filter (UKF)

The Unscented Kalman Filter addresses the approximation issues of Extended Kalman Filter [18]. The state distribution is again represented by a Gaussian Random Variable (GRV) through system dynamics, but is now specified using a minimal set of carefully chosen sample points. These sample points completely captures the true mean and covariance of the GRV, and when propagated through the true non-linear system, captures the posterior mean and covariance accurately to the 3rd order (Taylor series expansion) for any non-linearity [19].

The unscented transformation (UT) is a method for calculating the statistics of a random variable which undergoes a non-linear transformation. Consider propagating a random variable X (dimension L) through a non-linear function, $y = g(X)$. Assuming X has mean $\bar{X}$ and covariance P_X. To calculate the statistics of y, we form a matrix X of $2L+1$ sigma vectors X_i (with corresponding weights w_i, according to the following:

$$X_{k-1}^0 = X_k^a - 1 \tag{5}$$

$$-1 < w^0 < 1 \tag{6}$$

$$X_{k-1}^i = X_{k-1}^a + (\sqrt{\frac{n}{1 - w^0}} . P_{k-1})_i \tag{7}$$

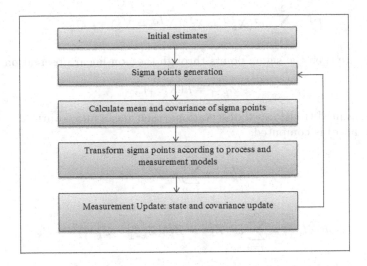

Fig. 1. Flow chart illustrating the operation of UKF

$$X_{k-1}^{i+n} = X_{k-1}^{a} + (\sqrt{\frac{n}{1-w^0}} \cdot P_{k-1})_i \tag{8}$$

for all i=1,...,n

$$w^j = \frac{1-w^0}{2n} \tag{9}$$

for all j=1,...,2n

where the weights must obey the condition:

$$\sum_{j=0}^{2n} w^j - 1 \tag{10}$$

and $(\sqrt{\frac{n}{1-w^0}} \cdot P_{k-1})$ is the row or column of the matrix square root of $(\frac{n}{1-w^0} \cdot P_{k-1})$. w_0 controls the position of sigma points: $w^0 \geq 0$ points tend to move further from the origin, $w^0 \leq 0$ points tend to be closer to the origin. A more general selection scheme for sigma points, called scaled unscented transformation.

4.1   UKF Forecast Step

Each sigma point is propagated through the non-linear process model:

$$X_k^{f,j} = f(X_{k-1}^j) \tag{11}$$

The transformed points are used to compute the mean and covariance of the forecast value of X_k:

$$X_k^f = \sum_{j=0}^{2n} w^j X_k^{f,j} \tag{12}$$

$$P_k^f = \sum_{j=0}^{2n} w^j (X_k^{f,j} - X_k^f)(X_k^{f,j} - X_k^f)^T + Q_{k-1} \tag{13}$$

We propagate then the sigma points through the non-linear observation model:

$$Z_{k-1}^{f,j} = h(X_{k-1}^j) \tag{14}$$

With the resulted transformed observations, their mean and covariance (innovation covariance) is computed:

$$Z_{k-1}^f = \sum_{j=0}^{2n} w^j Z_{k-1}^{f,j} \tag{15}$$

$$Cov(\tilde{Z}_{k-1}^f) = \sum_{j=0}^{2n} w^j (Z_k^{f,j} - Z_k^f)(Z_{k-1}^{f,j} - Z_{k-1}^f)^T + R_k \tag{16}$$

The cross covariance between $\tilde{x}_k^f$ and $\tilde{Z}_{k-1}^f$ is:

$$Cov(\tilde{X}_{k-1}^f, \tilde{Z}_{k-1}^f) = \sum_{j=0}^{2n} w^j (X_k^{f,j} - Z_k^f)(Z_{k-1}^{f,j} - Z_{k-1}^f)^T + R_k \tag{17}$$

4.2   Data Assimilation Step

We like to combine the information obtained in the forecast step with the new observation measured Z_k. Like in KF assume that the estimate has the following form:

$$X_k^a = X_k^f + K_k(Z_k - Z_{k-1}^f) \tag{18}$$

The gain K_k is given by:

$$K_k = Cov(\tilde{X}_k^f, \tilde{Z}_{k-1}^f)Cov^{-1}(\tilde{Z}_{k-1}^f) \tag{19}$$

The posterior covariance is updated after the following formula:

$$P_k = P_k^f - K_k Cov(\tilde{Z}_{k-1}^f)K_k^T \tag{20}$$

The operation of UKF is summarized as shown in Figure 1.

5   Experimental Results

Our analysis was conducted using very high-frequency time series on KLCI stock indices, for the period : January, 03, 2000 to January, 03, 2011 and KLSE tin index for the period : April, 1988 to April, 2013.

Before experimental simulations, we let $X_k = 0$. Similarly we need to choose an initial value for P_k, i.e. If we were absolutely certain that our initial state estimate $X_k = 0$ was correct, we would let $P_k = 0$. However given the uncertainty

Table 1. Comparison of UKF Smoothed Outputs using different R Values

DATE	KLCI	R=1	R=0.2	R=0.1	R=0.01	R=0.001
03/01/2000	**833.89**	416.95	694.88	758.00	825.55	**833.06**
04/01/2000	**832.8**	555.56	757.58	793.64	829.18	**832.93**
05/01/2000	**815.8**	659.66	778.13	801.26	824.72	**827.22**
06/01/2000	**818.43**	719.20	793.96	808.06	822.20	**823.70**
07/01/2000	**818.43**	757.36	803.20	811.96	820.79	**821.73**
10/01/2000	**818.43**	780.63	809.04	814.45	819.88	**820.46**
11/01/2000	**846.74**	805.91	823.42	826.76	830.11	**830.47**
12/01/2000	**869.62**	830.23	841.08	843.14	845.22	**845.44**
13/01/2000	**890.60**	853.29	859.99	861.26	862.55	**862.68**
14/01/2000	**928.24**	881.92	886.06	886.85	887.64	**887.73**
17/01/2000	**953.42**	909.23	911.79	912.28	912.77	**912.82**
18/01/2000	**949.62**	924.66	926.24	926.54	926.84	**926.88**
19/01/2000	**933.38**	927.99	928.97	929.15	929.34	**929.36**
20/01/2000	**950.86**	936.73	937.33	937.44	937.56	**937.57**
21/01/2000	**953.96**	943.31	943.68	943.75	943.82	**943.83**
24/01/2000	**954.50**	947.58	947.81	947.86	947.90	**947.91**
25/01/2000	**952.06**	949.29	949.44	949.46	949.49	**949.49**

in our initial estimate X_k , choosing $P_k=0$ would cause the filter to initially and always believe $X_k= 0$. As it turns out, the alternative choice is not critical. We could choose almost any initial estimate of X_k and the filter would eventually converge. The first simulation was conducted by setting the filter with $P_k=$ 1 and measurement variance R=0.001. The second simulation was conducted using measurement variance R=0.01, the third, fourth and fifth simulation was conducted using 0.1,0.2 and 1 respectively. In the UKF algorithm as in Figure 1, the values that are evaluated in the measurement update stage are also known as

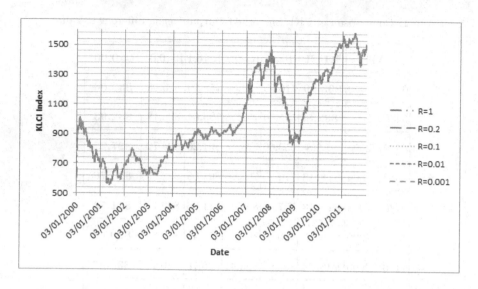

Fig. 2. Filtered and Smoothed KLCI Index using different R values

posterior values which are plotted as shown in Figure 2. Figure 3 shows only one value of R (0.001) which was selected having the minimum error rate using KLSE Tin index values. The overall performance was obtained when the measurement variance was set at 0.001 which outperformed all other measurement variance values as shown in Table 1. This was justified by performing quantitative analysis on all the posterior values to determine the best measuremnet variance using the most common used quantitative error anlaysis as shown in Table 2.

Table 2. Comparison of UKF Posterior Outputs in Terms of Error Rates

MEASUREMENT VARIANCE	RMSE	MAPE	MAD
R=0.001	**0.35**	**0.01**	**0.00011**
R=0.01	0.36	0.02	0.00012
R=0.1	0.41	0.03	0.00013
R=0.2	0.46	0.04	0.00015
R=1	0.71	0.07	0.0023

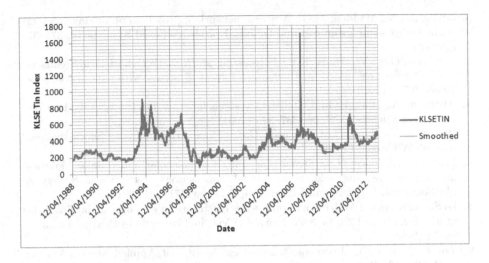

Fig. 3. Filtered and Smoothed KLSE Tin Index using R=0.001

6   Conclusion

This paper has presented an approach for filtering and smoothing noisy multivariate time series data which are characterized by non-linearity, noise, chaotic in nature and volatility thus making the process of financial planning cumbersome. Hence, the need for smoothing and noise reduction through the use of Unscented Kalman Filter (UKF) [19]. The goal was to determine if UKF could be exploited to smooth the direction of KLCI stock price movements using different measurement variance values. The filter has been applied extensively to the field of non-linear estimation for both state-estimation and parameter-estimation, this has been justified through simulations and comparative statistical analysis verifying the validity of UKF filter posterior values as shown in Table 1. The only drawback of this technique is that it is computationally intensive, and yet it can be quite effectively implemented by the parallel computation technique.

Acknowledgment. The authors would like to thank Universiti Teknologi PETRONAS (UTP) for the financial support received in conducting and publishing of this conference paper.

References

1. Scher, I., Koomey, J.G.: Is accurate forecasting of economic systems possible? Climatic Change 104(3-4), 473–479 (2011)
2. Atsalakis George, P.V.K.S.: Surveying stock market forecasting techniques Part II: soft computing methods. Experts Systems Applications 36, 5932–5941 (2009)

3. O'Connor, N., Madden, M.G.: A neural network approach to predicting stock exchange movements using external factors. Knowledge-Based Systems 19, 371–378 (2006)
4. Atsalakis, G.S., Valavanis, K.P.: Forecasting stock market short-term trends using a neuro-fuzzy based methodology. Expert Systems with Applications 36, 10696–10707 (2009)
5. Abraham, A., et al.: Hybrid intelligent systems for stock market analysis. In: Computational Science-ICCS, pp. 337–345 (2001)
6. Box, G.E., Jenkins, G.M., Reinsel, G.C.: Time series analysis: forecasting and control. Wiley.com (2013)
7. Little, M.A., Jones, N.S.: Signal processing for molecular and cellular biological physics: an emerging field. Philosophical Transactions of the Royal Society A: Mathematical, Physical and Engineering Sciences, 371–384 (2013)
8. Di Salvo, R., et al.: Multivariate time series clustering on geophysical data recorded at Mt. Etna from 1996 to 2003. Journal of Volcanology and Geothermal Research, 65–74 (2013)
9. Tiwari, Kumar, A.: Economic time series. Journal of Applied Statistics 40(6), 1384–1385 (2013)
10. Abdulkadir, S.J., et al.: Moisture Prediction in Maize using Three Term Back Propagation Neural Network. International Journal of Environmental Science and Development 3(2), 199–204 (2012)
11. Tsai, J.-F., et al.: Electroencephalography when meditation advances: a case-based time-series analysis. Cognitive Processing, pp. 1–6 (2013)
12. Ma, L., Khorasani, K.: A new strategy for adaptively constructing multilayer feedforward neural networks. Neurocomputing 51, 361–385 (2003)
13. De Oliveira, J.V., Pedrycz, W.: Advances in fuzzy clustering and its applications. Wiley Online Library (2007)
14. Chen, Y., et al.: Time-series forecasting using flexible neural tree model. Information Sciences 174, 219–235 (2005)
15. Giordano, F., et al.: Forecasting non-linear time series with neural network sieve bootstrap. Computational Statistics and Data Analysis 51, 3871–3884 (2007)
16. Chu, J.L., et al.: Seasonal forecast for local precipitation over northern Taiwan using statistical downscaling. J. Geophys. Res. 113, D12118 (2008)
17. Lin, J.-W., Chen, C.-W., Peng, C.-Y.: Kalman filter decision systems for debris flow hazard assessment. Natural Hazards 60(3), 1255–1266 (2012)
18. Li, L., Xia, Y.: Stochastic stability of the unscented Kalman filter with intermittent observations. Automatica, 978–981 (2012)
19. Dini, D.H., Mandic, D.P., Julier, S.J.: A widely linear complex unscented Kalman filter. IEEE Signal Processing Letters 18(11), 623–626 (2011)

Reconstructing Gene Regulatory Network Using Heterogeneous Biological Data

Farzana Kabir Ahmad and Nooraini Yusoff

Bio-Inspired Agent Systems, Artificial Inelligence Lab, School of Computing,
College of Arts and Sciences, Universiti Utara Malaysia, 06010 Sintok, Kedah Malaysia
{farzana58,nooraini}@uum.edu.my

Abstract. Gene regulatory network is a model of a network that describes the relationships among genes in a given condition. However, constructing gene regulatory network is a complicated task as high-throughput technologies generate large-scale of data compared to number of sample. In addition, the data involves a substantial amount of noise and false positive results that hinder the downstream analysis performance. To address these problems Bayesian network model has attracted the most attention. However, the key challenge in using Bayesian network to model GRN is related to its learning structure. Bayesian network structure learning is NP-hard and computationally complex. Therefore, this research aims to address the issue related to Bayesian network structure learning by proposing a low-order conditional independence method. In addition we revised the gene regulatory relationships by integrating biological heterogeneous dataset to extract transcription factors for regulator and target genes. The empirical results indicate that proposed method works better with biological knowledge processing with a precision of 83.3% in comparison to a network that rely on microarray only, which achieved correctness of 80.85%.

Keywords: Gene regulatory, Bayesian network, heterogeneous data.

1   Introduction

DNA microarray is one of the most fascinating and latest breakthrough technologies in molecular biology. This technology has been used to facilitate the quantitative studies of thousand of genes in order to answer various research questions. To date, this technology is widely employed to construct gene regulatory network (GRN). The construction of GRN using microarray data has enabled the measurement of global response of biological system to examine specific inventions. For example, scientists can look into large number of gene interactions that are perturbed during cancer progression.

GRN also known as cellular network is a set of molecular components that includes genes, proteins and other molecules, which collectively accomplish cellular functions as these molecules interact with each other [1]. However, in this study we only used microarray data to model gene network. The fundamental idea behind GRN

S. Ramanna et al. (Eds.): MIWAI 2013, LNCS 8271, pp. 97–107, 2013.
© Springer-Verlag Berlin Heidelberg 2013

analysis is to discover regulator genes by examining gene expression patterns. Notably, some genes regulate other genes, which mean that the amount of a gene expressed at a certain time could activate or inhibit the expression of another gene. Thus, the regulation of gene expression has an important role in cellular functions. Changes in the expression levels of particular genes across a whole process, such as the cell cycle or response to certain treatments, have provided information that allows reconstruction of cellular network using reverse engineering technique.

A large number of works have reported that GRN can possibly assist researchers in suggesting and evaluating innovative hypotheses in the context of genetic regulatory processes [2-3]. Such data-driven regulatory networks analysis ultimately would provide clearer understanding of the genetic regulatory processes, which are normally complex and intricate. Furthermore, it would bring significant implications in the biomedical fields and many other pharmaceutical industries. Thus, identifying GRNs and understanding regulatory processes at the genetic level has become an imperative goal in computational biology.

Various mathematical and computational methods have been used to model GRN from microarray data, including Boolean network, pair-wise comparison, differential equations estimation, Bayesian network [4-5] and other techniques. Amongst these, the Bayesian network model attracts the most attention and has become the prominent technique because it can capture linear, nonlinear, combinatorial, stochastic and casual relationships between variables. Compared to other methods, Bayesian network model establishes considerable relationships between all genes in the system. In addition, due to rich probabilistic semantics, this model is also capable of working with noisy data that is a common problem in microarray data. Furthermore, this technique allows for different implicit variable information to be added to the networks, which possibly enhances the interpretation of the gene regulation process. Thus, Bayesian network is used in this study to analyze gene regulatory processes and to model gene relationships for breast cancer metastasis.

The key challenge in using Bayesian network to model GRN is related to it's structure learning. Bayesian network structure learning is NP-hard and computationally complex, as the number of possible graphs increases super-exponentially with the number of genes and an exhaustive search is untraceable. This difficulty is a common problem in gene regulatory analysis because network is usually learnt from a relatively small number of measurements. The high dimensionalities of microarray data, which usually contain insufficient sample measurement plus large number of genes to examine, are the main causes of this problem.

The basic idea is to develop GRN by measuring the dependencies among nodes of the given data. Low-order conditional independence is used to examine the relationships between genes. Although the proposed method has increased the accuracy of inferred network as reported in [6], such gene network is solely based on the microarray data and is often insufficient for rigorous analysis. In many cases, microarray data is often daunted by noisy, incomplete data and misleading outliers, which can produce high number of false positive edges. Accordingly, an inferred GRN may contain some incorrect gene regulations that are unreliable from the biological point of view. Thus, integration of biological knowledge into gene network has become necessary to overcome the problem.

This study has used heterogeneous biological data to improve the structure learning of Bayesian network. The remainder of this paper is organized as follows. Section 2 describes some previous works, whose have utilized biological data to achieve better construction of GRN. Section 3 on the other hand, presents the proposed method. Section 4 meanwhile presents experimental results and discussion. Finally, Section 5 offers concluding and future direction remarks.

2 Previous Works

Recent years have witnessed the increasing amount of genomic data such as gene expression, single-nucleotide polymorphism (SNP) and proteomic which are available in public databases. This trend has triggered a new research direction whereby researchers now are motivated to combine various kinds of genomic data to reconstruct GRN. However, the integration of different data source is not simple as these data varies in term of sizes, formats and types. Furthermore, most of these data is partly independent and provide complementary information on the whole genome. Since, there is no complete GRN that are available for any species, the best option in hand is to integrate diverse biological data that presents fragmented information and seek a better explanation for the development at a system level.

Data integration has been defined as a data fusion process that not only includes various data sources but also provides biological meaning with the use of bioinformatics and computational tools. The overreaching goal of data integration is to obtain more accurate, precise and broader view of network than any of single dataset. Based on this concept several works have been done to seek for better explanation of GRN (as explained in Section 2.1. and Section 2.2). Generally there are two types of data integration in the field of GRN, namely homologous data integration and heterogeneous data integration. Homologous data integration mainly uses of similar data type for example combination of multiple microarray datasets from different studies to answer question raised by researchers. Meanwhile, heterogeneous data integration make used of different data types across or within studies to seek for better clarify of information provided by a single data type.

The main idea for homologous data integration is to increase the number of samples to address the issue of high-dimensional data. Most studies in homologous data integration have focused on comparing two or more related datasets to identify significant genes that can distinguish different group of samples (e.g. disease and normal samples). For an example, Rhodes *et al.* [7] have combined multiple microarray datasets to classify common transcription profiles that are universally activated in most cancer types.

Unlike homologous data integration, where it used similar data types, heterogeneous data integration mainly focuses on applying various data sources to ensure the reliability of results obtained. Among the popular data integration is gene expression and proteomic data. Protein is the end product of translation process and is also used as a trigger to initiate the expression of other genes. Therefore, the combination of these data type is reasonable to most of researchers. Thus, many previous works have estimated co-expressed relationship as a gene regulatory instead of looking at protein-protein interactions [8-9]. Besides that, large number of researchers also utilized transcription factor binding sites (TFBS) to verify the GRN. Like protein, TFBS is another complementary data to measure cellular state. Hence, more recent works have explored data integration of external knowledge to identify transcription factors and their target genes [10-11]. Transcription factors are very essential in regulating gene expression. Motivated by this fundamental concept, transcription factors have been used in this research to discover significant biological information from high-throughput data.

3   Methods

The Bayesian network is a graphical model that was introduced by Pearl and Wright in 1980s [12]. To deal with a large number of genes in microarray data, this research defines the Bayesian network, BN as: $BN = (G, P)$ where $G = (X, E(G))$ is a DAG with a set of variables X representing $\{X_i; i \in V\}$, and $E(G) \subseteq X_i * X_j$ (set of pairs that represents the dependent among v variables). The element E is an edge from node X_i to X_j, indicating X_i is a parent to X_j. On the other hand, P corresponds to joint distribution on the variables in the network. The $Pa(V)$ represents the parent for a set of vertex V and can be defined as:

$$Pa(X_i, G) = \{X_j, \text{ such that } (X_i, X_j) \in E(G); j \in V\} \qquad (1)$$

where $Pa(X_i, G)$ is the parent of X_i in the graph, G and having node X_j pointing toward X_i. Mathematically, the joint distribution of all node values in DAG can be decomposed as the product of the local distribution of each node and its' parents:

$$P(X_1, X_2, ..., X_n) = \prod_{i=1}^{n} P(X_i \mid parents(X_i)) \qquad (2)$$

The Markov Blanket is another important characteristic of the Bayesian network. The Markov Blanket of a variable is the set of variables that completely shield off this variable from the other variables. To such an importance aim and for the purpose of classification, this study focuses on the Markov Blanket to identify minimal set of variables that are required to predict the metastasis outcome

3.1   Structure Learning of Bayesian Network

There are generally two main approaches to construct the Bayesian network from data: (1) the score and searching approach and (2) the dependency analysis approach. The first approach involves measuring fitness of structure and searching for the best structure that describes the data. Several scoring methods have been applied, including the Bayesian scoring, entropy based and minimum description length. Once the scores are obtained, the next step is to use search methods, such as the heuristic search, to build the best-fit network structure. Although this approach have been used by many researcher in reconstructing GRN, it mainly suffered from computational complex search and disability to provide posterior distributions over all the parameters of the model that are needed to quantify uncertainty in the gene regulators.

The dependency analysis approach or constraint-based learning on the other hands, aims to identify from the data the dependencies to construct the network structure. In this study, we proposed to low-order conditional independence and its variants, full-order conditional independence, to construct a cellular network.

Full-order conditional independence is the exact set of edges between the successive variables X_j and X_i, given the remaining variables X_{Vj}, $V_j = V \setminus \{j\}$ and $X_{Vj} = \{X_k; \ k \in V_j\}$. It can be defined as:

$$\tilde{G} = \left(X, \ \left\{ \left(X_j, X_i \right); X_i \perp X_j \mid X_{V_j} \right\} i, j, \in V \right) \tag{3}$$

DAG $\tilde{G}$ is the smallest sub graph to which the probability distribution P has allows for a Bayesian network representation. Reverse discovery of DAG $\tilde{G}$ to model a cellular network requires determining each variable X_i and the set of variables X_j on which variable X_i is conditionally independent given the remaining variables X_{Vj}. Hence, by using the Equation 3, this approach has extended the principles of concentration graph that employed conditional independence to the Bayesian network case. However, by applying this approach, the curse of dimensionality is still a problem because the number of genes v is much greater than the number of measurements in n samples ($v >> n$) and conditional independence for each variables X_i given others remaining variable X_{Vj} is yet to be computed.

To reduce the high dimension of gene expression data, q^{th} order conditional independence, DAGs $G^{(q)}$ (whereby $q < v$) is estimated from DAG $\tilde{G}$. By doing so, the Bayesian network is extended based on the consideration of low-order conditional independence, and this is similar to the work of Wille and Buhlmann [13] for the GGM. DAGs $G^{(q)}$ is defined as below:

$$\forall q < v, \quad G^{(q)} = \left(X, \left\{ \left(X_j, X_i \right); \ \forall Q \subseteq V_j, |Q| \models q, X_i \perp X_j \mid X_Q \right\} i, j \in V \right) \quad (4)$$

$$\forall q < v, \quad \tilde{G} \subseteq G^{(q)} \quad (5)$$

DAGs $G^{(q)}$ is different from $\tilde{G}$ but it provides an alternative way of producing dependence relationship between variables, which is particularly suited for sparse network such as gene networks. However, DAGs $G^{(q)}$ is no longer associated with global relationship in the Bayesian network representation. Nevertheless, DAGs $G^{(q)}$ circumvents heavy statistical tasks and computation costly search in large number of variables. For additional technical details on this proposed method please refer to Ahmad et al. [6].

3.2   Revising Gene Regulatory Relationship with Integration of Transcription Factors

In a nutshell, two genes are regulating if transcription factor of regulator gene can bound at promoter region of target genes. Using such intrinsic biological feature, the regulatory relationships obtain by using the proposed Bayesian network model are verified. A number of necessary bioinformatics toolkits are wrapped to identify significant regulatory relationships. Three bioinformatics toolkits; (1) Ensembl, (2) TFSearch and (3) TRANSFAC, and their corresponding website are used in this study.

The name of dependent gene, T_h is entered in Ensembl and 1000 base pair upstream DNA sequence is then selected as a promoter region of target gene. This sequence is then copied and used as an input in TFSearch tool to find all possible transcription factors that can bind to a given promoter region. The regulator gene, T_g is examined using TRANFAC. This tool presents a list of transcription factors that are associated with regulator genes and DNA binding motif. If the transcription factor for both dependent gene and regulator gene match, then there is a dependency between X_i and X_j where $X_i \rightarrow X_j$.

3.3   Dataset Description

We tested this proposed method using a data set of 97 breast cancer microarray from van't Veer et al [14]. These cohorts of breast cancer patients are 55 years old or younger. We obtained this data from the Integrated Tumor Transcriptome Array and Clinical data Analysis database (ITTACA, 2006). Among the remaining 97 samples, 46 developed distant metastasis within 5 years and 51 remained metastasis free for at least 5 years. DNA microarray analysis was used by van't Veer to determine the expression levels of approximately 25,000 genes for each patient.

4   Experimental Results and Discussion

To obtain insights into the mechanism of gene regulation and how gene mutations act to turn on tumor development and metastasis progression in a cellular network context, the proposed method is executed on the breast cancer dataset producing a GRN as shown in Fig. 1.

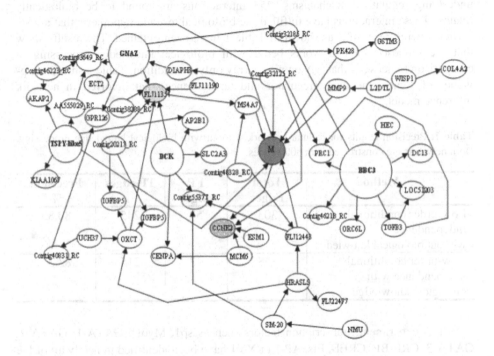

Fig. 1. The GRN for breast cancer metastasis using the low-order conditional independence method

This learned network revealed a group of genes which are primarily associated with causing metastasis, M. The larger nodes in the graph specify the genes when expressed at different levels lead to a major effect on the status of other genes (e.g., on or off). Meanwhile, the light-shaded nodes denote the highly regulated genes. Four genes that are found to regulate the expression levels of other genes are: BBC3, GNAZ, TSPY-like5 (TSPY5), and DCK. Two genes are highly regulated: FLJ11354 and CCNE2. This GRN involved 50 genes associated with metastasis, M, and 39 of them are annotated. Additionally, the p-value of the conditional independence test between the transcription regulatory genes and their co-expressed genes is given in [6].

Based on the experiment that has been carried out by using the proposed method, the relation between genes in the GRN is required to be verified. With the transcription factors of regulator gene and list of DNA binding sites of target gene at hand, as explained in Section 3.2, the regulatory relationship between G and H, can be examined. If the transcription factor of regulator gene TFg can bind to promoter site of target gene H, whereby $TF_h = TF_g$, then gene G and gene H could possibly has a relation. Together with this intrinsic biological features that play important role in the underlying regulatory mechanisms, 258 interactions are found to be biologically related. These interactions have fulfilled the biological test and hypothesis that are set earlier. Therefore, results as shown in Table 1 have been obtained. The results show that the proposed method works better with biological knowledge processing in comparison to network that rely on microarray only. In addition, Table 2 shows the p-values of some gene pairs (regulator and target genes) that involved in network inference model.

Table 1. Precision results for cellular network without/with biological knowledge processing. Both networks are constructed with 5000 genes.

Method	Total Edges	FP edges	TP edges	Precision %
Low-order conditional independence <u>without</u> biological knowledge	303	58	245	80.85
Low-order conditional independence <u>with</u> biological knowledge	258	43	215	83.33

In this experiment, transcription factors such as sp1, MyoD, GATA-1, GATA-2, GATA-3, CRE-BP; CREB, Ets, AP-1 or YY1 have been identified in nearly all of the interactions. Most of these transcription factors are discovered to be related to breast cancer metastasis, for instance, MyoD, AP-1 and sp1 have been identified by Mi *et al.* [15] to play important role in tumor progression, while Ets family of transcription factors are reported to be involved in cellular proliferation and apoptosis [16]. GATA transcription factor, particularly GATA-3 on the other hand has recently been identified as the key in controlling genes that involved in differentiation and proliferation of breast cancer [17]. Similar to the rest of transcription factors, CREB also has shown involvement in tumor initiation, progression and metastasis. It has been identified as proto-oncogene by Xiao *et al.* [18] and is found active in breast cancer, prostate cancer, lung cancer and leukemia cells. In addition, YY1 is discovered to play an essential role in tumorigenesis and is generally related to poor breast cancer prognosis [19].

Table 2. The prominent transcription factors for top ranked regulators

Regulator gene, G	Target gene, H	Transcription factors	P-value
BBC3	HEC	GATA-1; GATA-2; GATA-3; Oct-1;C/EBPalpha, C/EBPbeta	0.002178
BBC3	DC13	LyF-1; YY1; E2F; c-Ets-	0.001315
BBC3	Contig32185_RC	*	0.002261
BBC3	PRC1	c-Ets-; Oct-1; AP-1; c-Myb	0.002178
BBC3	ORC6L	AML-1a; USF-1; CP2; NF-Y	0.003732
BBC3	Contig46218_RC	*	0.021104
BBC3	LOC51203 (NUSAP1)	c-Ets-; LyF-1; HNF-3b; RORalp	0.003515
BBC3	TGFB3	E2F; AP-4; STATx; GATA-1; Sp1	0.013909
TGFB3	PRC1	SRY; c-ETS; GATA-1; SREBP-CRE-BP	3.60741E-06
TGFB3	LOC51203 (NUSAP1)	c-ETS; SRY; MyoD; GATA-1; AP-1	4.90607E-07
WISP1	COL4A2	AML-1a; Sp1; GATA-1; GATA-2; E2F;	0.595393
L2DTL	MMP9	GATA-1; GATA-3; NF-kap; AP-1; Sp1	3.53821E-05
PK428	GSTM3	GATA-1; SRY; HNF-3b; HNF-1; YY1	0.200820
GNAZ	PK428	USF; Sp1; MyoD; AP-2; YY1	0.007184
GNAZ	Contig32185_RC	*	0.020627
GNAZ	ECT2	Oct-1; LyF-1; MyoD; USF;	0.001813
GNAZ	Contig63649_RC	*	0.032360
GNAZ	FLJ11354	*	0.002381
GNAZ	GPR126	MyoD; CRE-BP; CREB; Oct-1; GATA-1	0.003125
TSPY-like 5	AP2B1	CREB; USF; GATA-1; GATA-2;GATA-3	0.042655

5   Conclusion and Future Remarks

This paper described the need to integrate diverse data integration for better interpretation of GRN model. Two types of data integration approaches have been comprehensively explained; (1) homologous data integration and (2) heterogeneous data integration. Since most GRN models are mainly implemented based on microarray data, issues like reliability and quality concern are also debated by many researchers. The best available alternative is to integrate different data to address this problem and obtain a better understanding of the underlying gene regulatory mechanisms. Furthermore, with the currently available and enormous public databases, this effort appears to be the most promising since it utilizes the independent and complementary information to answer research questions.

The use of transcription factors to identify relevant regulatory interactions is the key idea in this research. In achieving this, three main bioinformatics toolkits for instance Ensembl, TFSearch and TRANFAC have been used. Each of these tools is used to apprehend the concept of biological intrinsic features of transcription factor and promoter. Based on the experiments that were conducted, 258 out of 303 interactions are identified to be biologically relevant. Furthermore, the p-value of regulated genes are computed to gain significant and efficient statistical results. The empirical results indicated several transcription factors such as sp1, MyoD, GATA-1, GATA-2, GATA-3, CRE-BP; CREB, Ets, AP-1 or YY1 play essential role in breast cancer metastasis. In the future, many more different data types will be integrated to obtain more insightful view of GRN and further facilitate our understanding of cancer growth.

References

1. Yavari, F., Towhidkhah, F., Gharibzadeh, S.: Gene regulatory network modeling using Bayesian networks and cross correlation. Biomedical Engineering Conference, CIBEC, Cairo (2008)
2. Gevaert, O., Van Vooren, S., Moor, B.D.: A framework for elucidating regulatory networks based on prior information and expression data. In: Eklund, P., Mann, G.A., Ellis, G. (eds.) ICCS 1996. LNCS, vol. 1115, pp. 240–248. Springer, Heidelberg (1996)
3. Huang, Z., Li, J., Su, H., Watts, G.S., Chen, H.: Large-scale regulatory network analysis from microarray data: Modified Bayesian network learning and association rule mining. Decision Support Systems 43, 1207–1225 (2007)
4. Friedman, N., Linial, M., Nachman, I., Pe'er, D.: Using bayesian networks to analyze expression data. Journal of Computational Biology 7(3), 601–620 (2000)
5. Zou, M., Conzen, S.D.: A new dynamic Bayesian network (DBN) approach for identifying gene regulatory networks from time course microarray data. Bioinformatics 21(1), 71–79 (2005)
6. Ahmad, F.K., Deris, S., Othman, N.H.: The Inference of Breast Cancer Metastasis through Gene Regulatory Networks. Journal of Biomedical Informatics (JBI) 45(2), 350–362 (2012)

7. Rhodes, D.R., Yu, J., Shanker, K., Deshpande, N., Varambally, R., Ghosh, D., et al.: Large-scale meta-analysis of cancer microarray data identifies common transcriptional profiles of neoplastic transformation and progression. Proceedings of the National Academy of Sciences of the United States of America 101(25), 9309–9314 (2004)
8. Zhang, Y., Zha, H., Wang, J.Z., Chu, C.H.: Gene co-regulation vs. co-expression: The Pennsylvania State University, University Park, PA (2004)
9. Yeung, K.Y., Medvedovic, M., Bumgarner, R.E.: From co-expression to co-regulation: how many microarray experiments do we need? Genome Biology 5, R48 (2004)
10. Zhao, W., Serpedin, E., Dougherty, E.R.: Recovering genetic regulatory networks from chromatin immunoprecipitation and steady-state microarray data. Eurasip. J. Bioinform. Syst. Biol. (2008)
11. Kaleta, C., Göhler, A., Schuster, S., Jahreis, K., Guthke, R., Nikolajewa, S.: Integrative inference of gene-regulatory networks in Escherichia coli using information theoretic concepts and sequence analysis. BMC Systems Biology 4(116) (2010)
12. Pearl, J.: Probabilistic reasoning in intelligent systems: Networks of plausible inference. Morgan Kaufmann Publishers Inc., Francisco (1988)
13. Wille, A., Buhlmann, P.: Low-order conditional independence graphs for inferring genetic networks. Statistical Applications in Genetics and Molecular Biology 4(32) (2006)
14. van't Veer, L.J., Dai, H., van de Vijver, M.J., He, Y.D., Hart, A.A.M., Mao, M., Peterse, H.L., van de Kooy, K., Marton, M.J., Witteveen, A.T., Schreiber, G.J., Kerkhoven, R.M., Roberts, C., Linsley, P.S., Bernards, R., Friend, S.H.: Gene expression profiling predicts clinical outcome of breast cancer. Nature 415, 530–536 (2002)
15. Mi, Z., Guo, H., Wai, P.Y., Gao, C., Wei, J., Kuo, P.C.: Differential Osteopontin expression in phenotypically distinct subclones of murine breast cancer cells mediates metastatic behavior*. Journal of Biological Chemistry 279(45), 46659–46667 (2004)
16. Kato, T., Katabami, K., Takatsuki, H., Han, S.A., Takeuchi, K., Irimura, T., et al.: Characterization of the promoter for the mouse a3 integrin gene Involvement of the Ets-family of transcription factors in the promoter activity. Eur. J. Biochem. 269, 4524–4532 (2002)
17. Fang, S.H., Chen, Y., Weigel, R.J.: GATA-3 as a marker of hormone response in breast cancer. Journal of Surgical Research 157(2), 290–295 (2009)
18. Xiao, X., Li, B., Mitton, B., Ikeda, A., Sakamoto, K.: Targeting CREB for cancer therapy: friend or foe. Curr. Cancer Drug Targets 10(4), 384–391 (2010)
19. Gordon, S., Akopyan, G., Garban, H., Bonavida, B.: Transcription factor YY1: structure, function, and therapeutic implications in cancer biology. Oncogene 25, 1125–1142 (2006)

Design of a Multi-day Tour-and-Charging Scheduler for Electric Vehicles*

Junghoon Lee and Gyung-Leen Park**

Dept. of Computer Science and Statistics,
Jeju National University, Republic of Korea
{jhlee,glpark}@jejunu.ac.kr

Abstract. Aiming at alleviating the range anxiety problem in electric vehicles taking advantage of well-developed computational intelligence, this paper designs a multi-day tour-and-charging scheduler and measures its performance. Some tour spots have charging facilities for the vehicle battery to be charged during the tour. Our scheduler finds a multi-day visiting sequence, permitting different day-by-day start and end points. To exploit genetic algorithms for the extremely vast search space, a feasible schedule is encoded to an integer-valued vector having $(n+m-1)$ elements, where n is the number of places to visit and m is the number of tour days. The cost function evaluates the waiting time, namely, the time amount the tourist must wait for the battery to be charged enough to reach the next place. It also integrates the time budget constraint and quantizes the tour length. The performance measurement result obtained from a prototype implementation shows that our scheme achieves 100 % schedulability until 13 places for the 2-day trip and 17 places for the 3-day trip on given parameter setting.

Keywords: Electric vehicle, multi-day trip, tour-and-charging schedule, genetic algorithm, waiting time.

1 Introduction

Energy efficiency is the most important keyword in the future power grid [1]. According to the development and penetration of electric vehicles, or EVs in short, the smart grid extends its coverage to the transport system [2]. Not just efficient power consumption, EVs have many environmental benefits over their counterparts, namely, gasoline-powered vehicles, as it is not necessary to burn fossil fuels. Hence, many modern cities are trying to accelerate the large deployment of EVs and are building city-wide charging facilities. However, in spite of recent improvement in battery technologies, it is not yet quite sure that the cost

* This research was financially supported by the Ministry of Knowledge Economy (MKE), Korea Institute for Advancement of Technology (KIAT) through the Inter-ER Cooperation Projects.
** Corresponding author.

S. Ramanna et al. (Eds.): MIWAI 2013, LNCS 8271, pp. 108–118, 2013.

advantage in operating EVs outweighs their main drawback stemmed from range anxiety. If a user wants to drive more than the battery capacity, the EV must be charged somewhere on the route. It may take a few hours with slow chargers. Moreover, battery switch systems for EVs are rarely commercialized yet.

In the mean time, EV-based rent-a-car services are now appearing, not just in the form of a short-term hourly sharing model but also a long-term multi-day rental [3]. As their driving distance is highly likely to exceed the driving range of a fully-charged EV, drivers are facing the problem of deciding when and where to charge their EVs during their trips. Here, the visiting sequence is important, as drivers must wait and waste their time when they want to go to the next place but current battery remaining is not enough [4]. Moreover, in some places such as shopping malls and tour spots, EVs can be charged in parking areas while the drivers are taking their desired activities. How much an EV can fill electricity definitely depends on the stay time at a place. In case the number of visit places increases and charging facilities are limitedly available, this problem gets even more complex, and it is necessary to exploit sophisticated computer algorithms, mainly in artificial intelligence domains.

Generally, tourists select the set of tour places they want to visit while planning their tours. From tourist information, the stay time and the availability of chargers can be retrieved. Even if the scheduling problem is one of the most popular applications in computer sciences, it must be adapted for problem-specific constraints and scheduling goals. For example, the daily tour time, that is, the sum of driving and stay time at all places, is limited, while the waiting time must be kept as small as possible. Moreover, for multi-day trips, its search space grows too much, making it necessary to exploit suboptimal techniques such as genetic algorithms. Here, the modern well-organized communication infrastructure provides an efficient channel for such intelligent services to be delivered even to mobile users [5]. Moreover, the seamless interaction between many different objects over the Internet makes it possible for users easily specify their requirement and receive map-enriched information on diverse types of devices.

In this regard, this paper designs a multi-day tour-and-charging scheduler for EVs, aiming at reducing the inconvenience brought by their long charging time and short driving range [1]. Our research team has been developing a tour scheduler, mainly focusing on single day trips taking EVs. This effort includes waiting time estimation, genetic scheduler design, and orienteering problem modeling. The genetic scheduler will be extended for multi-day trips and its performance will be evaluated to check if it can be practically applied for an information service on EV-based rent-a-car systems. To this end, our scheme encodes each multi-day tour schedule to an integer vector, in which negative numbers are inserted to separate each day schedule. In addition, the relevant constraints are investigated and the corresponding fitness function is defined. The scheduling goal is to reduce the waiting time, for which tourist wait until their EV gets enough power to reach the next place along the route.

The rest of this paper is organized as follows: Section 2 reviews some related works. Section 3 describes the proposed scheme in detail, focusing on how to encode a schedule for multi-day tour to apply genetic operations. After Section 4 demonstrates and discusses the performance measurement results, Section 5 concludes this paper with a brief introduction of future work.

2   Related Work

As for an example of a multi-day trip scheduler, the City Trip Planner creates personalized tour routes and is currently providing services in world's most famous cities including New York [6]. Even though this service is not developed for EVs, it models tour scheduling as a team orienteering problem with time window. For a given set of user-selected spots, tour time at each spot, and tour length constraint, this service tries to maximize the sum of scores gained by visiting tour spots. The authors design a fast local search heuristic to respond to the user request within a reasonable time bound [7]. The search procedure iterates shake and insertion steps, removing and inserting some tour places, based on the estimated time window the tour spot can be placed. It automatically inserts those tour spots not selected but thought to be preferred by a tourist. It will be advantageous to insert the spots having charging facilities in the tour schedule generation for EVs.

Multi-day scheduling is quite similar to the multiple Traveling Salesman Problem (mTSP) in that each personal schedule can be mapped to a daily tour schedule [8]. In multi-day scheduling, the tourists do not have to return to the starting place. It has much more time complexity than the classic TSP, as the number of places to visit is much larger. Intuitively, to cope with this problem, a decomposition approach may cluster the places to visit into several groups [9]. In addition, it can be solved by exact solutions, heuristics, and transportations. For example, [10] first represents a visiting sequence by a chromosome having negative numbers (pseudo cities) to separate day-by-day schedules and runs evolutionary searches to achieve multi-objective goals. This encoding scheme will be exploited by this paper. In addition, mTSP is applied for vehicle routing problems which embrace multiple vehicles [11]. Here, vehicle specific constraints can be integrated in heuristic-based searches such as the Tabu scheme.

As for charging reservation over the route, [12] proposes a charging scheduling scheme with minimal waiting time on the large-scale network consisting of large number of charging stations cooperating with each other. The authors first define the waiting time as the sum of queuing time and charging time. Based on the theoretical analysis and observation that *the overall waiting time is minimized if charging demand of all charging stations is balanced*, a load distribution method is designed. Here, an EV can issue the next charging reservation request periodically. The EV-issued reservation request is forwarded to the charging stations within the driving range along the chain of the ad-hoc style communication network. Each station estimates the waiting time for the request and the result is sent backward to the issuer. The best one is selected and confirmed to both the EV and the charging station. As contrast to this model which just considers the next reservation, our scheme can build an entire tour schedule and possibly make a reservation.

In [13], authors have developed a waiting time estimation model for EV-based multi-destination tours. A genetic scheduler finds a visiting sequence having smaller waiting time just as the classical TSP reduces the tour distance, considering the charging facility availability, stay time, inter-destination distance. Moreover, to further reduce the waiting time, system-recommended tour spots, usually equipped with chargers, are added in the tour schedule [14]. This scheduler is a hybrid version of the ordinary TSP for user-selected spots and the selective TSP, or orienteering problem, for system-recommended spots. A chromosome includes all candidate recommended spots, while some of them will be deactivated with a tunable omission degree to exclude from the visit schedule. The scheduling goal is to reduce the waiting time and include as many spots as possible.

3 Tour Scheduler

3.1 Overview

Let $S = \{S_1, S_2, ..., S_n\}$ be the set of tour spots an EV driver wants to visit during the whole trip. They are all different, have their own stay time, and must be visited just once. The stay time will be affected by many factors such as user preference, weather, and the like. Considering the scheduling goal of reducing the waiting time, stay time will be set to the reasonable lower bound of the entire distribution. The stay time at S_i will be denoted by $T(S_i)$. In addition, we assume that the start and end points are given for every day, namely, $H = \{(H_i^s, H_i^e)\}$ for $1 \leq i \leq m$, where m is the number of tour days. Generally, $H_i^e = H_{i+1}^s$, as a tour team stays at a hotel and starts the next day trip from the same hotel. The start point of the first tour day, namely, H_1^s, and the end point of the last tour day, namely, H_m^e, will be pick-up and return stations of an EV rental service, respectively. Based on the road network, the distance between every pair of two spots is given for tour scheduling and $Dist(S_i, S_j)$ represents the distance between S_i and S_j.

The genetic algorithm is one of the most widely-used suboptimal search techniques, built upon the principle of natural selection and evolution [15]. Beginning from initial population consisting of given number of feasible solutions, genetic iterations improve the fitness of solutions generation by generation. Each generation is created from its parent generation by applying genetic operators such as selection, crossover, and mutation. To apply those operators, it is necessary to represent a feasible solution as an integer-valued vector. For example, the crossover operator swaps substrings from two parents, Hence, the first step of the multi-day tour-and-charging scheduler design is to encode a multi-day schedule. Basically, n user-selected tour spots are included in the schedule. Then, if the number of tour days is m, $(m-1)$ different negative numbers are added, creating a vector of $(n+m-1)$ discrete elements [10]. After all, a multi-day schedule can be handled just like a single-day schedule.

Figure 1 shows an example which builds a 3-day schedule for 9 places, namely, from S_1 to S_9. Hence, n is 9 and m is 3. This example also includes a chromosome for a visiting schedule, in which each daily schedule is separated by -1 and -2, resulting in 3 groups. Even if -2 comes first, it doesn't matter, as negative numbers are used just for separation of each daily schedule. Hence, the total length of a chromosome is 11. Each day has 3 places to visit, while the total trip begins and ends at rent-a-car station in the airport. The tourists stay at a hotel at the first and second days, respectively. The evaluation step calculates the predefined cost, such as the waiting time and total tour length, for each subschedule and hotel specification.

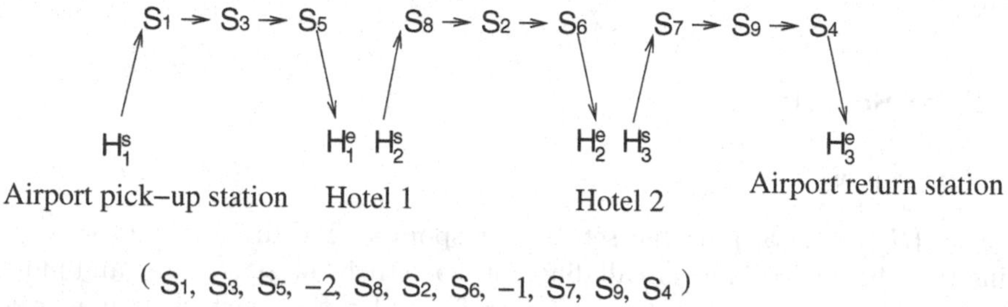

$$S_1 \rightarrow S_3 \rightarrow S_5 \qquad S_8 \rightarrow S_2 \rightarrow S_6 \qquad S_7 \rightarrow S_9 \rightarrow S_4$$

$$H_1^s \qquad\qquad H_1^e \; H_2^s \qquad\qquad H_2^e \; H_3^s \qquad\qquad H_3^e$$

Airport pick-up station   Hotel 1   Hotel 2   Airport return station

$$(\; S_1, \; S_3, \; S_5, \; -2, \; S_8, \; S_2, \; S_6, \; -1, \; S_7, \; S_9, \; S_4 \;)$$

Fig. 1. Effect of genetic iterations

3.2 Fitness Function

A fitness function, or cost function evaluates the quality of a schedule. The schedules in the population are sorted by fitness values. Straightforwardly, the higher the cost, the lower the fitness. Hence, they can be used interchangeably. To estimate the waiting time, our cost function defines two primitive operations, namely, *Move* and *Stay*. Along a tour route, two operations are invoked in turn. Here, both the waiting time and the tour time of a schedule, globally defined as W and L_i ($1 \leq i \leq n$), are estimated spot by spot. First, the *Move* operation accounts for driving between two spots. If battery remaining is enough, this operation just decreases battery remaining without changing the waiting time. Otherwise, the lacking battery amount and then corresponding charging time are calculated. The waiting time increases by the charging time. The tour time also increases by the charging time in addition to the driving time.

Next, the *Stay* operation traces battery charging at a tour spot. If a tour spot has no chargers, battery remaining does not increase. Otherwise, it linearly increases according to the stay time. Here, any battery charge and discharge model can be employed [16]. However, it should not exceed the maximum capacity, B_{Max}. Hence, the visiting sequence had better avoid two or more consequent spots having chargers and long stay time. For an example of 2-day trip specified by $\{(H_1^s, H_1^e), (H_2^s, H_2^e)\}$, and $\{S_1, S_2, -1, S_3, S_4\}$, the waiting time can be calculated by sequentially invoking $Move(Dist(H_1^s, S_1))$, $Stay(T(S_1))$, $Move(Dist(S_1, S_2))$, $Stay(T(S_2))$, and $Move(Dist(S_2, H_1^e))$ for the first day while $Move(Dist(H_2^s, S_3))$, $Stay(T(S_3))$, $Move(Dist(S_3, S_4))$, $Stay(T(S_4))$, and $Move(Dist(S_4, H_2^e))$ for the second day. The total waiting time is the sum of those two. At the same time, per-day tour length, L_i, and total tour length, $\sum L_i$, are calculated for the evaluation of fitness or cost of a tour schedule.

Users can set an upper bound on the daily tour length. Hence, each daily tour time is compared with this time budget, and if it is larger than the budget, the schedule is not valid. However, it is not discarded in the population as it can contribute to the improvement of fitness by reproduction. Instead, its cost function gives the largest value to it. Next, the waiting time is most critical to the user-side convenience, and it must be kept as small as possible. Besides, we can assign the tour length of each day as evenly as possible to make room for another activities or rest. Otherwise, we can also pack the tour schedule to as small number of tour days as possible, making a large free time chunk. It is hard to decide which one is better, but either requirement can be taken into account in the cost function. For the first case, the longest tour length is counted in the cost function. After all, the cost function F for a visiting sequence, x, is defined as follows:

$$
F(x) = \begin{cases} \infty & \text{for } \exists i \; L_i \geq T_b, \\ W \times 1000.0 + Max(L_i) & \text{for } 1 \leq i \leq m. \end{cases}
$$

where W is the waiting time and T_b is the time budget.

4   Experiment Result

This section measures the performance of our tour scheduling scheme via a proto-type implementation using Microsoft Visual Studio 2012. It employs the Roulette wheel selection method and randomly sets the initial population of chromosomes, or feasible solutions encoded by an integer vector. For better population diversity, it does not permit duplicated chromosomes in the contemporary population. As contrast, $(S_1, -1, S_2)$ and $(S_2, -1, S_1)$ are logically equivalent, having the same cost value. However, they are allowed to coexist in the population, as they can generate different offsprings. The inter-spot distance exponentially distributes with the average of 15.0 km. No two spots are separated by more than the driving range, which is the distance reachable with full battery capacity. In addition, the stay time also distributes exponentially with the average of 30 minutes, with its maximum limited to 2 hours.

The first experiment measures the effect of genetic iterations on performance metrics such as waiting time, longest daily tour time, and total tour time. As for genetic parameters, the population size is set to 32 and the number of iterations to 1,000. Its execution time is less than 1 second, regardless of the number of places to visit, for practical use. We select two sets of user-selected tour places, one for 16 places in 2 days and the other for 23 places in 3 days. As shown in Figure 2(a), the two cases begin from the cost of 1,000, which means no valid schedule is included in the initial population. Then, the 2-day case finds a feasible

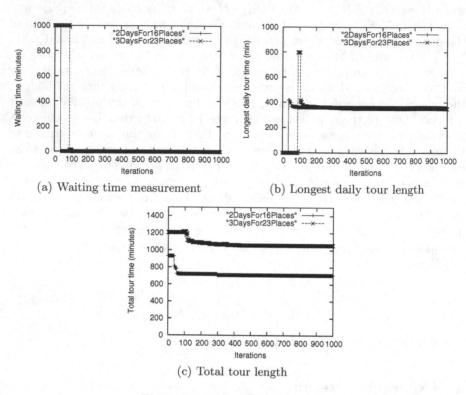

(a) Waiting time measurement   (b) Longest daily tour length

(c) Total tour length

Fig. 2. Effect of genetic iterations

solution after about 36 iterations, while the 3-day case after 106 iterations. For the 3-day case, the cost remains 10.0 temporarily, but goes down to 0 after 106 iterations. This result indicates that the time budget constraint is more critical to the cost function.

Next, Figure 2(b) plots the longest daily tour time for the two cases. Here, the value of 0 is meaningless as it takes place when there is no valid schedule. After the valid schedule is found, the daily tour time has non-zero values and then converges to a stable value. For the 2-day case, the longest daily tour becomes 416 minutes and then converges to 396 minutes. In addition, for the 3-day case it temporarily becomes 800 minutes and then converges to 351 minutes. For the case having larger number of places to visit, the longest daily tour time changes more often, or have more descending stages. Figure 2(c) plots the change in total tour length for above two cases. The total tour time in the 23-place case is definitely longer than that in the 16-place case. The initial value of the total tour time is 929 and 1,204 minutes, respectively. Then, each of them drops sharply when feasible schedules are found. After those points, the total tour length shows just minor improvement. During this iteration interval, the waiting time has already touched 0, and no more improvement is expected for it.

Now, the second experiment measures the schedulability of the proposed tour-and-charging scheduler according to the number of tour spots, average stay time, and the availability of chargers at each tour spot. The schedulability is the ratio of successfully finding a schedule having no waiting time out of 50 sets. Figure 3(a) plots the schedulability obtained by changing the number of tour spots from 10 to 20 for the 2-day trip and from 17 to 27 for the 3-day trip, respectively. For the 2-day case, up to 12 places, all selection sets are schedulable, and the schedulability begins to drop from 13 places, and touches zero on 19 places. Even 1 set is scheduled when the number of places is 20, it can be disregarded. Likewise, for the 3-day trip, all sets can be scheduled up to 17 places and no set can create a feasible schedule on 27 places.

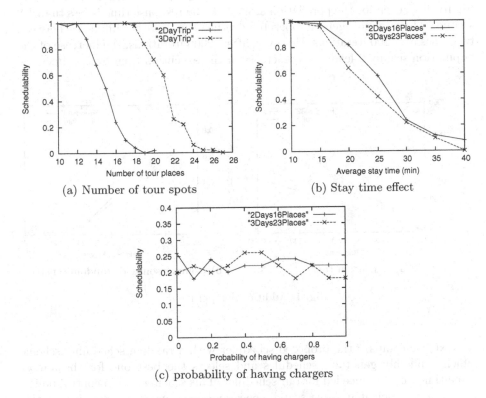

(a) Number of tour spots

(b) Stay time effect

(c) probability of having chargers

Fig. 3. Schedulablity measurement

Next, Figure 3(b) shows the effect of the average stay time to schedulability. With larger stay time, EVs can get more electricity during the trip. However, it makes both the total tour length and the longest daily tour too long. Hence, according to the increase in the average stay time, the schedulability gets poorer. Particularly, when the average stay time is 40 minutes, no set can be scheduled for the 2-day trip and just 5 can be scheduled for the 3-day trip. Finally, Figure

3(c) measures the effect of the availability of chargers. At first, we expect that schedulability will be enhanced with more chargers. However, as the time budget constraint dominates in tour set scheduling, its effect is not so significant. This parameter setting makes the schedulability fluctuate between 0.17 and 0.25, no one outperforming the other.

Additionally, Figure 4(a) plots the actual execution time of the implemented scheduler on an average performance PC, which consists of 2.4 GHz Intel(R) Core(TM)2 Duo CPU and 3 GM main memory, running Windows Vista operating system. The experiment takes a tour of 24 spots for 3 days. The population size is set to 32 and 64. The effect of population size to accuracy is already measured in Figure 2, so this experiment focuses on the execution speed according to the progress of genetic loops. The execution time of a genetic algorithm mainly depends on population size and the number of iteration. As shown in Figure 4(a), even for the population size of 64, the response time is less than 1.8 second. The common genetic operator overhead makes the difference between two cases 31.4 200 iterations. However, after 2,000 iterations, the increase of the population size by 2 lead to the increase in the execution time by 2.1 times.

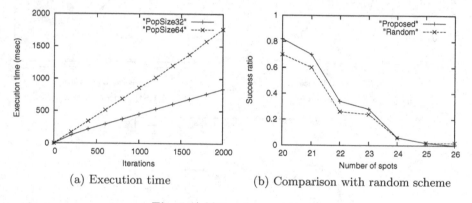

(a) Execution time   (b) Comparison with random scheme

Fig. 4. Additional experiment

Next, we compare the proposed scheduler with a random scheduling scheme which randomly generates schedules and selects the best one for the almost same time duration needed for our scheduler. This scheme has no control policy, but is quite efficient in large search space and gives us good reference. In this experiment, 3-day tour is assumed and the population size is set to 32. Figure 4(b) plots the success ratio for both schemes and shows that the proposed scheme outperforms by 12 % for the case of 20 tour spots. According to the increase in the number of selected spots, the difference gets smaller, as both schemes can hardly find a feasible schedule.

5   Conclusions

The smart grid is extending its coverage area to the transport system for better energy efficiency, particularly making an effort to prompt the penetration of EVs into our daily lives. They are suffering from short driving range, long charging time, and insufficient charging infrastructure. Those problems can be much serious when the trip length exceeds the driving range and the tour lasts for more than one day. It can be efficiently relieved by intelligent information technologies commonly available in these days. Accordingly, in this paper, we have designed a tour-and-charging scheduler for EVs, deciding when and where to charge EVs as well as the visiting sequence having the acceptable waiting time. Suboptimal techniques are unavoidably exploited to cope with large search space resulting from the extended number of places to visit for multi-day tours.

Our scheduler is developed based on the genetic algorithm, one of the most widely used suboptimal search schemes. To exploit genetic algorithms, a feasible schedule is encoded to an integer-valued vector having $(n+m\text{-}1)$ elements, where n is the number of places to visit and m is the number of tour days. $m\text{-}1$ negative number are inserted to separate each day schedule. The cost function mainly evaluates the waiting time, namely, the time amount the tourist must wait for the battery to be charged enough to reach the next station. It also integrates the time budget constraint and quantizes the daily and total tour length. The performance measurement result obtained from a prototype implementation shows that our scheme shows 100 % schedulability until 13 places for the 2-day trip and 17 places for the 3-day trip. In addition, for most cases, the reasonable answers are found in earlier stage of genetic iterations.

As future work, we are planning to apply our scheduler to the actual road network, specifically, in Jeju city, Republic of Korea. The distribution of tour places and terrain effect will give us more hints to improve or adapt the tour-and-charge scheduler proposed in this paper.

References

1. Ipakchi, A., Albuyeh, F.: Grid of the Future. IEEE Power & Energy Magazine, 52–62 (2009)
2. Goebel, C., Callaway, D.: Using ICT-Controlled Plug-in Electric Vehicles to Supply Grid Regulation in California at Different Renewable Integration Levels. IEEE Transactions on Smart Grid 4(2), 729–740 (2013)
3. Lue, A., Colorni, A., Nocerino, R., Paruscio, V.: Green Move: An Innovative Electric Vehicle-Sharing System. Procedia-Social and Behavioral Sciences 48, 2978–2987 (2012)
4. Bessler, S., Grønbæk, J.: Routing EV users towards an Optimal Charging Plan. In: International Battery, Hybrid and Fuel Cell Electric Vehicle Symposium (2012)
5. Ferreira, J., Pereira, P., Filipe, P., Afonso, J.: Recommender System for Drivers of Electric Vehicles. In: Proc. International Conference on Electronic Computer Technology, pp. 244–248 (2011)
6. Vansteenwegen, P., Souffriau, W., Berghe, G., Oudheusden, D.: The City Trip Planner: An Expert System for Tourists. Expert Systems with Applications 38, 6540–6546 (2011)

7. Vansteenwegen, P., Souffriau, W., Berghe, G., Oudheusden, D.: Iterated Local Search for the Team Orienteering Problem with Time Windows. Computers & Operations Research 36, 3281–3290 (2009)
8. Bektas, T.: The Multiple Traveling Salesman Problem: An Overview of Formulations and Solution Procedures. International Journal of Management Science 34, 209–219 (2006)
9. Lian, L., Castelain, E.: A Decomposition Approach to Solve a General Delivery Problem. Engineering Letters 18(1) (2010)
10. Shim, V., Tan, K., Tan, K.: A Hybrid Estimation of Distribution Algorithm for Solving the Multi-Objective Multiple Traveling Salesman Problem. In: IEEE World Congress on Computational Intelligence (2012)
11. Tarantilis, C., Zachariadis, E., Kiranoudis, C.: A Hybrid Metaheuristic Algorithm for the Integrated Vehicle Routing and Three-Dimensional Container-Loading Problem. IEEE Transactions on Intelligent Transportation Systems 10(2), 255–271 (2009)
12. Qin, H., Zhang, W.: Charging Scheduling with Minimal Waiting in a Network of Electric Vehicles and Charging Stations. In: ACM International Workshop on Vehicular Internetworking, pp. 51–60 (2011)
13. Lee, J., Kim, H., Park, G.: Integration of Battery Charging to Tour Schedule Generation for an EV-Based Rent-a-Car Business. In: Tan, Y., Shi, Y., Ji, Z. (eds.) ICSI 2012, Part II. LNCS, vol. 7332, pp. 399–406. Springer, Heidelberg (2012)
14. Lee, J., Park, G.: A Tour Recommendation Service for Electric Vehicles Based on a Hybrid Orienteering Model. In: ACM Symposium on Applied Computing, pp. 1652–1654 (2013)
15. Sivanandam, S., Deepa, S.: Introduction to Genetic Algorithms. Springer (2008)
16. Kim, H., Shin, K.: Scheduling of Battery Charge, Discharge, and Rest. In: IEEE Real-Time Systems Symposium, pp. 13–22 (2009)

Using a Normalized Score Centroid-Based Classifier to Classify Multi-label Herbal Formulae

Verayuth Lertnattee[1], Sinthop Chomya[1], and Virach Sornlertlamvanich[2]

[1] Faculty of Pharmacy, Silpakorn University, Sanamchandra Palace, Muang,
Nakorn Pathom 73000, Thailand
verayuths@hotmail.com, {verayuth,sinthop}@su.ac.th
[2] National Electronics and Computer Technology Center,
Pathumthani 12120 Thailand
{virach.sornlertlamvanich}@nectec.or.th

Abstract. The popularity of herbal medicines has greatly increased in worldwide countries over recent years. Herbal formula is a form of traditional medicine where herbs are combined to heal patient to heal faster and more efficiency. Herbal formulae can be divided into one or more therapeutic categories. The categories of a formula are usually based on decision from a group of experts. To support experts for classifying a formula, the normalized score centroid-based, is proposed for multi-label herbal formulae classification. The centroid-based classifier with more advanced term weight scheme is used. The normalized scores are calculated. The maximum number and cutoff point are set to adjust the decision for multi-label herbal formulae. The experiment is done using a mixed data set of herbal formulae collected from the Natural List of Essential Medicine and the list of common household remedies for traditional medicine. Moreover, a set of well-known commercial products are used for evaluating the effectiveness of the proposed method. From the results, the normalized score centroid-based classifier is an efficient method to classify multi-label herbal formulae. Its performance is depended on the set values of the maximum category and the cutoff point.

Keywords: Multi-label document, text classification, text categorization, herbal formula, centroid-based classifier.

1 Introduction

Origins of many traditional treatments in Thailand can be traced to India. The derivation has been diversified throughout many cultures since then [1]. Herbs are natural products that have been used safely for thousands of years to promote healing in patients. The popularity of herbal medicines has greatly increased in worldwide countries over recent years since the World Health Organization (WHO) suggested its member countries to use folk healing practices and herbal medicines as part of the basic public health projects [2]. They should be taken

S. Ramanna et al. (Eds.): MIWAI 2013, LNCS 8271, pp. 119–130, 2013.

with caution, and careful consideration of the dosage recommended. Traditional herbal formulae can be usually characterized by the use of several herbs. Various patterns of combinations from these herbs, can be applied on a disease. According to Thai traditional medicine, herbal formulae can be divided into one or more categories. The categories are usually based on indications of herbs in formulae. A combination of herbs may cause a formula has several categories. These categories may be arranged in flat and/or hierarchy. When the categories are arranged in flat, several main indications of the herbal formula can be applied to patients. In a complex situation, a formula is classified with one main category (or more) and a set of subcategories under the main category. With observation from human, it is hard to discover the combinational patterns of herbs in formulae. Nowadays, several data mining techniques, i.e., classification, clustering, association rules and, etc., have been developed and applied on several types of data. However, only few reliable research works have applied these techniques on herbal information. In this paper, we apply concept of text classification to categorize herbal formulae into therapeutic categories. However, many formulae have more than one category, even in a single component formula. Therefore, the normalized score centroid-based classifier is introduced to the multi-label herbal formulae. To the best of our knowledge, there is no research work contributes to classify multi-label herbal formulae. Performance of the proposed method is investigated on a set of herbal formulae found in the National List of Essential Drugs on both combine and single herbal formulae. Furthermore, a set of commercial traditional herbal formulae products is used for evaluating our proposed method. In the rest of this paper, section 2 presents herbal formulae. The concept of text categorization for herbal formulae is given in section 3. Section 4 introduces the multi-label herbal formula classification by the normalized score centroid-based classifier. The experimental settings are described in section 5. In section 6, a number of experimental results are given. A conclusion is made in section 7.

2   Herbal Formulae

As opposed to the western medicine, herbs are often used in formulae instead of being used singularly in larger amounts. Formulae allow us to blend herbs to enhance their positive effects (indications) and reduce or eliminate any negative effects (e.g., side effect and/or adverse effects) they may have. These formulae take a very long period of practice to master. Furthermore, the one benefit herbal medicine is that it allows practitioners to set formulae to match each patient and their signs and symptoms exactly. Instead of having a standard formula for a particular condition we can increase the clinical effectiveness of the herbs through this tailoring. For the patient this ideally means faster results with fewer side effects. In Thailand, some items of herbal medicines were placed on the National List of Essential Medicines (EM) in 1999, as part of an effort to promote the use of herbal medicines. Moreover, a list of common household traditional medicines was set. Two types of herbal medicine are placed. The first one is a set of composite well-known herbal formulae. The other one is a set of single herbs that can be used as single herbal formulae or can be combined to composite herbal formulae. According to the National List of Essential Medicine 2013 (the current version), two types of herbal medicines are placed, i.e., 1) Composite herbal medicines which are composed of several herbs, have been used traditionally and widely by the people for a long time. 2) Herbal medicines which have been developed from a single herb with evidence indicating its safety for use in humans. In the current version, eight therapeutic categories are placed, i.e., cardiovascular, gastrointestinal, gynecologic, antipyretic, respiratory, blood tonic, musculoskeletal and elementary balance. An herbal formula (single or composite), is belong to one or more than one category.

3   Multi-label Text Classification and Text Classifiers

With the increasing availability of online information, text classification turns into the important techniques by using machine learning. The objective of machine learning is to learn classifiers from examples which perform the category assignments automatically. This type of learning is induction-based supervised concept learning or just supervised learning. The supervised learning is the process of employing one or more computer learning techniques to automatically analyze and extract knowledge from data contained within a database [3]. Therefore, text classification falls within the machine learning paradigm and data mining. The definition of text classification is the activity of labeling natural language texts with thematic categories from a predefined set [4]. Several researches on text classification contributed to single-label classification. However, this paper focuses on multi-label herbal formulae classification. Classification techniques have been developed in a variety of learning techniques such as probabilistic models [5], example-based models (e.g., k-nearest neighbor) [6], linear models [7], support vector machine [8] and so on. Among these methods, a linear model called a centroid-based approach is attractive since it requires relatively less computation than other methods in both the learning and classification stages. Based on a so-called vector space model, the centroid-based method computes beforehand, for each class (category), an explicit profile (or class prototype), which is a centroid vector for all positive training documents of that category. Despite less computation time, centroid-based methods were shown in several literatures, including those in [9,7,10], to achieve relatively high classification accuracy. For the rest of this section, details of the centroid-based text classifier for herbal formulae and multi-label herbal formulae classification are given.

3.1   Centroid-Based Text Classifier for Herbal Formulae

In the centroid-based text categorization, a document (or a class) is represented by a vector using a vector space model with a bag of words (BOW) [11]. A set of words in the task of herbal formulae classification is the set of components to combine into a formula of traditional herbal medicine.

These components may be terms of crude drugs or natural sources for the crude drugs, i.e., plants, animals and elements. The simplest and popular one is applied term frequency (tf) and inverse document frequency (idf) in the form of $tf \times idf$ for representing a document. In this work, an herbal formula is used instead of a document. The tf means a weight of a crude drug in gram or milliliter in a formula. In a vector space model, given a set of herbal formulae $\mathcal{F} = \{f_1, f_2, ..., f_{|\mathcal{F}|}\}$, a formula f_j is represented by a formula vector $\mathbf{f}_j = \{w_{1j}, w_{2j}, ..., w_{|\mathcal{T}|j}\} = \{tf_{1j} \times idf_1, tf_{2j} \times idf_2, ..., tf_{|\mathcal{T}|j} \times idf_{|\mathcal{T}|}\}$, where w_{ij} is a weight assigned to a term (crude drug/natural product sources) t_i in a set of terms ($\mathcal{T}$) of the herbal formula. In this definition, tf_{ij} is term frequency of a term t_i (a component or a crude drug) in a formula f_j and idf_i is inverse document (formula) frequency, defined as $log(|\mathcal{F}|/df_i)$. The idf can be applied to eliminate the impact of frequent terms that exist in almost all documents. Here, $|\mathcal{F}|$ is the total number of formulae in a collection and df_i is the number of formulae which contain the term t_i. Besides term weighting, normalization is another important factor to represent a document or a class. A class prototype c_k is obtained by summing up all document vectors in $\mathcal{C}_k$ and then normalizing the result by its size. The formal description of a class prototype c_k is $\sum_{fj \in C_k} \mathbf{f}_j / \| \sum_{fj \in C_k} \mathbf{f}_j \|$, where $\mathcal{C}_k = \{f_j | f_j$ is a document belonging to the class $c_k\}$. The simple term weighting is $\overline{tf} \times idf$ where $\overline{tf}$ is average class term frequency of the crude drug. The formal description of $\overline{tf}$ is $\sum_{fj \in C_k} tf_{ijk}/|\mathcal{C}_k|$, where $|\mathcal{C}_k|$ is the number of formulae in a class c_k. The number of classes is $|\mathcal{C}|$, where C is the set of possible classes $\{c_1, c_2, ..., c_{|\mathcal{C}|}\}$. Term weighting described above can also be applied to a query or a test document. In general, the term weighting for a query is $tf \times idf$. Once a class prototype vector and a query vector have been constructed, the similarity between these two vectors can be calculated. The most popular one is cosine similarity [12]. This similarity can be calculated by the dot product between the class prototype vector and the normalized query vector. The test formula will be assigned to the class whose class prototype vector is the most similar to the vector of the test formula.

Some previous works, such as those in [7], attempted to apply some factors to improve performance of a centroid-based classifier with the basic term weighting of $tf \times idf$. In this paper, a centroid-based classifier with a standard deviation of term frequency for a term (sd), is used along with the standard $tf \times idf$.

3.2   Multi-label Herbal Formulae Classification

For a single-label text classification, a document is assigned only one category. Two approaches of classification are utilized to handle single-label classification, i.e., binary classification or multi-class classification. However, the problems in real work usually fall into the problem of multi-label text categorization, where each text document is assigned to one or more categories. Existing methods for multi-label classification can be divided into two main methods, i.e., problem transformation methods and algorithm adaptation methods [13]. Problem transformation methods can be defined as methods that transform the multi-label classification problem either into one or more single-label classification problems or regression problems. This is the same solution to solve the problem of single-label classification. However, the binary classification is based on the assumption of label independence. Therefore, during its transformation process, this method ignores label correlations that exist in the training data. Due to this information loss, predicted label sets from the binary classification are likely to contain either too few or too many labels, or labels that would never co-occur in practice [14]. With some limitations of binary classification, some extensions had been done such as [14,15] to provide better performance of classification. For algorithm adaptation methods, they can be defined as methods that extend or modify specific learning algorithms in order to handle multi-label data directly. The examples for these methods are shown in [16].

4   Classification of Multi-label Herbal Formulae by the Normalized Score Centroid-Based Classifier

In centroid-based classification, the similarity can be calculated by the dot product between these two vectors. Therefore, the test document will be assigned to the class whose class prototype vector is the most similar to the vector of the test document. In case of both vector are normalized into the unit length, the value of similarity of between the two vectors is in range of 0 and 1. For single-label classification, the maximum score is used.

The most popular one is cosine distance [11,12]. This similarity can be calculated by the dot product between these two vectors. Therefore, the test formula (f^t) will be assigned to the class C' whose class prototype vector is the most similar to the query vector (f^t) of the test formula.

$$C' = \arg\max_{C_k \in C} \ sim(f^t, c_k) \tag{1}$$

$$= \arg\max_{C_k \in C} \ \cos(f^t, c_k) \tag{2}$$

$$= \arg\max_{C_k \in C} \ \frac{f^t \cdot c_k}{\|f^t\|\|c_k\|} \tag{3}$$

$$= \arg\max_{C_k \in C} \ f^t \cdot c_k \tag{4}$$

Here, as stated before, $\|c_k\|$ is equal to 1 since the class prototype vector has been normalized. Moreover, the normalization of the test formula has no effect on ranking. Therefore, the test formula is assigned to the class when the dot product of the test formula vector and the class prototype vector achieves its highest value.

In this paper, a multi-label formula is also taken into account. A ranking categorization is applied for this work. Given a formula f_j, a system ranks the categories in according to their estimated similarity to a formula f_j . A ranked list of possible therapeutic categories will be taken into account. The problem of multi-label multiclass classification is how many categories belong to this formula. In this paper, two parameters are applied, i.e., the maximum number of predicted categories and the cutoff point. The normalized score centroid-based classifier (NS-CB) is introduced. The score for each category is divided by the maximum score. The normalized score of the first category in a new ranked list is 1. The categories belong to the formula, selected when the order of their score are less than or equal to the maximum number of predicted categories and the normalized score of the category is greater than or equal to the cutoff point. For example, when the maximum number of predicted categories is fixed to 3 and the cutoff point is 0.25, a list of normalized scores is shown in Figure 1.

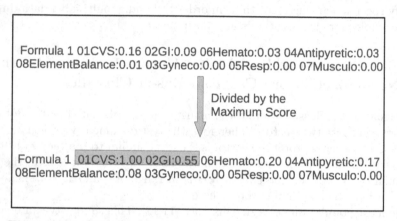

Fig. 1. A set of therapeutic categories is selected by normalized score

In Figure 1, the predicted therapeutic categories of the herbal formula 1 are 01CVS (cardiovascular) and 02GI (gastrointestinal) with the normalized scores of 1.00 and 0.55. The third category (06Hemato, i.e., blood tonic) is not selected due to the fact that its normalized score value is less than 0.25.

5   Experimental Settings and Evaluation

The experimental settings for multi-label herbal formulae and evaluation on experiments are described in this section.

5.1   Experimental Settings of Herbal Formulae

To evaluate the concept of text categorization for classifying multi-label herbal formulae, a set of 58 composite herbal formulae from the National List Essential Medicine 2011 and 230 single herbal formulae from a list of common household remedies for traditional medicines 1999, was used as a training set. The total number of herbal formulae was 288. The minimum number of therapeutic categories was 1 on both types of herbal formulae. The maximum numbers of therapeutic categories were 2 and 5 on composite and single herbal formulae, respectively. The distribution of herbal formulae on each therapeutic categories is shown it Table 1. The category-id (CAT ID), therapeutic categories' name (CAT Name) and the numbers of herbal formulae on each therapeutic category (# Formulae) are presented. The numbers of formulae for 1, 2, 3, 4 and 5 categories are 244, 37, 6, 0 and 1, respectively.

Table 1. The distribution of the numbers of herbal formulae on each category

CAT ID	CAT Name	# Formulae
01	Cardiovascular	38
02	Gastroinstestinal	96
03	Gynecologic	43
04	Antipyretic	85
05	Respiratory	20
06	Blood Tonic	12
07	Musculoskelatal	30
08	Elementary Balance	18

A unigram model was applied in all experiments. The term weighting schemes of , $\overline{tf} \times idf$ and $\overline{tf} \times idf/\sqrt{sd}$, were used for prototype vectors. The default term weighting scheme for a test herbal formulae was $tf \times idf$. The prototype and test formula vectors were normalized by their length. The cosine similarity was used. The value of score was in range from 0 to 1.

5.2 Evaluation

To assign a suitable category to a test document, we can apply the score between the class vector and the test document vector. For finding the best classifier on each dimension, it can be evaluated using measures similar to the traditional measures for evaluating a ranking-based retrieval system called precision (P), recall (R) and F_1 [4]. Two types of F_1 are used as performance indices, i.e., per-class effectiveness and all-class effectiveness. The per-class effectiveness of a classifier, is the average F_1 from ten trials for each class. The all-class performance of a classifier is calculated by averaging the local measures on a data set. Two types of all-class measures are usually used. For the macro-average, an equal weight is given to the performance on every class, regardless of how large the class is. When the size of each class is also affected to the overall performance, the micro-average is applied. In this work, both the macro-average and micro-average of the F_1 were used.

6 Experimental Results

Three experiments were conducted to evaluate the proposed method. In the first experiment, two term weighting schemes were used to constructed prototype vectors. The better performance one was selected in the next experiment. The value of maximum category and the cutoff point were set on the selected centroid-based classifier from the first experiment. In the last experiment, a set of commercial herbal medicinal products was used to evaluate the proposed method.

6.1 Finding the Appropriated Term Weight of Centroid-Based Classifiers

In order to evaluate the good term weight, two weighting schemes were used, i.e., $\overline{tf} \times idf$ (CB1) and $\overline{tf} \times idf/\sqrt{sd}$ (CB2), on a data set of multi-label herbal formulae. The experiment was performed using on formulae as a training set and a test set. The results of F_1, macro-average F_1 and micro-average F_1 on each class and on each classifier are shown in Table 2.

Table 2. Performance of text classifiers by F1 on therapeutic classes

CAT ID	CAT Name	F1	
		CB1	CB2
01	Cardiovascular	0.67	0.71
02	Gastroinstestinal	0.66	0.65
03	Gynecologic	0.59	0.67
04	Antipyretic	0.82	0.84
05	Respiratory	0.68	0.61
06	Blood Tonic	0.60	0.65
07	Musculoskelatal	0.59	0.61
08	Elementary Balance	0.53	0.51
Macro Average F1		0.64	**0.66**
Micro Average F1		0.68	**0.69**

From the results, some observations could be found. With the default decision of the multiclass classifiers, the highest score was used. Only single category is assigned. The overall performance is fair. The $\overline{tf} \times idf/\sqrt{sd}$ (CB2) was the best classifier among the two. The category of Antipyretic was gained the highest F_1. The CB2 was selected to adjust the two parameters in the next step.

6.2 Adjusting the Maximum Category and the Cutoff Point

In this experiment, the values of the maximum category and the cutoff point were set for the CB2 The normalized score was applied in this experiment. Therefore, CB2 is tuned to NS-CB2. For the number of maximum category, it was set to 3. Due to detail described in Section 5, the value of 3 could cover 287 of 288 formulae. The cutoff points were set to 80%, 25% and 0%. The cutoff points at 80% (NS-CB2Top3_80) and 25% (NS-CB2Top3_25) mean we consider only top three therapeutic category (by score) and accept the therapeutic categories which their normalized score ≥ 0.80 and 0.25, respectively. The cutoff point at 0%

Table 3. The results of the classifiers with maximum category and cutoff point

CAT ID	CAT Name	F1		
		SC-CB2Top3_80	SC-CB2Top3_25	SC-CB2Top3
01	Cardiovascular	0.74	0.87	0.93
02	Gastroinstestinal	0.73	0.87	0.95
03	Gynecologic	0.72	0.93	0.96
04	Antipyretic	0.87	0.91	0.97
05	Respiratory	0.61	0.75	0.84
06	Blood Tonic	0.74	0.87	0.90
07	Musculoskelatal	0.70	0.86	0.98
08	Elementary Balance	0.56	0.76	0.81
Macro Average F1		0.71	0.85	**0.92**
Micro Average F1		0.75	0.87	**0.94**

(NS-CB2Top3) means, all three therapeutic categories are accepted. The results from three different cutoff points are shown in Table 3.

From the results, some conclusions could be described. Among the three sets of a classifier, the NS-CB2Top3 gained the best performance. However, over-predicted categories may be occurred. For example, a formula with two categories may be suggested three categories. Although the NS-CB2Top3_25 achieved the lower performance, the numbers of categories suggested may be more appropriated.

6.3   Evaluation on the Commercial Products

In this experiment, the twelve well-known commercial products in herbal medicine were used as a test formulae (F01, F02, ..., F12). The ten formulae had more than one therapeutic categories. However, the eight therapeutic categories were taken into account. In order to evaluate the effectiveness of the algorithm, The NS-CB2top3_25 and NS-CB2top3 were selected to perform in this experiment. The results of the NS-CB2top3_25 and NS-CB2top3 are shown in Table 4 and 5, respectively. For each formula, the number of correct predicted categories/the number of total categories is shown in the last column (Predict/Actual). the category ids in bold type and in italic type mean the correctly and incorrectly predicted categories, respectively. The other actual category ids are presented in regular type.

From the results, some observations were made. For the NS-CB2Top3_25, it was optimum of the F01, F05 and F07. These formulate belonged to two categories and the NS-CB2Top3_25 classified predicted them two categories and both therapeutic categories were correctly classified. Only two categories may not be enough to predicted all therapeutic categories of formulae, e.g, F06. The NS-CB3Top3 often suggested more categories than a formula should have. Although three categories over predicted some categories, this value may be not cover all categories for the formulae F08, F10 and F11. The predicted categories by NS-CB2Top3_25 which were not correct, provided us interesting information.

Table 4. The result of NS-CB2Top3_25 on commercial products

Formula ID	Predicted Categories			Other Actual CAT	Predict/ Actual
	First CAT	Second CAT	Third CAT		
F01	01	02	-	-	2/2
F02	01	02	06	-	2/2
F03	04	01	06	-	2/2
F04	04	01	-	-	1/1
F05	04	02	-	-	2/2
F06	02	05	07	-	2/2
F07	02	07	-	-	2/2
F08	02	08	01	03	1/2
F09	03	06	08	-	1/1
F10	06	01	02	03	2/3
F11	01	03	06	08	2/3
F12	05	03	02	-	1/1

Table 5. The result of NS-CB2Top3 on commercial products

Formula ID	Predicted Categories			Other Actual CAT	Predict/ Actual
	First CAT	Second CAT	Third CAT		
F01	01	02	06	-	2/2
F02	01	02	06	-	2/2
F03	04	01	08	-	2/2
F04	04	01	02	-	1/1
F05	04	02	01	-	2/2
F06	02	05	07	-	2/2
F07	02	07	01	-	2/2
F08	02	08	01	03	1/2
F09	03	06	08	-	1/1
F10	06	01	02	03	2/3
F11	01	03	06	08	2/3
F12	05	03	02	-	1/1

These may be minor therapeutic categories in normal dose. It may be effective when higher dose is applied. On the other hand, it may cause some side effects. For example, the F08 may produce some side effects on gastrointestinal tract. For overall performance on both methods, the number of correct predicted categories/the number of total categories for all formulae was 89.96% (20/23*100).

7   Conclusion and Future Work

In this paper, the normalized score centroid-based, was proposed for multi-label herbal formulae classification. The centroid-based classifier with more advanced term weight scheme was used. The normalized scores were calculated. The maximum number and cutoff point were set to adjust the decision for multi-label herbal formulae. The experiment was done using a mixed data set of herbal formulae collected from the Natural List of Essential Medicine and the list of common household remedies for traditional medicine. Moreover, a set of well-known commercial products were used for evaluating the effectiveness of the proposed method. From the results, the normalized score centroid-based classifier was an efficient method to classify a formula into one category or multiple categories. Its performance depended on the set values of the maximum category and the cutoff point. We plan to find the way to automatically detect the appropriate number of categories for each formula. We left this for our future work.

Acknowledgments. This work was supported in part by the National Science and Technology Development Agency (NSTDA) under project number P-09-00159.

References

1. Lovell-Smith, H.D.: In defence of ayurvedic medicine. The New Zealand Medical Journal 119, 1–3 (2006)
2. Aziz, Z., Peng, T.N.: Herbal medicines: prevalence and predictors of use among malaysian adults. Complementary Therapies in Medicine 44, 44–50 (2009)
3. Roiger, R., Geatz, M.: Data Mining: A Tutorial Based Primer. Addison-Wesley, Boston (2002)
4. Sebastiani, F.: Machine learning in automated text categorization. ACM Computing Surveys 34, 1–47 (2002)
5. Nigam, K., McCallum, A.K., Thrun, S., Mitchell, T.M.: Text classification from labeled and unlabeled documents using em. Machine Learning 39, 103–134 (2000)
6. Duwairi, R., Al-Zubaidi, R.: A hierarchical k-nn classifier for textual data. The International Arab Journal of Information Technology 8, 251–259 (2011)
7. Lertnattee, V., Theeramunkong, T.: Effect of term distributions on centroid-based text categorization. Information Sciences 158, 89–115 (2004)
8. Joachims, T.: Learning to Classify Text using Support Vector Machines. Kluwer Academic Publishers, Dordrecht (2002)
9. Han, E.-H., Karypis, G., Kumar, V.: Text categorization using weight-adjusted k-nearest neighbor classification. In: Cheung, D., Williams, G.J., Li, Q. (eds.) PAKDD 2001. LNCS (LNAI), vol. 2035, pp. 53–65. Springer, Heidelberg (2001)
10. Schapire, R.E., Singer, Y., Singhal, A.: Boosting and Rocchio applied to text filtering. In: Croft, W.B., Moffat, A., Van Rijsbergen, C.J., Wilkinson, R., Zobel, J. (eds.) Proceedings of SIGIR-98, 21st ACM International Conference on Research and Development in Information Retrieval, pp. 215–223. ACM Press, New York (1998)
11. Salton, G., Buckley, C.: Term-weighting approaches in automatic text retrieval. Information Processing and Management 24, 513–523 (1988)
12. Singhal, A., Salton, G., Buckley, C.: Length normalization in degraded text collections. Technical Report TR95-1507 (1995)
13. Tsoumakas, G., Katakis, I.: Multi-label classification: An overview. International Journal Data Warehousing and Mining 3, 1–13 (2007)

14. Read, J., Pfahringer, B., Holmes, G., Frank, E.: Classifier chains for multi-label classification. Machine Learning 85, 333–359 (2011)
15. Fujino, A., Isozaki, H., Suzuki, J.: Multi-label text categorization with model combination based on f1-score maximization. In: Proceeding of The 3rd International Joint Conference on Natural Language Processing, pp. 823–828 (2008)
16. Hua, L.: Research on multi-classification and multi-label in text categorization. In: Proceeding of International Conference on Intelligent Human-Machine Systems and Cybernetics, pp. 86–89 (2009)

On the Computation of Choquet Optimal Solutions in Multicriteria Decision Contexts

Thibaut Lust[1] and Antoine Rolland[2]

[1] LIP6-CNRS Université Pierre et Marie Curie, Paris, France
thibaut.lust@lip6.fr
[2] Université de Lyon (ERIC), France
antoine.rolland@univ-lyon2.fr

Abstract. We study in this paper the computation of Choquet optimal solutions in decision contexts involving multiple criteria or multiple agents. Choquet optimal solutions are solutions that optimize a Choquet integral, one of the most powerful tools in multicriteria decision making. We develop a new property that characterizes the Choquet optimal solutions. From this property, a general method to generate these solutions in the case of several criteria is proposed. We apply the method to different Pareto non-dominated sets coming from different knapsack instances with a number of criteria included between two and seven. We show that the method is effective for a number of criteria lower than five or for high size Pareto non-dominated sets. We also observe that the percentage of Choquet optimal solutions increase with the number of criteria.

Keywords: Choquet integral, Multicriteria decision making, Multiagent optimization, Fuzzy measure, Multiobjective optimization.

1 Introduction

The Choquet integral [1] is one of the most powerful tools in multicriteria decision making [2, 3]. A Choquet integral can be seen as an integral on a non-additive measure (or capacity or fuzzy measure). It presents extremely wide expressive capabilities and can model many specific aggregation operators, including, but not limited to, the weighted sum, the minimum, the maximum, all the statistic quantiles, the ordered weighted averaging operator [4], the weighted ordered weighted averaging operator [5], etc.

However, this high expressiveness capability has a price: while the definition of a simple weighted sum operator with p criteria requires $p - 1$ parameters, the definition of the Choquet integral with p criteria requires setting of $2^p - 2$ values, which can be a problem even for low values of p.

Many approaches have been studied to identify the parameters of the Choquet integral [6]. Generally, questions are asked to the decision maker and the information obtained is represented as linear constraints over the set of parameters. An optimization problem is then solved in order to find a set of parameters which minimizes the error according to the information given by the decision maker.

S. Ramanna et al. (Eds.): MIWAI 2013, LNCS 8271, pp. 131–142, 2013.
© Springer-Verlag Berlin Heidelberg 2013

The approach considered in this paper is quite different: we will not try to identify the parameters of the Choquet integral but we will compute the solutions that are potentially optimal for at least one parameter set of the Choquet integral. Therefore, the parameters of the Choquet integral will not have to be determined. Instead, a set of solutions of smaller size comparing to the set of Pareto optimal solutions (which can be very huge in the case of multiobjective or multiagent problems) will be presented to the decision-maker. Each solution proposed will have interesting properties since they optimize at least one Choquet integral. Also, by computing all the Choquet optimal solutions, all the solutions that optimize one of the operators that the Choquet integral can model (weighted sum, ordered weighted averaging operator, etc.) will be generated.

We will present in the paper a new property that characterizes the Choquet optimal solutions. From this property, a general method to generate the Choquet optimal solutions is proposed. The method can be applied in different decision contexts involving multiple criteria or agents. The first application is in multicriteria decision making [7]: different alternatives are proposed to a decision maker and each alternative is evaluated according to a set of p criteria. No alternative Pareto dominates another and therefore no alternative can be a priori rejected. However, if we plan to use the Choquet integral in order to select the best alternative according to the preferences of the decision maker, we can first generate the solutions that are potentially optimal for at least one Choquet integral. This can be done in the absence of the decision maker. At the end, a smaller set comparing to the Pareto optimal set is proposed.

Another context in which the method can be applied is in decision contexts involving multiple agents, like multiagent knapsack problems [8], paper assignment problems [9], marriage problems in social networks [10], etc. In these problems, each agent has its own cost function and the aim is to generate a solution which is fair according to all the agents. Since the Choquet integral can model fairness operators like the max-min operator, the ordered weighted averaging and the weighted ordered weighted averaging operators, we can first compute all the potentially Choquet optimal solutions of these multiagent problems in order to generate a first set of candidate solutions.

The last application of the method is in multiobjective combinatorial optimization (MOCO) problems, which model situations where a decision-maker has to optimize several objectives simultaneously. These situations often come from a problem with a combinatorial number of solutions, for example spanning tree, shortest path, knapsack, traveling salesman tour, etc. [11]. To solve a MOCO problem, three different approaches are usually followed. In the *a posteriori* approach, all the Pareto optimal solutions are first generated. Once this has been done, the decision-maker is free to choose among all solutions the one that corresponds the best to his/her preferences. Another possibility, called the *a priori* approach, is to first ask the decision-maker what are his/her preferences among all the objectives and to compute an aggregation function [3] with specified parameters. The aggregation function is then optimized and at the end, only one solution is generally proposed to the decision-maker. A last possibility is to *interact* with the decision-maker along the process of generation of the solutions [12]. In this *interactive* approach, we ask the decision-maker to establish his/her preferences among different solutions, in order to guide the search, and to finally obtain a solution that suits him/her.

We propose here a new approach, between the *a posteriori* approach and the *a priori* approach, that consists in trying to find the set of solutions that are potentially optimal for at least one set of parameters of an aggregation function, and more specifically in this paper the Choquet integral.

Some papers already deal with the optimization of the Choquet integral of MOCO problems [13–15] but only when the Choquet integral is completely defined by the decision-maker. To our knowledge, the development of a method to generate the whole set of Choquet optimal solutions has not yet been studied, except the recent work of [16], where Lust and Rolland study the particular case of biobjective combinatorial optimization problems. They characterize the Choquet optimal solutions through a property and they define a method to generate all the Choquet optimal solutions. They apply the method to the biobjective knapsack problem and the biobjective minimum spanning tree.

We focus here on the general problem where the number of criteria can be more than two. We present a new property that characterizes the Choquet optimal set and develop a method based on this property to generate the Choquet optimal set, containing the solutions that are optimal solutions of Choquet integrals. We analyze the computational property of the method and we propose results for Pareto non-dominated sets. We will show that the Choquet integral becomes more expressive (can attain more Pareto optimal solutions) than the weighted sum, especially if the number of objectives increase.

The paper is placed in the context of MOCO problems and is organized as follows. In the next section, we first recall the definition of a MOCO problem and the Choquet integral. In section 3 we expose a property that characterizes the Choquet optimal set. In Section 4, we experiment the method on different instances of the multiobjective knapsack problem.

2   Aggregation Operators

In this section, we first introduce the formalism of a MOCO problem, and then present the weighted sum as it is the most popular aggregation operator. We then introduce the Choquet integral.

2.1   Multiobjective Combinatorial Optimization Problems

A multiobjective (linear) combinatorial optimization (MOCO) problem is generally defined as follows:

$$\text{``max''}_x f(x) = Cx = (f_1(x), f_2(x), \ldots, f_p(x))$$
$$\text{subject to } Ax \le b$$
$$x \in \{0, 1\}^n$$

$$x \in \{0,1\}^n \qquad \longrightarrow \qquad n \text{ variables}, i = 1, \ldots, n$$
$$C \in \mathbb{R}^{p \times n} \qquad \longrightarrow \qquad p \text{ objective functions}, k = 1, \ldots, p$$
$$A \in \mathbb{R}^{r \times n} \text{ and } b \in \mathbb{R}^{r \times 1} \qquad \longrightarrow \qquad r \text{ constraints}, j = 1, \ldots, r$$

A feasible solution x is a vector of n variables, having to satisfy the r constraints of the problem. Therefore, the feasible set in decision space is given by $\mathcal{X} = \{x \in \{0,1\}^n : Ax \leq b\}$. The image of the feasible set is given by $\mathcal{Y} = f(\mathcal{X}) = \{f(x) : x \in \mathcal{X}\} \subset \mathbb{R}^p$. An element of the set $\mathcal{Y}$ is called a cost-vector or a point.

Let us recall the concept of Pareto efficiency. We consider that all the objectives have to be maximized and we design by $\mathcal{P}$ the set of objectives $\{1, \ldots, p\}$.

Definition 1. *The Pareto dominance relation (P-dominance for short) is defined, for all $y^1, y^2 \in \mathbb{R}^p$, by:*

$$y^1 \succ_P y^2 \iff [\forall k \in \mathcal{P}, y_k^1 \geq y_k^2 \text{ and } y^1 \neq y^2]$$

Definition 2. *The strict Pareto dominance relation (sP-dominance for short) is defined as follows:*

$$y^1 \succ_{sP} y^2 \iff [\forall k \in \mathcal{P}, y_k^1 > y_k^2]$$

Within a feasible set $\mathcal{X}$, any element x^1 is said to be *P-dominated* when $f(x^2) \succ_P f(x^1)$ for some x^2 in $\mathcal{X}$, *P-optimal* (or *P-efficient*) if there is no $x^2 \in \mathcal{X}$ such that $f(x^2) \succ_P f(x^1)$ and *weakly P-optimal* if there is no $x^2 \in \mathcal{X}$ such that $f(x^2) \succ_{sP} f(x^1)$. The *P-optimal set* denoted by $\mathcal{X}_P$ contains all the P-optimal solutions. The image $f(x)$ in the objective space of a P-optimal solution x is called a P-non-dominated point. The image of the P-optimal set in $\mathcal{Y}$, equal to $f(\mathcal{X}_P)$, is called the Pareto front, and is denoted by $\mathcal{Y}_P$.

2.2   Weighted Sum

The most popular aggregation operator is the weighted sum (WS), where non-negative importance weights $\lambda_i (i = 1, \ldots, p)$ are allocated to the objectives.

Definition 3. *Given a vector $y \in \mathbb{R}^p$ and a weight set $\lambda \in \mathbb{R}^p$ (with $\lambda_i \geq 0$ and $\sum_{i=1}^p \lambda_i = 1$), the WS $f_\lambda^{ws}(y)$ of y is equal to:*

$$f_\lambda^{ws}(y) = \sum_{i=1}^p \lambda_i y_i$$

Note that there exist P-optimal solutions that do not optimize a WS, and they are generally called *non-supported* P-optimal solutions [11].

2.3   Choquet Integral

The Choquet integral has been introduced by Choquet [1] in 1953 and has been intensively studied, especially in the field of multicriteria decision analysis, by several authors (see [7, 2, 3] for a brief review).

We first define the notion of capacity, on which the Choquet integral is based.

Definition 4. *A capacity is a set function* $v \colon 2^{\mathcal{P}} \to [0,1]$ *such that:*

- $v(\emptyset) = 0$, $v(\mathcal{P}) = 1$ *(boundary conditions)*
- $\forall \mathcal{A}, \mathcal{B} \in 2^{\mathcal{P}}$ *such that* $\mathcal{A} \subseteq \mathcal{B}, v(\mathcal{A}) \leq v(\mathcal{B})$ *(monotonicity conditions)*

Therefore, for each subset of objectives $\mathcal{A} \subseteq \mathcal{P}$, $v(\mathcal{A})$ represents the importance of the coalition $\mathcal{A}$.

Definition 5. *The Choquet integral of a vector* $y \in \mathbb{R}^p$ *with respect to a capacity* v *is defined by:*

$$f_v^C(y) = \sum_{i=1}^{p} \left(v(Y_i^{\uparrow}) - v(Y_{i+1}^{\uparrow}) \right) y_i^{\uparrow}$$

$$= \sum_{i=1}^{p} (y_i^{\uparrow} - y_{i-1}^{\uparrow}) v(Y_i^{\uparrow})$$

where $y^{\uparrow} = (y_1^{\uparrow}, \ldots, y_p^{\uparrow})$ *is a permutation of the components of* y *such that* $0 = y_0^{\uparrow} \leq y_1^{\uparrow} \leq \ldots \leq y_p^{\uparrow}$ *and* $Y_i^{\uparrow} = \{j \in \mathcal{P}, y_j \geq y_i^{\uparrow}\} = \{i^{\uparrow}, (i+1)^{\uparrow}, \ldots, p^{\uparrow}\}$ *for* $i \leq p$ *and* $Y_{(p+1)}^{\uparrow} = \emptyset$.

We can notice that the Choquet integral is an increasing function of its arguments.

We can also define the Choquet integral through the Möbius representation [17] of the capacity. Any set function $v \colon 2^{\mathcal{P}} \to [0,1]$ can be uniquely expressed in terms of its Möbius representation by:

$$v(\mathcal{A}) = \sum_{\mathcal{B} \subseteq \mathcal{A}} m_v(\mathcal{B}) \quad \forall \mathcal{A} \subseteq \mathcal{P}$$

where the set function $m_v \colon 2^{\mathcal{P}} \to \mathbb{R}$ is called the Möbius transform or Möbius representation of v and is given by

$$m_v(\mathcal{A}) = \sum_{\mathcal{B} \subseteq \mathcal{A}} (-1)^{(a-b)} v(\mathcal{B}) \quad \forall \mathcal{A} \subseteq \mathcal{P}$$

where a and b are the cardinals of $\mathcal{A}$ and $\mathcal{B}$.

A set of 2^p coefficients $m_v(\mathcal{A})$ $(\mathcal{A} \subseteq \mathcal{P})$ corresponds to a capacity if it satisfies the boundary and monotonicity conditions [18]:

1. $m_v(\emptyset) = 0$, $\displaystyle\sum_{\mathcal{A} \subseteq \mathcal{P}} m_v(\mathcal{A}) = 1$

2. $\displaystyle\sum_{\mathcal{B} \subseteq \mathcal{A}, \, i \in \mathcal{B}} m_v(\mathcal{B}) \geq 0 \quad \forall \mathcal{A} \subseteq \mathcal{P}, i \in \mathcal{P}$

We can now write the Choquet integral with the use of Möbius coefficients. The Choquet integral of a vector $y \in \mathbb{R}^p$ with respect to a capacity v is defined as follows:

$$f_v^C(y) = \sum_{\mathcal{A} \subseteq \mathcal{P}} m_v(\mathcal{A}) \min_{i \in \mathcal{A}} y_i$$

3   Characterization of Choquet Optimal Solutions

We present in this section a characterization of the Choquet optimal solutions based on WS-optimal solutions, that is solutions that optimize a weighted sum. The set of Choquet optimal solutions of a MOCO problem with p objectives is called $\mathcal{X}_C$, and contains at least one solution $x \in \mathcal{X}$ optimal for each possible Choquet integral, that is $\forall v \in \mathcal{V}, \exists x_c \in \mathcal{X}_C \,|\, f_v^C(f(x_c)) \geq f_v^C(f(x)) \; \forall x \in \mathcal{X}$, where $\mathcal{V}$ represents the set of capacity functions defined over p objectives. Note that each Choquet optimal solution is at least weakly P-optimal [16].

In [16], Lust and Rolland studied the particular case of two objectives and they showed that $\mathcal{X}_C$ could be obtained by generating all WS-optimal solutions in each subspace of the objectives separated by the bisector ($f_1(x) \geq f_2(x)$ or $f_2(x) \geq f_1(x)$), and by adding a particular point M with $M_1 = M_2 = \max\limits_{x \in \mathcal{X}} \min(f_1(x), f_2(x))$. We show here how this property can be generalized to more than two objectives.

We will work with the image of $\mathcal{X}_C$ in the objective space, $\mathcal{Y}_C$, equal to $f(\mathcal{X}_C)$. To each point $y_c \in \mathcal{Y}_C$ corresponds thus at least one solution x_c in $\mathcal{X}_C$.

Let σ be a permutation on $\mathcal{P}$. Let O_σ be the subset of points $y \in \mathbb{R}^p$ such that $y \in O_\sigma \iff y_{\sigma_1} \geq y_{\sigma_2} \geq \ldots \geq y_{\sigma_p}$.

Let p_{O_σ} be the following application:

$$p_{O_\sigma} : \mathbb{R}^p \to \mathbb{R}^p, (p_{O_\sigma}(y))_{\sigma_i} = (\min(y_{\sigma_1}, \ldots, y_{\sigma_i})), \forall i \in \mathcal{P}$$

For example, if $p = 3$, for the permutation (2,3,1), we have:

$$p_{O_\sigma}(y) = \big(\min(y_2, y_3, y_1), \min(y_2), \min(y_2, y_3)\big)$$

We denote by $\mathcal{P}_{O_\sigma}$ the set containing the points obtained by applying the application $p_{O_\sigma}(y)$ to all the points $y \in \mathcal{Y}$. As $(p_{O_\sigma}(y))_{\sigma_1} \geq (p_{O_\sigma}(y))_{\sigma_2} \geq \ldots \geq (p_{O_\sigma}(y))_{\sigma_p}$, we have $\mathcal{P}_{O_\sigma} \subseteq O_\sigma$.

3.1   Characterization Theorem

We propose a new characterization of the Choquet optimal set.

Theorem 1

$$\mathcal{Y}_C \cap O_\sigma = \mathcal{Y} \cap WS(\mathcal{P}_{O_\sigma})$$

where $WS(\mathcal{P}_{O_\sigma})$ designs the set of WS-optimal points of the set $\mathcal{P}_{O_\sigma}$.

This theorem characterizes the solutions which can be Choquet optimal in the set of feasible solutions as being, in each subspace of the objective space $\mathcal{Y}$ where $y_{\sigma_1} \geq y_{\sigma_2} \geq \ldots \geq y_{\sigma_p}$, the solutions that have an image corresponding to a WS-optimal point in the space composed of the original subspace plus the projection of all the other points following the application p_{O_σ}.

Proof

In the following, we will denote O_σ as simply O for the sake of simplicity, and we will consider, without loss of generality, that the permutation σ is equal to $(1, 2, \ldots, p)$, that is $y \in O \Leftrightarrow y_1 \geq y_2 \geq \cdots \geq y_p$.

We know that $\mathcal{Y}_C \subseteq \mathcal{Y}$ and then $\mathcal{Y}_C \cap O \subseteq \mathcal{Y} \cap O$. We also know that $WS(\mathcal{P}_O) \subseteq O$ and then $\mathcal{Y}_C \cap WS(\mathcal{P}_O) \subseteq \mathcal{Y} \cap O$. Let y be in $\mathcal{Y}_C \cap O$.

- $[y \in \mathcal{Y}_C \cap O \Rightarrow y \in \mathcal{Y} \cap WS(\mathcal{P}_O)]$

  Let us write the Choquet integral of $y \in O$ related to a capacity v, with $Y_i = \{1, \ldots, i\}$ and $Y_0 = \emptyset$:

$$f_v^C(y) = \sum_{i=1}^{p} \big(v(Y_i) - v(Y_{i-1}) \big) y_i$$

$$= \sum_{i=1}^{p} \lambda_i y_i$$

As v is monotonic for the inclusion, $\lambda_i = v(Y_i) - v(Y_{i-1})$ is always positive. We have also $\sum_{i=1}^{p} \lambda_i = \sum_{i=1}^{p} (v(Y_i) - v(Y_{i-1})) = v(\mathcal{P}) - v(\emptyset) = 1$.

Let $z \in \mathcal{Y}$. As $y \in \mathcal{Y}_C$, we have $f_v^C(y) \geq f_v^C(z)$. We also have $\forall i \in \mathcal{P}$, $z_i \geq \min\{z_1, z_2, \ldots, z_i\} = (p_O(z))_i$, and as the Choquet integral is an increasing function of its arguments, we have $f_v^C(z) \geq f_v^C(p_O(z))$. And since $p_O(z) \in O$, we have $f_v^C(p_O(z)) = \sum_{i=1}^{p} \lambda_i p_O(z)_i$. Therefore we have $\forall z \in \mathcal{Y}$:

$$\sum_{i \in \mathcal{P}} \lambda_i y_i = f_v^C(y) \geq f_v^C(z) \geq f_v^C(p_O(z)) = \sum_{i \in \mathcal{P}} \lambda_i p_O(z)_i$$

where $\lambda_i \geq 0 \ \forall i \in \mathcal{P}$ and $\sum_{i=1}^{p} \lambda_i = 1$. So $y \in WS(\mathcal{P}_O)$ and as $y \in \mathcal{Y}$, $y \in \mathcal{Y} \cap WS(\mathcal{P}_O)$.

- $[y \in WS(\mathcal{P}_O) \cap \mathcal{Y} \Rightarrow y \in \mathcal{Y}_C \cap O]$

  Let $y \in WS(\mathcal{P}_O) \cap \mathcal{Y}$. Then there are $\lambda_1, \ldots, \lambda_p \geq 0$ such that $\sum_{i=1}^{p} \lambda_i = 1$ and

$$\forall z \in \mathcal{Y}, \ \sum_{i \in \mathcal{P}} \lambda_i y_i \geq \sum_{i \in \mathcal{P}} \lambda_i p_O(z)_i$$

By definition, $(p_O(z))_i = \min(z_1, \ldots, z_i), \forall i \in \mathcal{P}$.

Let $\mathcal{A} \subseteq \mathcal{P}$. Let us define a set function m such that $m(\mathcal{A}) = \lambda_i$ if $\mathcal{A} = \{1, \ldots, i\}$ and $m(\mathcal{A}) = 0$ if not.

Then

$$\sum_{i \in \mathcal{P}} \lambda_i (p_O(z))_i = \sum_{i \in \mathcal{P}} \lambda_i \min(z_1, \ldots, z_i)$$

$$= \sum_{\mathcal{A} \subseteq \mathcal{P}} m(\mathcal{A}) \min_{i \in \mathcal{A}} z_i$$

Let us remind that the set function m corresponds to a capacity if:

1. $m(\emptyset) = 0$, $\displaystyle\sum_{\mathcal{A} \subseteq \mathcal{P}} m(\mathcal{A}) = 1$

2. $\displaystyle\sum_{\mathcal{B} \subseteq \mathcal{A}, \; i \in \mathcal{B}} m(\mathcal{B}) \geq 0 \quad \forall \mathcal{A} \subseteq \mathcal{P}, i \in \mathcal{P}$

All these conditions are satisfied:

- $m(\emptyset) = 0$ by definition
- $\displaystyle\sum_{\mathcal{A} \subseteq \mathcal{P}} m(\mathcal{A}) = \sum_{i=1}^{p} \lambda_i = 1$
- all $m(\mathcal{B})$ are non-negative as $\lambda_i \geq 0$

So we have a set of Möbius coefficients such that $\forall z \in \mathcal{Y}$,

$$f_v^C(y) = \sum_{\mathcal{A} \subseteq \mathcal{P}} m(\mathcal{A}) \min_{i \in \mathcal{A}} y_i$$

$$= \sum_{i \in \mathcal{P}} \lambda_i y_i$$

$$\geq \sum_{i \in \mathcal{P}} \lambda_i p_O(z)_i$$

$$\geq \sum_{\mathcal{A} \subseteq \mathcal{P}} m(\mathcal{A}) \min_{i \in \mathcal{A}} z_i$$

$$\geq f_v^C(z)$$

$\square$

4   Generation of Choquet Optimal Solutions

4.1   Algorithm for Generating $\mathcal{X}_C$

We present in this section an algorithm to generate the set $\mathcal{X}_C$ containing all the Choquet optimal solutions of a MOCO problem. The algorithm straightly follows from Theorem 1.

For all the permutations σ on $\mathcal{P}$, we have to:

1. Determine the projections with the application p_{O_σ}
2. Solve a WS problem

The projections are defined with the application

$$(p_{O_\sigma}(y))_{\sigma_i} = (\min(y_{\sigma_1}, \ldots, y_{\sigma_i})), \forall i \in \mathcal{P}$$

for each $y \in \mathcal{Y}$.

However, among these projections, only the P-non-dominated points are interesting (since if a point is P-dominated, its WS is inferior to the WS of at least another point). Therefore, to determine the projections, the following MOCO problem (called P_σ) has to be solved:

$$\text{``max''} p(x) = \underset{x \in \mathcal{X} \setminus \mathcal{X}_\sigma}{\text{``max''}} (f_{\sigma_1}(x), \min(f_{\sigma_1}(x), f_{\sigma_2}(x)), \ldots, \min(f_{\sigma_1}(x), f_{\sigma_2}(x), \ldots, f_{\sigma_p}(x)))$$

where $\mathcal{X}_\sigma$ is the set such that $x \in \mathcal{X}_\sigma \iff f_{\sigma_1}(x) \geq f_{\sigma_2}(x) \geq \ldots \geq f_{\sigma_p}(x)$.

Once the projections have been defined, a WS problem has to be solved, in $\mathcal{X}_\sigma$, and by adding the P-non-dominated points obtained from P_σ.

We give the main lines of the method in Algorithm 1.

Algorithm 1. Generation of $\mathcal{X}_C$

Parameters ↓: a MOCO problem
Parameters ↑: the set $\mathcal{X}_C$
Let σ be a permutation on $\mathcal{P}$, and Σ the set of permutations
Let $\mathcal{X}_\sigma$ be the set such that $x \in \mathcal{X}_\sigma \iff f_{\sigma_1}(x) \geq f_{\sigma_2}(x) \geq \ldots \geq f_{\sigma_p}(x)$
$\mathcal{X}_C \leftarrow \{\}$
for all $\sigma \in \Sigma$ **do**
--| Determination of the projections:
   Solve the following MOCO problem, called P_σ, in $x \in \mathcal{X} \setminus \mathcal{X}_\sigma$:
   $\text{``max''} p(x) = \underset{x \in \mathcal{X} \setminus \mathcal{X}_\sigma}{\text{``max''}} (f_{\sigma_1}(x), \min(f_{\sigma_1}(x), f_{\sigma_2}(x)), \ldots, \min(f_{\sigma_1}(x), f_{\sigma_2}(x), \ldots, f_{\sigma_p}(x)))$
   Let $\mathcal{Y}_{P_\sigma}$ the Pareto non-dominated points obtained from solving (P_σ)
   Solve the WS problem (called WS_σ) $\underset{x \in \mathcal{X}_\sigma}{\max} f_\lambda^{ws}(f(x))$ with the additional points of $\mathcal{Y}_{P_\sigma}$.
   Let $\mathcal{X}_{ws_\sigma}$ the solutions obtained from (WS_σ)
   $\mathcal{X}_C \leftarrow \mathcal{X}_C \cup \mathcal{X}_{ws_\sigma}$
end for

4.2 Experiments

We present results for defined Pareto fronts, that is, a Pareto front is given, and the aim is to determine, among the P-non-dominated points, the Choquet optimal points.

To generate Pareto fronts with different numbers of objectives, we have applied a heuristic to several multiobjective knapsack instances. We have used knapsack instances with random profits. The heuristic is an adaptation of the one presented in [19]. Note that the aim is not to generate the best possible approximation of the Pareto front of these instances, but only to generate a set of P-non-dominated points. The results are given in Table 1, for $p = 2, \ldots, 7$, and for 3000 points.

We respectively indicate the number of criteria, the number of WS-optimal points, the number of Choquet optimal points, the proportion of WS-optimal points under the total number of points, the proportion of Choquet optimal

Table 1. Random multiobjective knapsack instances (3000 points)

# Crit	# WS	# C	% WS	% C	% C not WS
2	123	128	4.10	4.27	3.91
3	184	240	6.13	8.00	23.33
4	240	380	8.00	12.67	36.84
5	282	485	9.40	16.17	41.86
6	408	676	13.6	22.53	39.64
7	528	1016	17.6	33.87	48.03

points under the total number of points and the proportion of Choquet optimal points that are not WS-optimal.

We see that if the number of Choquet optimal points and the number of Choquet optimal points that are not WS-optimal points are very small for $p = 2$, these number grows rapidly with the number of criteria: for $p = 7$, we have that 33.87% of the P-non-dominated points are Choquet optimal points, and 48.03% of them are not WS-optimal. We see thus that when the number of criteria increases, the Choquet integral allows to attain considerably more P-non-dominated points than the WS.

In table 2, we indicate the CPU times needed to generate the Choquet optimal points of these sets (on a Intel Core i7-3820 at 3.6GHz). We see that if the CPU times are reasonable for $p \leq 5$, they become rapidly high for $p = 6$ (more than 4 minutes) or $p = 7$ (more than 1 hour). We also compare the CPU times obtained with the Algorithm 1 with a method based on a linear program: for each point of the Pareto front, we check if there exists a capacity v such that the Choquet integral of this point is better that all the other points. We see that this method is more effective once $p \geq 6$.

Table 2. CPU Random multiobjective knapsack instances (3000 points)

# Crit	CPU(s) Algorithm 1	CPU(s) LP
2	19.91	46.32
3	12.26	32.22
4	19.89	44.76
5	33.43	66.80
6	291.58	88.78
7	4694.23	110.56

In table 3, we compare the CPU times obtained by both methods, for $p = 5$, according to the number of solutions (between 100 and 3000). We see that until the number of solutions is equal to 2000, the method based on the linear program is more effective. It is only for sets with at least 2000 solutions that the method based on the Algorithm 1 becomes faster, since it is only for high size sets that we can take the most of enumerating all the permutations.

Table 3. CPU Random multiobjective knapsack instances (5 criteria)

# Sol	CPU (s) LP	CPU (s) Algorithm 1
100	0.54	3.11
250	0.85	3.49
500	2.01	5.34
1000	6.78	9.95
2000	24.72	20.57
3000	73.65	33.43

5 Conclusion

We have introduced in this paper a new characterization of the Choquet optimal solutions in multicriteria decision contexts, and more specifically for multiobjective combinatorial optimization problems. We have also presented an algorithm to obtain these solutions based on this characterization. The experimentations showed that increasing the number of objectives increase the expressiveness of the Choquet integral comparing to the WS (more P-non-dominated points can be attained). This work about generating all Choquet-optimal solutions opens many new perspectives:

- Following [20], it will be interesting to study and to define what brings exactly and concretely (for a decision maker) the Choquet optimal solutions that are not WS optimal solutions, given that they are harder to compute.
- We have shown that it can be very time-consuming to apply the general method developed in this paper due to the increase of parameters of the Choquet integral. Therefore, dedicated methods to compute all the Choquet optimal solutions of specific problems could be studied.
- It will be interesting to study if we can adapt the characterization of the Choquet optimal solutions to more restrictive set of capacities, such that k-additive capacities [21].

References

1. Choquet, G.: Theory of capacities. Annales de l'Institut Fourier (5), 131–295 (1953)
2. Grabisch, M., Labreuche, C.: A decade of application of the Choquet and Sugeno integrals in multi-criteria decision aid. Annals OR 175(1), 247–286 (2010)
3. Grabisch, M., Marichal, J.C., Mesiar, R., Pap, E.: Aggregation functions. Encyclopedia of Mathematics and its Applications, vol. 127. Cambridge University Press, Cambridge (2009)
4. Yager, R.: On ordered weighted averaging aggregation operators in multicriteria decision making. IEEE Trans. Systems, Man and Cybern. 18, 183–190 (1998)
5. Torra, V.: The weighted OWA operator. Int. J. Intell. Syst. (12), 153–166 (1997)
6. Grabisch, M., Kojadinovic, I., Meyer, P.: A review of methods for capacity identification in choquet integral based multi-attribute utility theory: Applications of the kappalab R package. European Journal of Operational Research 186(2), 766–785 (2008)

7. Grabisch, M.: The application of fuzzy integrals in multicriteria decision making. European Journal of Operational Research 89, 445–456 (1996)
8. Nicosia, G., Pacifici, A., Pferschy, U.: On multi-agent knapsack problems. In: Proceedings of the 8th Cologne-Twente Workshop on Graphs and Combinatorial Optimization, pp. 44–47 (2009)
9. Goldsmith, J., Sloan, R.: The conference paper assignment problem. In: AAAI Workshop on Preference Handling for Artificial Intelligence (2007)
10. Gent, I., Irving, R., Manlove, D., Prosser, P., Smith, B.: A constraint programming approach to the stable marriage problem. In: Walsh, T. (ed.) CP 2001. LNCS, vol. 2239, pp. 225–239. Springer, Heidelberg (2001)
11. Ehrgott, M.: Multicriteria Optimization, 2nd edn. Springer, Berlin (2005)
12. Ojalehto, V., Miettinen, K., Mäkelä, M.: Interactive software for multiobjective optimization: IND-NIMBUS. WSEAS Transactions on Computers 6, 87–94 (2007)
13. Fouchal, H., Gandibleux, X., Lehuédé, F.: Preferred solutions computed with a label setting algorithm based on Choquet integral for multi-objective shortest paths. In: 2011 IEEE Symposium on Computational Intelligence in Multicriteria Decision-Making (MDCM), Paris, France, pp. 143–150. IEEE (2011)
14. Galand, L., Perny, P., Spanjaard, O.: Choquet-based optimisation in multiobjective shortest path and spanning tree problems. European Journal of Operational Research 204(2), 303–315 (2010)
15. Galand, L., Perny, P., Spanjaard, O.: A branch and bound algorithm for Choquet optimization in multicriteria problems. In: Proceedings Lecture Notes in Economics and Mathematical Systems, vol. 634, pp. 355–365 (2011)
16. Lust, T., Rolland, A.: Choquet optimal set in biobjective combinatorial optimization. Computers & Operations Research 40(10), 2260–2269 (2013)
17. Rota, G.C.: On the foundations of combinatorial theory I. Theory of Mobius functions. Z. Wahrscheinlichkeitstheorie und Verw. Gebiete 2, 340–368 (1964)
18. Chateauneuf, A., Jaffray, J.Y.: Some characterizations of lower probabilities and other monotone capacities through the use of mobius inversion. Mathematical Social Sciences 17(3), 263–283 (1989)
19. Lust, T., Teghem, J.: The multiobjective multidimensional knapsack problem: a survey and a new approach. International Transactions in Operational Research 19(4), 495–520 (2012)
20. Meyer, P., Pirlot, M.: On the expressiveness of the additive value function and the Choquet integral models. In: DA2PL 2012: from Multiple Criteria Decision Aid to Preference Learning, pp. 48–56 (2012)
21. Grabisch, M.: k-order additive discrete fuzzy measures and their representation. Fuzzy Sets and Systems (92), 167–189 (1997)

Test-Cost-Sensitive Attribute Reduction
in Decision-Theoretic Rough Sets

Xi'ao Ma[1], Guoyin Wang[2], Hong Yu[2], and Feng Hu[2]

[1] School of Information Science and Technology,
Southwest Jiaotong University, Chengdu, 610031, China
[2] Chongqing Key Laboratory of Computational Intelligence,
Chongqing University of Posts and Telecommunications, Chongqing, 400065, China
`maxiao73559@163.com, wanggy.cq@hotmail.com,`
`{yuhong,hufeng}@cqupt.edu.cn`

Abstract. Decision-theoretic rough sets (DTRS) can be seen as a kind of misclassification cost-sensitive learning model. In DTRS, attribute reduction is the process of minimizing misclassification costs. However in parctice, data are not free, and there are test costs to obtain feature values of objects. Hence, the process of attribute reduction should help minimizing both of misclassification costs and test costs. In this paper, the minimal test cost attribute reduct (MTCAR) problem is defined in DTRS. The objective of attribute reduction is to minimize misclassification costs and test costs. A genetic algorithm (GA) is used to solve this problem. Experiments on UCI data sets are performed to validate the effectiveness of GA to solve MTCAR problem.

Keywords: Decision-theoretic rough sets, misclassification costs, test costs, attribute reduct, cost-sensitive.

1 Introduction

The rough set theory first described by Pawlak [11] is a valid mathematical tool to handle imprecise, uncertain and vague information. Rough set-based data analysis methods have been successfully applied in bioinformatics, knowledge discovery, mechine learning, web and data mining, signal and image processing, cluster analysis, and so on [12].

Pawlaks rough set models assume the classification must be fully correct or certain. Hence, it cannot effectively deal with data sets which have noisy data and latent useful knowledge. Various probabilistic rough sets (PRS) have been presented in order to solve the problem [2,6,13,19]. Decision-theoretic rough set (DTRS) proposed by Yao [16] is a typical probabilistic extension of rough set, which is based on Bayesian decision procedure to calculate probabilistic thresholds by minimizing the loss. Many PRS can be explicitly derived from DTRS by considering various classes of loss functions [16].

Attribute reduction is the key problem in rough set theory [3,15]. In PRS, attribute reduction can be seen as minimal subset of attributes preserving a

S. Ramanna et al. (Eds.): MIWAI 2013, LNCS 8271, pp. 143–152, 2013.
© Springer-Verlag Berlin Heidelberg 2013

particular property of the given information table [17]. For example, Nguyen and Slezak [10] defined the notions of α reduct and α relative reduct for decision tables. Mi et al. [7] proposed the concept of β lower distribution reduct and β upper distribution reduct based on the variable precision rough set model. Li et al. [5] investigated the monotonicity of the positive region, and presented a definition of the attribute reduct based on positive region expanding. Zhao et al. [18] examined the definition of the attribute reduct and pointed out three problems of the existing definition for the attribute reduct. We may regard these methods as cost-insensitive attribute reduct methods.

However, in many real world applications, many problems, such as medical diagnosis, credit card fraud, intrusion detection, are often cost-sensitive. In cost-sensitive learning model, misclassification costs and test costs are the most active and important research areas [14]. Misclassification costs are the loss incurred for taking incorrectness action. Test costs are what we pay for collecting data items [8]. The process of attribute reduction should help in minimizing misclassification costs and test costs.

In fact, DTRS can be seen as a kind of cost-sensitive learning model [4]. In DTRS, decision costs is a very imporant concept, attribute reduction can be interpreted as selecting the minimal attribute sets that can minimize the decision cost [4]. Hence, Jia et al. [4] presented minimum cost attribute reduct. However, they concern only on misclassification costs, and does not consider test costs. Since data are not free in the real world, the objective of attribute reduction should aim at finding the cheapest attribute subsets which minimize the decision cost. Min et al. [8,9] defined the minimal test cost reduct problem, which is to select a set of tests satisfying a minimal test cost criterion, but they do not take into account misclassification costs.

In this paper, the minimal test cost attribute reduct (MTCAR) problem is introduced into DTRS, both misclassification costs and test costs are considered. The objective of attribution reduction is to select an attribute subset with a minimal test cost as well as minimize decision cost. A genetic algorithm (GA) to MTCAR problem is proposed. Experiment results on UCI data sets [1] shows the effectiveness of the proposed GA.

2 Preliminary Knowledge

In this section, we recall the basic notions related to decision-theoretic rough set models and minimum cost attribute reduction problem. [4,16].

2.1 Decision-Theoretic Rough Set Models

A decision system is a five-tuple $S = (U, C, D, V, f)$, where U is a finite nonempty set of objects called universe, C is a the set of conditional attributes, D is the set of decision attributes, $V = \bigcup_{a \in C \cup D} V_a$, where V_a is a nonempty set of values of attribute $a \in C \cup D$, called the domain of a, $f : U \rightarrow V_a$ is an information function that maps an object in U to exactly one value in V_a such as $\forall a \in C \cup D$, $x \in U$, $f(x, a) \in V_a$.

Each subset of attributes $R \subseteq C \cup D$ determines a binary indistinguishable relation $IND(R)$ as follows:

$$IND(R) = \{(x, y) \in U \times U | \forall a \in R, f(x, a) = f(y, a)\}. \tag{1}$$

It can be easily shown that $IND(R)$ is an equivalence relation on the set U. For $R \subseteq C \cup D$, the equivalence relation $IND(R)$ partitions U into some equivalence classes, which is denoted by $U/IND(R) = \{[x]_R | u \in U\}$, for simplicity, $U/IND(R)$ will be replaced by U/R, where $[x]_R$ is an equivalence class determined by x with respect to R, i.e., $[x]_R = \{y \in U | (x, y) \in IND(R)\}$.

Let $\Omega = \{X, \neg X\}$ denotes the set of states indicating that an object is in X and not in X, respectively. Let $A = \{a_P, a_B, a_N\}$ be the set of actions, where a_P, a_B, a_N represent the three actions to classify an object into $POS(X)$, $BND(X)$ and $NEG(X)$, respectively. The loss function regarding the risk or cost of actions in different states is given in Table 1:

Table 1. Loss function matrix

	a_P	a_B	a_N
X	λ_{PP}	λ_{BP}	λ_{NP}
$\neg X$	λ_{PN}	λ_{BN}	λ_{NN}

In Table 1, λ_{PP}, λ_{BP} and λ_{NP} denote the loss incurred for taking action a_P, a_B and a_N, respectively, when an object belongs to X and λ_{PN}, λ_{BN} and λ_{NN} denote the loss incurred for taking action a_P, a_B and a_N when an object does not belongs to X.

For an object with description $[x]_B$, the expected loss associated with taking the individual actions can be expressed as:

$$R(a_P | [x]_B) = \lambda_{PP} p(X | [x]_B) + \lambda_{PN} p(\neg X | [x]_B);$$
$$R(a_B | [x]_B) = \lambda_{BP} p(X | [x]_B) + \lambda_{BN} p(\neg X | [x]_B);$$
$$R(a_N | [x]_B) = \lambda_{NP} p(X | [x]_B) + \lambda_{NN} p(\neg X | [x]_B). \tag{2}$$

Consider a special kind of loss functions with:

$$(c0)\ \lambda_{PP} \le \lambda_{BP} < \lambda_{NP};$$
$$\lambda_{NN} \le \lambda_{BN} < \lambda_{PN}.$$

The Bayesian decision procedure leads to the following minimum-risk decisions:

$$(p1)\ \text{If } p(X | [x]_B) \ge \alpha,\ \text{decide } x \in POS(X);$$
$$(B1)\ \text{If } \beta < p(X | [x]_B) < \alpha,\ \text{decide } x \in BND(X);$$
$$(N1)\ \text{If } p(X | [x]_B) \le \beta,\ \text{decide } x \in NEG(X).$$

where the parameters α, β, and γ are defined as:

$$\alpha = \frac{(\lambda_{PN} - \lambda_{BN})}{(\lambda_{PN} - \lambda_{BN}) + (\lambda_{BP} - \lambda_{PP})};$$

$$\beta = \frac{(\lambda_{BN} - \lambda_{NN})}{(\lambda_{BN} - \lambda_{NN}) + (\lambda_{NP} - \lambda_{BP})}. \tag{3}$$

By using the thresholds, the three regions of the partition π_D are defined as follows:

$$POS_B^{(\alpha,\beta)}(\pi_D) = \{x \in U | p(D_{\max}([x]_B) | [x]_B) \geq \alpha\};$$
$$BND_B^{(\alpha,\beta)}(\pi_D) = \{x \in U | \beta < p(D_{\max}([x]_B) | [x]_B) < \alpha\};$$
$$NEG_B^{(\alpha,\beta)}(\pi_D) = \{x \in U | p(D_{\max}([x]_B) | [x]_B) \leq \beta\}. \tag{4}$$

where $p(D_{\max}([x]_B)) = argmax_{D_i \in \pi_D}\{\frac{|[x]_B \cap D_i|}{|[x]_B|}\}$.

2.2 Minimum Cost Attribute Reduction

In the decision-theoretic rough set model, based on the Bayesian decision procedure, the purpose of making decision is to minimize the decision cost. Hence, decision cost is a very key concept. Jia et al. [4] proposed an optimization representation of decision-theoretic rough set model by considering the minimization of the decision cost.

Given a decision system $S = (U, C, D, V, f)$, the decision cost of an attribute set $B \subseteq C$ is composed of the cost of the positive rules, the cost of the boundary rules and the cost of the negative rules. Consider the special case where the cost of the correct classification is zero, namely, $\lambda_{PP} = \lambda_{NN} = 0$, the decision cost can be expressed as [4]:

$$\begin{aligned}
\mathbf{COST_B} = &\sum_{x_i \in POS_B^{(\alpha,\beta)}(D)} (1 - p_i) \cdot \lambda_{PN} \\
&+ \sum_{x_j \in BND_B^{(\alpha,\beta)}(D)} (p_j \cdot \lambda_{BP} + (1 - p_j) \cdot \lambda_{BN}) \\
&+ \sum_{x_k \in NEG_B^{(\alpha,\beta)}(D)} p_k \cdot \lambda_{NP}.
\end{aligned} \tag{5}$$

The decision-theoretic rough set model can be seen as a kind of cost-sensitive learning model [4]. The objective of attribution reduction is to minimize the decision cost. Hence, It is easily found that finding a reduct is actually the procedure of solving the above optimization problem. The minimum cost attribute reduction is defined as follows.

Definition 1. [4] *Given a decision system* $S = (U, C, D, V, f)$, $B \subseteq C$ *is a minimum decision cost attribute reduct iff*

(1) $R = \arg\min_{R \subseteq C}\{\mathbf{COST_R}\}$,
(2) $\forall R' \subseteq R$, $\mathbf{COST_{R'}} > \mathbf{COST_R}$.

In this definition, condition (1) is called jointly sufficient conditions and condition (2) is called individually necessary conditions. Condition (1) guarantees that the cost induced from the reduct is minimal, and condition (2) guarantees that the reduct is minimal.

3   Test-Cost-Sensitive Attribute Reduction in DTRS

Minimum cost attribute reduction can be seen as a kind of misclassification-cost-sensitive attribute reduction method. The objective of attribution reduction is to minimize the misclassification cost, it does not consider test cost. However, data are not free in the real world, we need to pay test cost for collecting a data item of an object. Hence, the objective of attribution reduction should be to select an attribute subset with a minimal test cost and preserve minimal decision cost in order to save money, time, or other resources.

The test-cost-independent decision system is the simplest and general model pertinent to the minimal test cost reduct problem, and is defined as follows.

Definition 2. [8] *A test-cost-independent decision system* S *is the six-tuple:*

$$S = (U, C, D, V, f, c) \tag{6}$$

where U, C, D, V *and* f *have the same meanings as decision system, and* $c : C \to \mathbb{R}^+ \cup \{0\}$ *is an attribute cost function.*

A minimal test cost reduct problem in DTRS can be defined as follows.

Definition 3. *The minimal test cost attribute reduct* (*MTCAR*) *problem.*
Input: $S = (U, At = C \cup D)$, *cost functions* λ_{PP}, λ_{BP}, λ_{NP}, λ_{PN}, λ_{BN}, λ_{NN};
Output: $R \subseteq C$;
Optimization objective: (1) *min* $\mathbf{COST_R}$; (2) *min* $c(R)$.

It is worth reminding that two objectives are not equally important. They are the primary and the secondary objectives, respectively. The second objective is considered after the primary one is achieved. This definition can help us to design swarm intelligent algorithm to the minimal test cost reduct problem.

In fact, in DTRS if all attributes have same test cost which equal to a constant, then the minimal test cost reduct problem coincides to the minimum cost attribute reduct problem.

When the costs of making the wrong decision are greater than zero, there are some interesting results in DTRS in the following.

Theorem 1. *Given a test-cost-independent decision system* $S = (U, C, D, V, f, c)$. $\mathbf{COST_C} = 0$ *iff* $POS_C(D) = U$.

Proof. Let $U/C = \{X_1, X_2, \cdots, X_n\}$.

$$\mathbf{COST}_C = \sum_{x_i \in POS_C^{(\alpha,\beta)}(D)} (1 - p_i) \cdot \lambda_{PN} + \sum_{x_j \in BND_C^{(\alpha,\beta)}(D)} (p_j \cdot \lambda_{BP} + (1 - p_j) \cdot \lambda_{BN})$$

$$+ \sum_{x_k \in NEG_C^{(\alpha,\beta)}(D)} p_k \cdot \lambda_{NP}$$

$$= \sum_{X_i \in POS_C^{(\alpha,\beta)}(D)} (|X_i| - |D_{\max}(X_i)|) \cdot \lambda_{PN}$$

$$+ \sum_{X_j \in BND_C^{(\alpha,\beta)}(D)} (|D_{\max}(X_j)| \cdot \lambda_{BP} + (|X_j| - |D_{\max}(X_j)|) \cdot \lambda_{BN})$$

$$+ \sum_{X_k \in NEG_C^{(\alpha,\beta)}(D)} |D_{\max}(X_k)| \cdot \lambda_{NP} = 0.$$

$\Leftrightarrow \forall X_i \in POS_C^{(\alpha,\beta)}(D)$, we have $|X_i| = |D_{\max}(X_i)|$ and $BND_C^{(\alpha,\beta)}(D) = \emptyset$ and $NEG_C^{(\alpha,\beta)}(D) = \emptyset$, namely, $POS_C(D) = U$.

Thus, Theorem 1 holds.

According to Theorem 1, we have the following theorem.

Theorem 2. *Given a test-cost-independent decision system $S = (U, C, D, V, f, c)$. If $\mathbf{COST}_C = 0$, then R is a minimum decision cost attribute reduct iff R is a decision-relative reduct, namely,*

(1) $POS_R(D) = POS_C(D)$,
(2) $\forall a \in R, POS_{R-\{a\}}(D) \subset POS_C(D)$.

Proof. It is obvious according to Theorem 1 and the non-negative of decision cost.

Theorem 2 shows that the minimal test cost reduct problem in DTRS is equivalent to the minimal test cost reduct problem in the classical rough set [8], when a test-cost-independent decision system is consistent one.

Note that loss functions and test costs are positive real numbers, but we can always convert the positive real number into the positive integer. Hence, we only consider loss functions and test cost are positive integers in the following discussion.

4   The Algorithm

In this section, we develop genetic algorithm to test-cost-sensitive attribute reduction in DTRS.

In our algorithm, an attribute subset is represented as a chromosome by a binary string of length $|C|$. A 1 in a gene represents the corresponding attribute is present and a 0 in a gene represents the corresponding attribute is not present.

It is generally known that the fitness function is one of the key factors of success of genetic algorithm. According to Definition 3, the fitness function of a chromosome R is defined as follow.

$$f(ch_R) = \mathbf{COST}_{\max} - \mathbf{COST}_R + \frac{c(C) - c(R)}{c(C)} \tag{7}$$

where $\mathbf{COST}_{\max} = |U| \cdot (\lambda_{PN} + \lambda_{BP} + \lambda_{BN} + \lambda_{NP})$.

The above fitness function can guarantee the optimality equivalence between the MTCAR problem and the fitness maximization problem, which is proved as follows.

Theorem 3. *If R is an optimal solution to the fitness maximization problem, then R is also an optimal solution to the MTCAR problem.*

Proof. For $\forall B \subseteq C$ satisfying $f(ch_R) > f(ch_B)$. We must prove that $\mathbf{COST}_R < \mathbf{COST}_B$ or ($\mathbf{COST}_R = \mathbf{COST}_B$ and $c(B) > c(R)$). The latter is obvious. The former must be proven for all the three possible cases of $c(B) < c(R)$, $c(B) = c(R)$, $c(B) > c(R)$.

When $c(B) < c(R)$, we have $\frac{c(C)-c(B)}{c(C)} > \frac{c(C)-c(R)}{c(C)}$. Since $f(ch_R) > f(ch_B)$, then $\mathbf{COST}_{\max} - \mathbf{COST}_R > \mathbf{COST}_{\max} - \mathbf{COST}_B$, namely, $\mathbf{COST}_R < \mathbf{COST}_B$.

When $c(B) = c(R)$, we have $\frac{c(C)-c(B)}{c(C)} = \frac{c(C)-c(R)}{c(C)}$. Since $f(ch_R) > f(ch_B)$, then $\mathbf{COST}_{\max} - \mathbf{COST}_R > \mathbf{COST}_{\max} - \mathbf{COST}_B$, namely, $\mathbf{COST}_R < \mathbf{COST}_B$.

When $c(B) > c(R)$, let $U/C = \{X_1, X_2, \cdots, X_n\}$. Obviously, $0 \leq \frac{c(C)-c(B)}{c(C)} \leq 1$ and $0 \leq \frac{c(C)-c(R)}{c(C)} \leq 1$. Since

$$\begin{aligned}
\mathbf{COST}_B =\ & \sum_{x_i \in POS_B C^{(\alpha,\beta)}(D)} (1 - p_i) \cdot \lambda_{PN} + \sum_{x_j \in BND_B^{(\alpha,\beta)}(D)} (p_j \cdot \lambda_{BP} + (1 - p_j) \cdot \lambda_{BN}) \\
& + \sum_{x_k \in NEG_B^{(\alpha,\beta)}(D)} p_k \cdot \lambda_{NP} \\
=\ & \sum_{X_i \in POS_B^{(\alpha,\beta)}(D)} (|X_i| - |D_{\max}(X_i)|) \cdot \lambda_{PN} \\
& + \sum_{X_j \in BND_B^{(\alpha,\beta)}(D)} (|D_{\max}(X_j)| \cdot \lambda_{BP} + (|X_j| - |D_{\max}(X_j)|) \cdot \lambda_{BN}) \\
& + \sum_{X_k \in NEG_B^{(\alpha,\beta)}(D)} |D_{\max}(X_k)| \cdot \lambda_{NP},
\end{aligned}$$

apparently, we have $\mathbf{COST}_B \in \mathbb{N}$. Similarly, we have $\mathbf{COST}_R \in \mathbb{N}$, when $\mathbb{N}$ is the set of natural numbers. Again since $f(ch_R) > f(ch_B)$, then

$$\mathbf{COST}_{\max} - \mathbf{COST}_R + \frac{c(C) - c(R)}{c(C)} > \mathbf{COST}_{\max} - \mathbf{COST}_B + \frac{c(C) - c(B)}{c(C)}.$$

Hence $\mathbf{COST}_R < \mathbf{COST}_B$.

This completes the proof.

Then roulette wheel selection is employed in the selection operation, chromosomes which have larger fitness values have larger probability to select. The uniform crossover and uniform mutation are used to breed next generation.

The genetic algorithm is described in Algorithm 1.

Algorithm 1. A genetic algorithm to minimal test cost attribution reduction.

Input:
    A test-cost-independent decision system $S = (U, C, D, V, f, c)$;
Output:
    A minimal test cost reduct R.
1: Initialization: randomly initialize N individuals in a population, where N is population size.
2: Evaluate population: evaluate the fitness of each individual.
3: **while** not terminate **do**
4:    Selection: select $M/2$ pairs of individuals by roulette wheel selection, where $M \geq N$.
5:    Crossover: perform uniform crossover on the selected $M/2$ pairs of individuals to generate M temporary individuals.
6:    Mutate: perform uniform mutate on M temporary individuals.
7:    Selection: select N individuals from M temporary individuals by roulette wheel selection to generate next population.
8:    Evaluate population: evaluate the fitness of each individual.
9: **end while**
10: output R that has the maximum fitness value from the current population.

---

5   Experimental Results

In this section, we present an experiments to verify the effectiveness of Algorithm 1 to the minimal test cost attribute reduction in DTRS, namely, we compare test costs of reducts between the minimum cost attribute reduction that takes into account test costs (Algorithm 1) and the minimum cost attribute reduction that does not consider test costs (By GA in [4], denoted by Algorithm 2).

Eight data sets from the UCI Machine Learning Repository [1] are used in the empirical study, where the missing values are filled with mean values for numerical attributes and mode values for discrete attributes. All numerical attributes are discretized using equal-frequency discretization techniques.

In the experiments, the loss function is presented in Table 2. For each data set, 10 different group of test costs are randomly generated by normal distribution (parameter $\alpha = 8$) [8]. Their values are integers ranging from 1 to 100. The average value and standard deviation of test costs of reducts are recorded.

Genetic algorithm parameter settings are as follows:
- Population size: 100
- Number of generation: 100
- Probability of crossover: 0.7
- Probability of mutation: 0.3

The experimental results are shown in Table 3, where $|U|$ and $|C|$ are the cardinality of the universe and the conditional attribute set, respectively. TCAA represents test costs of all attributes, TCAlgorithm-1 represents test costs of reducts obtained by Algorithm 1 and TCAlgorithm-2 represents test costs of reducts obtained by Algorithm 2.

Table 2. Loss function matrix for experiments

	a_P	a_B	a_N
X	0	10	30
$\neg X$	60	30	0

Table 3. Results of the minimal test cost attribute reduction in DTRS

Data sets	$|U|$	$|C|$	TCAA	TCAlgorithm-1	TCAlgorithm-2
German	1000	20	1023.8 ± 47.4	**637.7 ± 47.6**	691.7 ± 76.8
Hill-Valley	1212	100	5037.6 ± 41.0	**2143.9 ± 158.9**	2372.1 ± 122.1
Horse-colic	386	22	918.0 ± 31.0	**437.0 ± 41.8**	452.5 ± 54.9
Ionosphere	351	34	1705.6 ± 57.8	**614.8 ± 34.9**	654.7 ± 54.0
Lymphography	148	18	918.0 ± 31.0	**437.0 ± 41.8**	452.5 ± 54.9
SPECTF Heart	267	44	2249.0 ± 49.7	**621.7 ± 59.4**	658.9 ± 37.5
Wdbc	569	30	1492.7 ± 30.5	**503.2 ± 34.4**	533.3 ± 30.1
Wine	178	13	645.8 ± 52.3	**247.8 ± 29.5**	276.4 ± 38.5

From the experimental results, we can see that two algorithms can reduce significantly total test costs (TCAA). However, test costs of reducts derived by Algorithm 1 are smaller than test costs of reducts derived by Algorithm 2 on all data sets. This is because Algorithm 2 concerns on the length of reducts instead of test costs of reducts, and Algorithm 1 focus on test costs of reducts. Thus in DTRS, Algorithm 1 can solve MTCAR problem more effectively.

6   Conclusion

Attribute reduction plays a large role in data analysis. In DTRS, attribute reduction can be achieved by minimizing decision cost. Actually, DTRS is a misclassification cost-sensitive learning model. The minimum cost attribute reduct focus only on misclassification costs, and does not consider test costs. However in the real world, test cost is also a important type of cost, because data are not free. In this paper, the minimal test cost attribute reduction problem is introduced into DTRS. The purpose of attribute reduction is to select attribute subset which can minimize misclassification costs and test costs. A GA to MTCAR problem has been proposed. The experimental results show that the proposed GA is able to deal with MTCAR problem effectively.

Acknowledgments. This work is supported the Natural Science Foundation of China under Grant No. 61379114 61073146, 61272060, the Key Natural Science Foundation of Chongqing of China under Grant Nos. CSTC2013jjB40003 and Natural Science Foundation Project of CQ CSTC under Grant No. cstc2013jcyjA 40063.

References

1. Ucirvine machine learning repository, http://archive.ics.uci.edu/ml/
2. Herbert, J.P., Yao, J.T.: Game-theoretic rough sets. Fundamenta Informaticae 108, 267–286 (2011)
3. Hu, Q.H., Liu, J.F., Yu, D.R.: Mixed feature selection based on granulation and approximation. Knowledge-Based Systems 21, 294–304 (2008)
4. Jia, X.Y., Liao, W.H., Tang, Z.M., Shang, L.: Minimum cost attribute reduction in decision-theoreticroughsetmodels. Information Sciences 219, 151–167 (2012)
5. Li, H.X., Zhou, X.Z., Zhao, J.B., Liu, D.: Attribute reduction in decision-theoretic rough set model: a further investigation. In: Yao, J., Ramanna, S., Wang, G., Suraj, Z. (eds.) RSKT 2011. LNCS, vol. 6954, pp. 466–475. Springer, Heidelberg (2011)
6. Liu, D., Li, T.R., Ruan, D.: Probabilistic model criteria with decision-theoretic rough sets. Information Sciences 181, 173–178 (2011)
7. Mi, J.S., Wu, W.Z., Zhang, W.X.: Approaches to knowledge reduction based on variable precision rough set model. Information Sciences 159, 255–272 (2004)
8. Min, F., He, H.P., Qian, Y.H., Zhu, W.: Test-cost-sensitive attribute reduction. Information Sciences 181, 4928–4942 (2011)
9. Min, F., Hu, Q.H., Zhu, W.: Feature selection with test cost constraint. International Journal of Approximate Reasoning (2013)
10. Nguyen, H.S., Ślęzak, D.: Approximate reducts and association rules correspondence and complexity results. In: Zhong, N., Skowron, A., Ohsuga, S. (eds.) RSFD-GrC 1999. LNCS (LNAI), vol. 1711, pp. 137–145. Springer, Heidelberg (1999)
11. Pawlak, Z.: Rough sets. International Journal of Computer and Information Science 11, 341–356 (1982)
12. Shen, Q., Jensen, R.: Rough sets, their extensions and applications. International Journal of Automation and Computing 4, 217–228 (2007)
13. Ślęzak, D.: Rough sets and bayes factor. In: Peters, J.F., Skowron, A. (eds.) Transactions on Rough Sets III. LNCS, vol. 3400, pp. 202–229. Springer, Heidelberg (2005)
14. Turney, P.: Types of cost in inductive concept learning. In: Proceedings of the ICML 2000 Workshop on Cost-Sensitive Learning, Stanford, CA, pp. 15–21 (2000)
15. Wu, W.Z.: Attribute reduction based on evidence theory in incomplete decision systems. Information Sciences 178, 1355–1371 (2008)
16. Yao, Y.Y.: Decision-theoretic rough set models. In: Yao, J.T., Lingras, P., Wu, W.-Z., Szczuka, M.S., Cercone, N.J., Ślęzak, D. (eds.) RSKT 2007. LNCS (LNAI), vol. 4481, pp. 1–12. Springer, Heidelberg (2007)
17. Yao, Y.Y., Zhao, Y.: Attribute reduction in decision-theoretic rough set model. Information Sciences 178, 3356–3373 (2008)
18. Zhao, Y., Wong, S.K.M., Yao, Y.Y.: A note on attribute reduction in the decision-theoretic rough set model. In: Chan, C.-C., Grzymala-Busse, J.W., Ziarko, W.P. (eds.) RSCTC 2008. LNCS (LNAI), vol. 5306, pp. 61–70. Springer, Heidelberg (2008)
19. Ziarko, W.: Probabilistic approach to rough sets. International Journal of Approximate Reasoning 49, 272–284 (2008)

Spatial Anisotropic Interpolation Approach for Text Removal from an Image

Morusupalli Raghava, Arun Agarwal, and Ch. Raghavendra Rao

School of Computer and Information Sciences, University of Hyderabad,
Hyderabad, Andhra Pradesh, India - 500046
raghava.m@cvr.ac.in, {aruncs,crrcs}@uohyd.ernet.in

Abstract. We propose a Spatial Anisotropic Interpolation (SAI) based, Design and Analysis of Computer Experiment (DACE) model for inpainting the gaps that are induced by the removal of text from images. The spatial correlation among the design data points is exploited, leading to a model which produces estimates with zero variance at all design points. Incorporating such a feature turns the model to serve as a surrogate for predicting the response at desired points where experiment is not carried out. This property has been tuned for the purpose of gap filling in images also called as Image Inpainting, while treating the pixel values as responses. The proposed methodology restores the structural as well as textural characteristics of input image. Experiments are carried out with this methodology and results are demonstrated using quality metrics such as SSIM and PSNR.

Keywords: Random Field, Kriging, Anisotropic Interpolation, Spatial Correlation.

1 Introduction

The low level computer vision problem, namely removal of inbuilt artifacts of imaging systems such as 'date in photographs', 'subtitles in movie frames' and 'location information in satellite images' induces gaps as shown in Figure 1(b) and hence results in discontinuities in the image. Such an image serves little purpose in providing information for understanding. Therefore, it is essential to fill these induced gaps (inpaint region) that gel with existing portions (source region) of the image. Such a filling activity is referred to in literature as inpainting - a challenging task for image processing community over the past decade.

This problem was attempted by many authors [1-8] visualizing it as a problem of information flow, matching, total variation minimization, texture interpolation and sparse representations. But, these existing models are complex in terms of model design, costly with respect to associated numerical method and show limited capabilities to mention a few. The approach presented in [1] can only preserve linear structures but, cannot synthesize textures. The model in [5] cannot handle structures but interpolates textures. An image is a composition of spatial characteristics such as structures and periodic variations that represent textures.

S. Ramanna et al. (Eds.): MIWAI 2013, LNCS 8271, pp. 153–164, 2013.

(a)   (b)

Fig. 1. (a) Image with Text superposed. (b) Holes in Image after removing Text.

Hence the inpainting algorithm must possess potential to inherit structures and interpolate the textures simultaneously [7, 8]. The classical image representation and analysis approaches, in general are based on edges or regions which are realized by directional operators. This aspect suggests that the capability of inpainting method is further enhanced if it accommodates anisotropy property. Thus developing a spatial interpolating process for predicting pixels in inpaint region, based on source region pixels with minimum variance, structure preserving and texture synthesizing capabilities is the focus of this paper. Further, this paper addresses the following:

- In cases where inpaint region is uniformly spread across the image, then the image is divided into non-overlapping tiles and inpainting is carried out using the proposed model.
- When source region is considerably larger than inpaint region, an appropriate scheme for sampling of source region is incorporated in the model to minimize the computational costs.

Recently, surrogate models [18] have been widely used in literature as knowledge representation metaphors for any stochastic process in engineering design. Kriging is the most sought after surrogate used by geologists to handle such spatial interpolation issues.

In this paper, we propose an adaptive surrogate model building methodology through a proper framework called Design and Analysis of Computer Experiments (DACE) that abet text removal, an instance of inpainting domain problem. Experiments are conducted on a surrogate to predict the responses at locations of inpaint regions. In order to accomplish the effective design, choosing a proper design space and parameter learning are essential. We adopt Maximum Likelihood Estimation (MLE) algorithm for initial parameter learning.

This paper is organized as follows: Section 2 presents the literature survey on inpainting problem. Section 3 provides a brief overview of DACE model, discusses its applicability and the proper customization to inpainting problem. Section 4 develops Spatial Anisotropic Inpainting procedure (SAI). Section 5 provides illustrative analysis and the contrastive results accrued by the proposed algorithm compared to existing algorithms. Section 6 gives the conclusions.

2   The Inpainting Problem

At its simplest manifestation, inpainting is an image restoration problem [1] wherein the underlying goal is to reconstruct or predict the missing regions within an image in such a way that the changes carried out on an image are not clearly noticeable to anyone who observes the modified image. In general, inpainting algorithms divide the given image domain D into two disjoint sub-regions, source region Φ denoting the set of all locations where the intensity value u_0 is known and inpaint region $\Omega(D - \Phi)$ over which the intensity values need to be estimated. Further, they attempt to fill Ω which is consistent with source region. Several researchers proposed algorithms which are broadly categorized as flow, variational, statistical and similarity based.

The operational principle involved in flow based models is to propagate the Laplacian of intensity values that are available in neighboring source region into inpaint region. This is carried out along the interpolated structures and is governed by a higher order Partial Differential Equations (PDE). Hence, these algorithms preserve structures. But, in many situations where the local information does not entirely characterize the missing information, the resulting reconstructed information in the inpaint region will not be visually consistent with the rest of the image. Also, these methods can handle only linear structures and give good results for narrow inpaint regions. To overcome these limitations the pixel value estimation is done such that 'total variation'(TV) is minimized [2]. These models are further improvised by incorporating the curvature term [3] which has made the resulting model to reconstruct Non-linear structures also. In recent works the TV minimization is implemented through application of Split-Bregman methods [4] and applied for inpainting. It is clear to note that the 'Text' is a composition of various strokes and hence contributes to structural aspects. On the other hand, images also can have textures. In order to interpolate the texture effectively, algorithms rely on 'texture interpolation techniques' [5] either through synthesizing it from some other suitable image or by picking up the best matching patch within the same image. The results of these methods are further enhanced by defining priority based filling order [6] so that the structures also get propagated. The other category of algorithms [7, 8] tries to combine both afore mentioned aspects to simultaneously recover the texture and structure. These algorithms separate the texture and structure components using filter banks, process each component individually in a different functional space and try to combine the results obtained. Recently, Morphological Component Analysis [9] based methods are developed to separate texture and cartoon (smooth structure) components.

Another school of thought for solving inpainting problem is drawn from a set of computer vision techniques namely, Graphical Models that couple probability theory with graph theory. Markov Random Field [10-13] is an undirected graphical model which provides a unified framework for representing pixel intensity as a random variable on an efficient data structure and offers algorithms to assign appropriate label to each node to solve the inference problem. In literature, such inference algorithms are broadly classified into two categories, 'message passing' algorithms and 'move making' algorithms. Prior to invoking these inference algorithms, the initial parameters and neighborhood structure of the model are learnt from the input data itself [14, 15] by employing expert agents (FoE).

3   Text Removal by Dace Model

A metamodel [16] combines the capabilities of numerical methods with a stochastic process. It provides an abstraction of latent physical phenomenon. This abstraction is realized either through interpolation or approximation. Such metamodels include Response Surface, Spline, Radial Basis Functions, Kriging, Neural Networks etc. The utility of a chosen metamodel is that it acts as a surrogate to the underlying actual experiment. The immediate advantage is, through proper surrogate models, it is not required to conduct the experiment physically to get the responses at all design sites, but they can be predicted with minimum error in Mean-Square sense.

3.1   Modeling and Prediction with DACE

DACE [17-18] offers a 'surrogate computer model' feature for any computer experiment. This model is endowed with Kriging approximation, for data resulting from computer experiment. Here, a computer experiment corresponds to a collection of pairs. Each pair includes an input location and the associated response from different runs of a computer model. Both the input and the response from the computer model are likely to be in higher dimensions.

Consider an image I of size $a \times b$ having source region Φ and inpaint region Ω. Then the source region pixel locations are represented as a set of design sites $S = [s_1, ..., s_m]^T$; $s_i \in R^n$ and the set of corresponding pixel values is represented as $Y = [y_1, ..., y_m]^T$; $y_i \in R^q$. With regard to image inpainting problem we have $n = 2$ and $q = 1$ (in case of gray image) or $q = 3$ (for color image). Then the DACE fits a Kriging model to the design sites as

$$\hat{y}(x) = z(x) + \sum_{j=1}^{p} \beta_j \varphi_j \tag{1}$$

The first term $z(x)$, $x \in R^n$ on the right hand side in Equation (1) is again a random field which models the residual, with mean zero and covariance $\sigma^2 R(\theta, s_i, s_l)$, for $i, l = 1...m$ and θ is the parameter of spatial correlation R. The latter term denotes the trend which is modeled through a known basis function φ with unknown coefficients β_j. For example, if the known basis is a set of polynomials, then the corresponding trend is a regression polynomial. Further, the trend plays a vital role in selecting an appropriate kriging model for the interpolation problem at hand.

Widely used kriging models are Simple, Ordinary and Universal. In case of simple kriging, the trend has a constant mean throughout the design space. In ordinary kriging the trend is constant (piecewise) within local neighborhood whereas in universal kriging it is modeled as a regression polynomial (it can be of a higher order). For text removal inpainting problem the universal kriging is selected. This aspect is substantiated in the upcoming discussions.

Now, the objective is to tune the unknown regression coefficients β_j, such that the minimum residual at an untried point, during prediction step is possible. Thus kriging accommodates even the error term in the modeling process.

To simplify the discussion, consider only single response at a given site and restrict the nature of predictor to be linear as, $\hat{y}(x) = C^T Y$ with $C \in R^m$, $q = 1$. Then, the error at the untried site x can be expressed in terms of MSE as

$$E[(\hat{y}(x) - y(x))^2] \qquad (2)$$

In order to tune the predictor as Best Linear Unbiased (BLUP)[17], i.e. which do not favor any design site x , it is required to impose a constraint on MSE as

$$E\left[C^T(x)Y\right] = E\left[y(x)\right] \qquad (3)$$

Upon introducing the design matrix $\varphi = \{\varphi(s_1)...\varphi(s_m)\}^T$, Correlation Matrix between every pair of design sites as $R = \{R(s_i, s_l)\}$ for $1 \leq i, l \leq m$ and Correlation Matrix between every design site s and an untried location x as $r(x) = [R(x, s_1), ...R(x, s_m)]^T$, now Equation (2) can be simplified to

$$MSE(\hat{y}(x)) = \sigma^2(1 + C^T(x)RC(x) - 2C^T(x)r(x)) \qquad (4)$$

Subject to constraint in Equation (3).

Then the corresponding Lagrange equation of the minimization function with constraint is expressed as

$$L(C, \lambda) = \sigma^2(1 + C^T(x)RC(x) - 2C^T(x)r(x)) - \lambda^T(x)(\varphi^T C(x) - \varphi(x)) \qquad (5)$$

The minimization problem is solved by equating the gradient of Equation (5) with respect to C, to zero which results in a system of equations

$$\begin{bmatrix} 0 & \varphi^T \\ \varphi & R \end{bmatrix} \begin{bmatrix} \lambda \\ C \end{bmatrix} = \begin{bmatrix} \varphi \\ r \end{bmatrix} \qquad (6)$$

The solution of Equation (6) for $C(x)$ results in the predictor

$$\hat{y}(x) = \varphi^T \hat{\beta} + r^T R^{-1}(Y - \varphi\hat{\beta}) \qquad (7)$$

with $\hat{\beta}$ being the generalized Least-Square solution for the regression problem $\varphi\beta = Y$ with respect to R which is evaluated as

$$\hat{\beta} = (\varphi^T R^{-1}\varphi)^{-1}\varphi^T R^{-1}Y \qquad (8)$$

3.2   Applicability of Kriging to Text Removal Problem

Text removal problem is fundamentally ill-posed and possesses certain features. One of these features includes locality *i.e.* during inpainting it is sufficient to propagate the information that is available in the vicinity and preserve the trend observed in the source region. Due to this property the Universal Kriging model is suitable and is sufficient to develop it only on non-overlapping sub-images of appropriate size but not on the entire image. Hence the proposed model is called as adaptive DACE. The second important feature is anisotropy, which insists that the information flowing into inpaint region should vary with the direction. It is prudent to observe that the Kriging model building involves spatial correlation for minimizing the 'prediction error' which is sensitive to distance. On the other hand, correlation is also capable of describing the directional differences in spatial domain which will fetch anisotropy property into the model.

4 Proposed Algorithm

4.1 Model Selection

DACE provides a wide range of correlation and regression models. Each correlation model involves a certain number of parameters. For example, in case of exponential correlation model

$$R(s_i, s_l) = \exp(-\theta \mid s_i - s_l \mid) \tag{9}$$

has θ as its parameter. According to Equation (9) small value of θ implies that even farther pixels could have high correlation among themselves and influence the prediction at intermediate locations. If such a parameter is learned properly then the correlation deteriorates appropriately and helps in getting better predictions. In this paper parameter learning is accomplished through well-known MLE algorithm. This algorithm is invoked on Latin Hyper Cube Sample (LHS)[19] which divides the design space into rows and columns and evenly samples the design sites by accommodating every row and column of design space. The correlation model can be extended to multi-dimensional data, by taking the product of all correlation models, wherein each one is selected along a dimension in the form Equation (9). Then the product model is expressed as

$$R(s, x) = \Pi_{d=1}^{n} \exp(-\theta^d \mid s^d - x^d \mid) \tag{10}$$

In Equation (10) variable d represents the dimension number and n represents the number of dimensions. The next step in model building is about modeling the regression which captures the trend in the sub-image. In this paper a quadratic polynomial is selected as the regression model.

Prior to fitting the DACE model, a correct design space should be chosen. Though there are sophisticated sampling methods available in literature, they are not suitable to inpainting problem. In this paper a precise sampling mechanism is chosen. According to this method, the design space is modeled to span 'uniformly spaced grid points' and the user can subsample the design sites at a selected scale $1, \frac{1}{2}, \frac{1}{3}$ or $\frac{1}{4}$. Then the sampled locations are normalized to lie in the unit square $[0 \ 1]^2$ and maintained on a linear array.

Another essential aspect of random field learning is related to learning its structure. In this paper, to simplify this issue, 8-Neighborhood structure is considered. Final pertinent issue is, its memory requirement. Kriging is a memory hungry algorithm as it involves computing the Inverse of Correlation Matrix. To overcome this problem the input image and the corresponding mask are divided into non-overlapping patches of size 50×50 and kriging model is constructed for every patch. This tiling approach is justified as the inpainting process should capture local trends to improve the quality of results. The size of the patch is selected in a subjective manner.

4.2   Algorithm Design (SAI)

Let I be input image of size $a \times b$ and the corresponding mask is M in which each inpaint region location is given the value 1 and source region location is assigned a value 0. Then extract all non-overlapping sub-images of size $k_1 \times k_2$ (in our experiments 50×50) from image I, and corresponding sub-masks form the mask M. Then Algorithm SAI is applied on each sub-image as listed below, which involves only one index variable i so that the notations appear to be simple.

Algorithm: SAI
> **Input:** Set of all tiles I_i and M_i extracted from I and M
> **Output:** Inpainted image
> 1. Choose corresponding sub-images (size 50×50) I_i and M_i from I and M respectively and extract the source region Φ_i and inpaint region Ω_i.
> 2. Subsample source region pixels by using LHS.
> 3. Choose appropriate correlation model R (exponential, in our experiments)
> 4. Apply ML on samples of step 2 for correlation parameter (θ) learning.
> 5. Select a regression model φ of interest (quadratic polynomial, in our experiments).
> 6. Choose a subsample independent of step 2 from source region Φ_i as defined in section 4.1.
> 7. Fit DACE model for chosen sub-sample, φ, R and θ.
> 8. Extract the inpaint locations, normalize them and put them on a linear array.
> 9. Use DACE model to predict the responses at inpaint locations maintained in step 8.
> 10. Retain the values obtained in step 9 into inpaint region of I_i.
> 11. Repeat steps 1 to 10 for each tile until all the tiles are inpainted.

The Algorithm SAI is applied on sub-images for which the ratio ρ between the size of inpaint region and source region is within the range $0 \le \rho \le 0.4$.

5   Results and Discussion

Experiments are conducted on a wide set of images with structural and texture elements on which text is imposed. Addition of text to original image in experiment 2 and experiment 3 is achieved by using tools developed by [4]. The proposed model has been exhaustively run while utilizing different correlation kernels and regression polynomials of different orders. The results presented below involve exponential correlation model and quadratic regression. In order to compare the results, experiments are conducted by utilizing tools [4], [14].

Experiment: 1

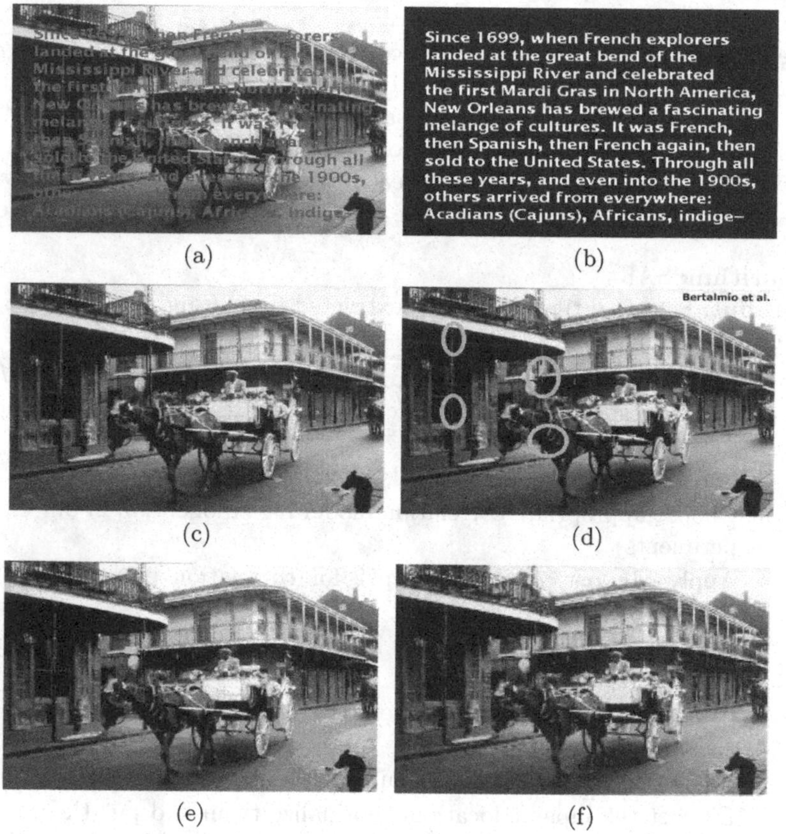

(a)   (b)

(c)   (d)

(e)   (f)

Fig. 2. (a) Presents the input image of size 297 X 438 with text overlaid. (b) Shows the corresponding Mask (c) Gives the result of proposed SAI model. (d) Shows the result of Bertalmio method[1]. (e) Presents result of Split-Bregman method [4]. (f) Presents the output of FoE model [12].

The results presented here are compared through visual inspection. It is evident to notice that the proposed method SAI successfully preserves all linear structures, (refer to Figure 2(c)) like poles and the window frames whereas all other methods fail to preserve these features. For demonstration purpose, such broken structures are highlighted in Figure 2(d) within pale green ellipses. The same disability is seen in results of [4] and [12] as shown in Figure 2(e) and Figure 2(f). In this experiment, metrics are not derived as there is no access to original image prior to overlaying text on it.

The results of this experiment establishes that even if the pole, a linear structural element in Figure 3(a) is totally obscured by characters 'l' and 'w' at different locations, still SAI is able to perfectly preserve the pole (Figure 3(d)). In addition to this, SAI preserved the portion of the wheel (compare Figure 3(b) with Figure 3(e)) with Non-linear structures. Figure 3(c) and Figure 3(f) presents the strength of SAI to retain the back of the knee and heel, even though very little information is available.

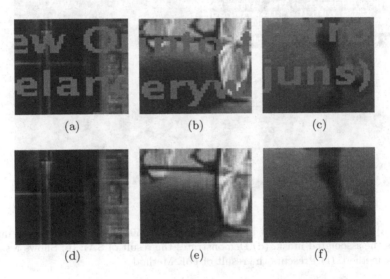

(a)   (b)   (c)

(d)   (e)   (f)

Fig. 3. Presents a close look at results of Experiment 1

In the upcoming experiments the performance of SAI against algorithm [4] is compared by deriving either Peak Signal to Noise Ratio (PSNR) or Structural Similarity Index Measure (SSIM) [20], which ever is relevant to the content in the input image. These metrics are regarded to be appropriate to inpainting problem by image processing community. Image restoration algorithm is analyzed to be effective if the resulting PSNR is between 30db and 50db where as SSIM is in the range [0 1] and is effective if its value is very close to 1.

Experiment: 2 The input image in this experiment has Non-linear structures as well as textures that are spread across the image. The result of SAI in Figure 4(d) speaks about its strength in terms of curvature preserving capability at peaks of the mountain. The textures are interpolated perfectly even though the inpainting is carried out on non-overlapping tiles. The results of SAI and

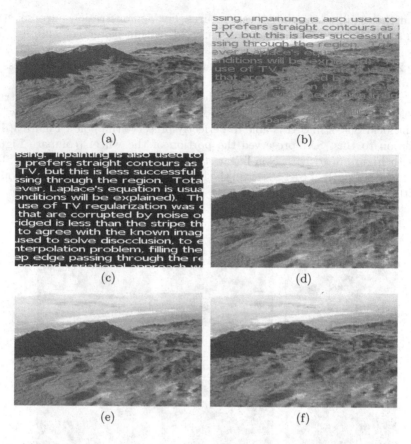

Fig. 4. (a) Presents the original image of size 250×250. (b) Gives the input image. (c) Gives the associated mask. (d) Demonstrates the result of SAI. (e) Shows result of Bregman method. (f) Presents the result of FoE Method.

Bregman method are compared by extracting the PSNR, a texture related metric. The PSNR realized by SAI is 38.87 db whereas Bregman method is 39.0 db.

Experiment: 3 In this experiment an image which has linear structures is taken up. In Figure 5(a) both vertical and horizontal edges are obscured by the text. A close look at the building appearing on the right hand side in Figure 5(d) reveals that crossovers on it are perfectly retained by SAI. In addition to that, the circular logo on the rear glass of parked vehicle is completely reconstructed as shown in Figure 5(d), which is not possible in Bregman method or FoE method (see Figures 5(e) and 5(f)). SSIM, a structure related metric is extracted from the results of SAI and Bregman experiments. SAI achieved a distinguishable SSIM 0.938 to that of Bregman method which achieved SSIM 0.918.

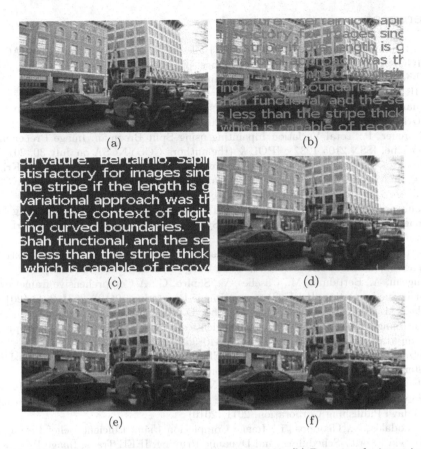

Fig. 5. (a) Presents, the original image of size 213×163. (b) Presents the input image. (c) Shows the associated mask. (d) Presents the result of SAI. (e) Shows the result of Bregman method. (f) Shows the result of FoE Method.

6   Conclusions and Future Work

In this paper we have solved the problem of removing text from the image by treating it as an instance of spatial interpolation and by utilizing adaptive DACE. The anisotropy property which is associated with this problem is realized by spatial correlation among the pixels of the image. The results suggest that SAI not only interpolates the texture but also preserves the linear structure effectively for smaller inpaint regions provided a proper design sampling is done and model parameters are learnt. The future work aims at extending the random field capabilities so that the resulting model becomes able to handle, even larger inpaint regions to solve image completion problem.

References

1. Bertalmio, M., Sapiro, G., Caselles, V., Ballester, C.: Image inpainting. In: ACM SIGGRAPH 2000 Conference Proceedings, pp. 417–424 (2000)
2. Chan, T.F., Shen, J.: Non-texture inpainting by curvature driven diffusion. JV-CIR 12(4), 436–449 (2001)
3. Chan, T.F., Kang, S.H., Shen, J.: Euler's elastica and curvature based inpainting. SIAM J. App. Math. 63(2), 564–592 (2002)
4. Getreuer, P.: Total Variation Inpainting using Split Bregman. Image Processing On Line, ISSN 2105 1232 c IPOL & (the authors CCBYNCSA on July 30, 2012)
5. Chandra, S., Petrou, M., Piroddi, R.: Texture interpolation Using Ordinary Kriging. In: Marques, J.S., Pérez de la Blanca, N., Pina, P. (eds.) IbPRIA 2005. LNCS, vol. 3523, pp. 183–190. Springer, Heidelberg (2005)
6. Tauber Z., Li Z N., Drew.: Review and Preview: Disocclusion by inpainting for image-based rendering. IEEE Trans. Systems, Man, and Cybernetics, Part C: Applications and Reviews 37(4) (2007)
7. Bertalmio, M., Vesa, L., Sapiro, G., Osher, S.: Simultaneous structure and texture image inpainting. IEEE Transactions on Image Processing 8, 882 (2003)
8. Bugeau, A., Bertalmio, M., Caselles, V., Sapiro, G.: A Comprehensive framework for image inpainting. IEEE Trans. on Image Processing 19(10), 2634–2645 (2010)
9. Elad, M., Starck, J., Querre, P., Donoho, D.L.: Simultaneous cartoon and texture image inpainting using morphological component analysis (MCA). Applied Computational Harmonic Analysis 19, 340–358 (2005)
10. Felzenszwalb, P.F., Huttenlocher, D.P.: Efficient belief propagation for early vision. International Journal of Computer Vision 70(1), 41–54 (2006)
11. Chen, S., Tong, H., Wang, Z., Liu, S., Li, M., Zhang, B.: Improved generalized belief propagation for vision processing. In: Mathematical Problems in Engineering. Hindawi Publishing Corporation, 2011 (2010)
12. Komodakis, N., Georgios, T.: Image Completion Using Efficient Belief Propagation via Priority Scheduling and Dynamic Pruning. IEEE Trans. Image Process., 2649–2661 (2007)
13. Richard, S.: A Comparative Study of Energy Minimization Methods for Markov Random Fields with Smoothness-Based Priors. IEEE Transactions on Pattern Analysis and Machine Intelligence 30 (June 2008)
14. Roth, S., Black, M.J.: Fields of experts. International Journal of Computer Vision 82(2), 205–209 (2009)
15. Wu, R., Srikant, R., Ni, J.: Learning graph structures in discrete Markov random fields. In: INFOCOM Workshops, pp. 214–219 (2012)
16. Santner, T., Williams, B., Notz, W.: The Design and Analysis of Computer Experiments. Springer (2003)
17. Lophaven, N., Nielsen, H.B., Jacob, S.: IMM. Informatics and Mathematical Modeling. Technical University of Denmark. DACE A Matlab Kriging Toolbox (2002)
18. Sacks, J., William, J.W., Mitchell, T.J., Henry, P.W.: Design and Analysis of Computer Experiments. H.P, Statistical Science 4(4), 409–423 (1989)
19. Martin, J., Simpson, T.: Use of kriging models to approximate deterministic computer models. AIAA Journal 43(4), 853–863 (2005)
20. Wang, Z., Bovik, A.C., Sheikh, H.R., Simoncelli, E.P.: Image quality assessment: From error visibility to structural similarity. IEEE Transactions on Image Processing 13(4), 600–612 (2004)

A Geometric Evaluation of Self-Organizing Map and Application to City Data Analysis

Shigehiro Ohara, Keisuke Yamazaki, and Sumio Watanabe

Department of Computational Intelligence and Systems Science,
Tokyo Institute of Technology,
4259 Nagatsuta-chou, Midori-Ku, Yokohama, Kanagawa, Japan, 226-8502
{ohhara,k-yam}@math.dis.titech.ac.jp, swatanab@dis.titech.ac.jp

Abstract. Kohonen's Self-Organizinig Map (SOM) is useful to make a low dimensional manifold which is embedded into a high dimensional data space, hence it is now being used in visualization, analysis, and knowledge discovery of complex data. However, for a given set of training samples, a trained SOM is not uniquely determined because of many local minima, resulting that we can not evaluate whether it is appropriately embedded or not. In this paper, we propose a new method how to evaluate a trained SOM from the viewpoint of geometric naturalness. The new criterion is defined by the average correspondence gap between different trained SOMs. We show the effectiveness of the proposed method by experimental results for artificial data, and then introduce its application to a real-world problem, analysis of cities and villages in Japan.
keywordsSelf-Organizing Map, Corresponding Gap, Square error.

1 Introduction

Statistical learning machines such as multi-layer perceptrons, Boltzmann machines, and competitive winner-takes-all learning models are now being applied to computational intelligence, bioinformatics, and data mining [7]. It was proved that such machines have capability to attain high learning performance by adapting hidden parts to complex environments [2]. However, artificial neural networks are statistically nonregular because the maps taking parameters to statistical models are not one-to-one and their Fisher information matrices are singular. A new mathematical method is necessary to evaluate such hierarchical learning machines [9].

Kohonen's Self-Organizing Map (SOM) [5], which is useful in visualization, analysis, and knowledge discovery of high dimensional data, is also a nonregular statistical model, because it contains hidden parts. To evaluate trained SOMs, statistical tools and empirical studies such as bootstrap methods were proposed [1][3], and estimating generalization error was studied [8]. Topographic idex using an external measure was also introduced [4]. However, in SOM training, there are many local minima which have globally different structures but almost same training and generalization errors, resulting that evaluation of them has been difficult.

S. Ramanna et al. (Eds.): MIWAI 2013, LNCS 8271, pp. 165–174, 2013.
© Springer-Verlag Berlin Heidelberg 2013

In this paper, in order to evaluate a trained SOM, we propose a new criterion which measures the geometric naturalness of the embedded SOM in the high dimensional data space. This criterion is defined as follows. Firstly, many different trained SOMs are collected by changing initial conditions. Secondly, the corresponding matrices [6] between all pairs in trained SOMs are calculated. Then from the corresponding matrix, we define the correspondence gap. And lastly, for a trained SOM, a new criterion is defined by the empirical average of correspondence gaps between a trained SOM and all other collected SOMs.

The effectiveness of the proposed criterion is shown by experiments for artificial data and real world data for analysis of cities and villages in Japan. It is experimentally shown that the proposed method can choose the more non-twisted SOM than the square error criterion. Also it is shown that, by using the proposed criterion, we can compare several SOM training algorithms.

This paper consists of 5 sections. In the second section, we introduce two different training algorithms of SOM, both of which were developed by Kohonen [5]. In the third section, we propose a new evaluation method of a SOM and explain its geometric meaning. In the fourth section, experimental results about artificial data and applications to city and village data analysis are reported. And, in the last section, we discuss and conclude the paper.

2   Self-Organizing Map

2.1   SOM Learning Algorithms

Let $\mathbb{R}^M$ be the data space. We study a two-dimensional Self-Organizing Map (SOM) which are embedded in $\mathbb{R}^M$. In other words, a SOM is defined by a two-dimensional manifold which is contained in $\mathbb{R}^M$. The set of training samples is denoted by

$$\{x_n \in \mathbb{R}^M; n = 1, 2, ..., N\},$$

where N is the number of data. A SOM $(K \times K)$ is represented by the set of its node points,

$$\{Y(j, k) \in \mathbb{R}^M; j, k = 1, 2, ..., K\}.$$

For each (j, k), its neighborhood is necessary. For example, the set

$$\{(j - 1, k), (j, k - 1), (j, k), (j + 1, k), (j, k + 1)\}$$

is defined as a neighborhood of (j, k). In figures of the present paper, Fig.2,3,4,5, two nodes in the neighborhood are connected by a line.

For a given training set, a SOM is optimized so that it can approximate the lower dimensional manifold made of the training samples. The basic SOM training algorithm is given by the following.

SOM Training Algorithm I. (T. Kohonen,[5])
(1) Initialize $\{Y(j, k)\}$ randomly.
(2) The set $S(j, k)$ is defined by

$$S(j, k) = \{n; \|x_n - Y(j, k)\| = \min_{(p,q)} \|x_n - Y(p, q)\|\}.$$

By the definition, $S(j,k)$ is the subset made of suffices $\{n\}$ such that the closest node point from x_n is $Y(j,k)$.

(3) For each $S(j,k)$, the partition number of data N_{jk} and the average vector $\overline{X}(j,k)$ in $S(j,k)$ are calculated, by the following equations,

$$N_{jk} = \sum_{n \in S(j,k)} 1,$$

$$\overline{X}(j,k) = \frac{1}{N_{jk}} \sum_{n \in S(j,k)} x_n.$$

Then, by the definition, $\sum_{j,k} N_{jk} = N$.

(4) Let $\alpha > 0$ is a positive constant value. Each node point $Y(j,k)$ and its neighborhood $Y(p,q)$ are updated by the following equations respectively.

$$Y(j,k) := Y(j,k) + \alpha N_{jk}(\overline{X}(j,k) - Y(j,k)), \tag{1}$$

$$Y(p,q) := Y(p,q) + (\alpha/2)N_{pq}(\overline{X}(j,k) - Y(p,q)).$$

Where $0 \le \alpha N \le 1$.

(5) The procedures (2),(3), and (4) are repeated.

This method is referred to as Algorithm I. By using this recursive procedure, a trained SOM is obtained which strongly depends on the initial $\{Y(j,k)\}$. The table which consists of the partion numbers $\{N_{jk}\}$ is called a partition matrix.

It is desirable that a trained SOM represents the manifold made of data. However, some node points are often overfitted for an outlier datum, leading to such a resulting that a trained SOM often becomes twisted. To prevent such a phenomenon, Kohonen also gave an improved algorithm.

SOM Training Algorithm II. (T. Kohonen,[5])
This algorithm is the same as the basic Algorithm I, except that eq.(1) is replaced by

$$Y(j,k) := Y(j,k) + \alpha N_{jk}\Big\{\overline{X}(j,k) - Y(j,k)$$
$$-(1/4)\sum_{p,q}(\overline{X}(j,k) - Y(p,q))\Big\}.$$

This method is called the Algorithm II. By the effect of the additional term, it is expected that a trained SOM by the Algorithm II becomes smoother. In this paper, we study an evaluation criterion to compare these different algorithms.

2.2   Evaluation Criterion

In order to evaluate a trained SOM, the square error E is sometimes employed,

$$E = \frac{1}{2N} \sum_{n=1}^{N} \min_{(j,k)} \|x_n - Y(j,k)\|^2.$$

The smaller square error means that the SOM is fitted for the data. However, the square error does not show the geometric naturalness of the SOM. Since this value can be understood as the training error, the generalization error has been proposed to evaluate the trained SOM for unseen data [8]. Even the generalization error can not estimate whether the trained SOM is appropriately embedded in the data space.

3   Proposed Method

In this chapter, we propose a new criterion which enables us to measure the geometric naturalness of a trained SOM.

Let $Y = \{Y(j,k)\}$ and $Z = \{Z(p,q)\}$ be two different trained SOMs for the same training data. The correspondence matrix $\{A(j,k)\}$ is defined by the following.

(1) $\{A(j,k)\}$ is a matrix whose (j,k) component is a pair of natural numbers $(A_1(j,k), A_2(j,k))$.

(2) Let $Z(p,q)$ be the nearest node point to $Y(j,k)$ in the set $\{Z(p,q)\}$. Then $A(j,k)$ is defined by

$$A(j,k) = (A_1(j,k), A_2(j,k)) = (p,q).$$

Note that A is defined by using a pair of two trained SOMs (Y, Z).

By using the correspondence matrix A, the correspondence gap $G(Y,Z)$ is defined by

$$G(Y,Z) = \frac{1}{K^2} \sum_{j=1}^{K-1} \sum_{k=1}^{K-1} \sum_{d=1}^{2} \left| A_d(j+1, k+1) - A_d(j, k+1) \right.$$

$$\left. - A_d(j+1, k) + A_d(j, k) \right|^2.$$

Then, $G(Y,Z) = 0$, if and only if, for $d = 1, 2$,

$$A_d(j+1, k+1) - A_d(j, k+1) = A_d(j+1, k) - A_d(j, k).$$

In other words, if the direction from $(j, k+1)$ to $(j+1, k+1)$ is equal to that from (j, k) to $(j+1, k)$, then $G(Y,Z) = 0$. Therefore, the larger $G(Y,Z)$ indicates that the difference of $\|V(k+1) - V(k)\|$ is larger, where

$$V(k) = A_d(j+1, k) - A_d(j, k).$$

From the mathematical point of view, this value $G(Y,Z)$ calculates the difference of affine connection, therefore, $G(Y,Z)$ shows the geometric naturalness of the correpondence matrix. For a given set of many trained SOMs $Y_1, Y_2..., Y_L$, the average correspondence gap G for Z is defined by

$$G = \frac{1}{L} \sum_{\ell=1}^{L} G(Y_\ell, Z).$$

The small G shows that Z is geometrically naturally embedded into the data space.

4   Experiments

4.1   Artificial Data

To clarify the effectiveness of the proposed method, we conducted an experiment using artificial data. A uniform distribution on the positive subset of the sphere is defined,

$$\{(x, y, z) \; ; \; x^2 + y^2 + z^2 = 1, \quad x, y, z > 0\}$$

and 1000 training samples are independently taken from this distribution. In both algorithms, $\alpha = 0.00025$ was used, and the number of iteration epoch was 25000. By using the same training samples, 100 trained SOMs were obtained by using different initial values for both training Algorithms I and II.

In Fig.1, (1) and (2) respectively show the histograms of the square error E and the corresponding gap G by the Algorithm I. In Fig.1, (3) and (4) those by the Algorithm II. By the Algorithm I, E is made smaller than the Algorithm II, but G becomes larger. By the Algorithm II, E is larger and G smaller. Note that the variances of E and G by Algorithm I are larger than those by Algorithm II, hence Algorithm II is more stable than I. If we would adopt the square error E

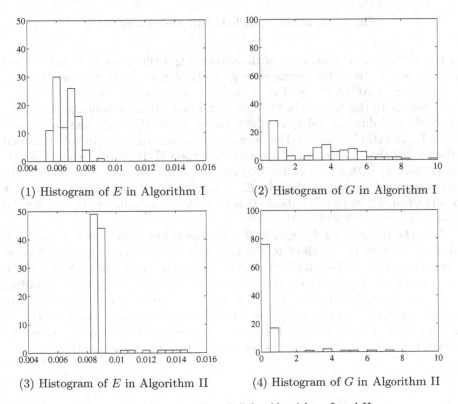

(1) Histogram of E in Algorithm I   (2) Histogram of G in Algorithm I

(3) Histogram of E in Algorithm II   (4) Histogram of G in Algorithm II

Fig. 1. Histograms of E and G for Algorithms I and II

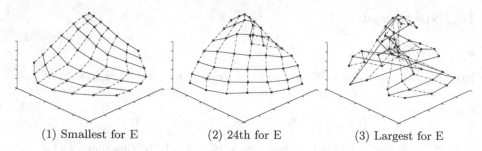

| (1) Smallest for E | (2) 24th for E | (3) Largest for E |

Fig. 2. Algorithm I: Smallest, 24th, and Largest for E

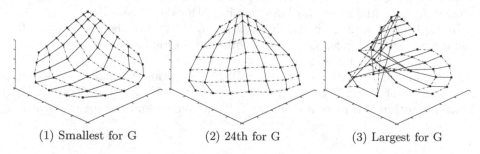

| (1) Smallest for G | (2) 24th for G | (3) Largest for G |

Fig. 3. Algorithm I: Smallest, 24th, and Largest for G

as the evaluation criterion of algorithms, then Algorithm I would be better than II. However, by using the corresponding gap G, Algorithm II is better than I.

The trained 100 SOMs by Algorithm I were sorted according to E or G from the smallest to the largest. In Fig.2, (1), (2), and (3) show respectively figures of SOMs in 3-dimensional space about the smallest, the 24th, and the largest E. In Fig.3, (1), (2), and (3) show respectively figures of SOMs in 3-dimensional space about the smallest, the 24th, and the largest G. By sorting trained SOMs by G, first 34 SOMs were nontwisted and the other 66 ones were twisted. On the other hand, by sorting them by E, such geometrical natural order could not be obtained. These results indicate that G is the better criterion than E for the geometric naturalness of SOM.

Also the trained 100 different SOMs by Algorithm II were sorted according to E or G from the smallest to the largest. In Fig.4, (1), (2), and (3) show respectively the smallest, the 50th, and the largest according to E. In Fig.5, (1), (2), and (3) show respectively the smallest, the 50th, and the largest according to G. By Algorithm II, more than half of trained SOMs were nontwisted. By comparing Fig.4 with Fig.2, and Fig.5 with Fig.3, it is shown that Algorithm II is more stable than Algorithm I. However, it seems that the figure that has the smallest E or G by Algorithm I is better than that by Algorithm II.

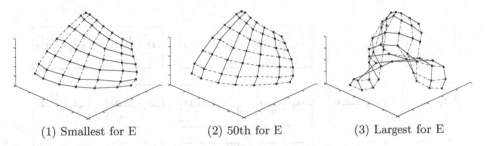

| (1) Smallest for E | (2) 50th for E | (3) Largest for E |

Fig. 4. Algorithm II: Best, 50th, and Worst for E

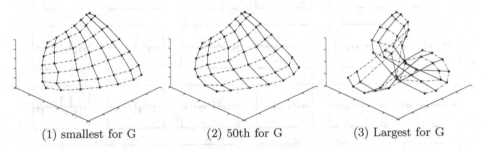

| (1) smallest for G | (2) 50th for G | (3) Largest for G |

Fig. 5. Algorithm II: Best, 50th, and Worst for G

4.2   Data of Cities and Villages in Japan

A SOM was applied to analysis of cities and villages in Japan. The number of training samples is 1700 and each datum is represented by the eleven-dimensional vector, (1) population, (2) young population (younger than 15 years old), (3) working population (15 - 65 years old), (4) older population (older than 65 years old), (5) number of births, (6) number of fatalities, (7) number of moving-in, (8) number of moving-out, (9) daytime population, (10) number of marriage, (11) number of divorce.

Every datum was normalized and the average of each dimension was made to be zero.

The 100 different trained SOMs made of 6×6 node points were constructed by Algorithm II, and the SOM that made G smallest was chosen. The partition matrices of the smallest E and G is respectively denoted by N_{E}. and N_{G}. They were given as

$$
N_E = \begin{pmatrix} 81 & 36 & 44 & 61 & 38 & 61 \\ 21 & 47 & 60 & 37 & 30 & 35 \\ 75 & 80 & 68 & 54 & 45 & 40 \\ 89 & 78 & 67 & 36 & 30 & 47 \\ 71 & 83 & 55 & 47 & 27 & 26 \\ 114 & 49 & 44 & 48 & 13 & 33 \end{pmatrix}, \quad N_G = \begin{pmatrix} 65 & 39 & 52 & 67 & 41 & 55 \\ 36 & 47 & 36 & 42 & 49 & 44 \\ 50 & 69 & 51 & 52 & 69 & 73 \\ 71 & 38 & 48 & 53 & 64 & 80 \\ 40 & 40 & 46 & 39 & 70 & 61 \\ 51 & 26 & 31 & 37 & 38 & 100 \end{pmatrix}.
$$

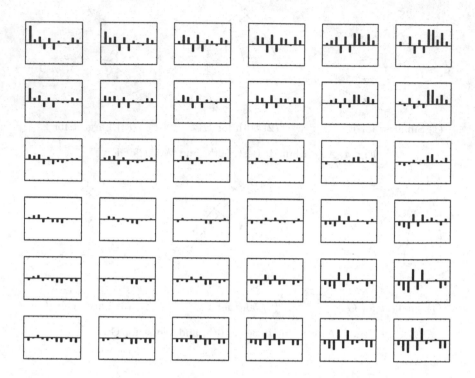

Fig. 6. Trained SOM for Cities and Villages

The standard deviation of elements in N_E is 21.9, whereas that in N_G is 15.7. Hence by minimization of the corresponding gap, we can choose the better trained SOM than by minimization of the square error.

Fig.6 shows the SOM trained by Algorithm II that makes G smallest. In Fig.6, (top, left) is a city where the number of people is larger than the average, whereas (bottom, right) is a village where the number of old people is larger than the average. In Fig.6, (top, right) is a city where the numbers of moving-in and moving-out is very large. From the trained SOM, we can understand that the map made of cities and villages in Japan can be characterized by the two viewpoints, young-old and turn-in-out. One viewpoint is urban and rural, another is busy and calm. In an urban city, the number of young and working peple is large, whereas in a rural village, the number of old people is large. In a busy city, the number of marriage and divorce is large, whereas in calm a city, the number of marriage and divorce is nearly equal to the average.

5   Discussion and Conclusion

5.1   Fluctuation of Ensemble

In this paper, we studied an evaluation criterion of trained SOMs based on geometric naturalness. The proposed criterion needs an emsenble of local minima, which were randomly generated by the fluctuation of the initial values of SOM traning algorithm. Therefore the proposed criterion depends on the method how to determine the initial values. To make the more objective criterion which is independent of initial values, the local minima should be chosen by some concrete probability distribution such as a Boltzmann distribution with a Hamilton function, for which the error function E would be appropriate. The method how to determine the inverse temperature of the Boltzmann distribution is the problem for the future study.

5.2   Possible Hybrid Algorithm

We compared two basic SOM training algorithms proposed by Kohonen. Nowadays, several methods how to improve SOM training are being developed, which can be also evaluated by the proposed criterion.

In experiments, it was shown that Algorithm II was more stable than Algorithm I, whereas the best SOM by Algorithm I was better than that by Algorithm II. In other words, there is a trade-off structure between stability and optimality. From this fact, a hybrid combination algorithm in which Algorithm I and II are respectively used firstly and lastly would be the more appropriate for SOM training. It is the future study how to apply a new evaluation criterion G to controlling of the best combination of two algorithms.

5.3   Generalization to Higher Dimension

In this paper, we studied two-dimensional SOM which is embedded into a higher dimensional Euclidean space. The method used in this paper can be generalized for the higher dimensional SOMs. For example, for three-dimensional SOM consists of points (i_1, i_2, i_3), the correspondence $\{A(i_1, i_2, i_3)\}$ is represented by three natural numbers $(A_1(i_1, i_2, i_3), A_2(i_1, i_2, i_3), A_3(i_1, i_2, i_3))$. Then the correspondence gap improved as

$$G(Y, Z) = \frac{1}{K^3} \sum_{i_1=1}^{K-1} \sum_{i_2=1}^{K-1} \sum_{i_3=1}^{K-1} \sum_{d=1}^{3} g_d(i_1, i_2, i_3),$$

where

$$g_1(i_1, i_2, i_3) = \left| A_1(i_1 + 1, i_2 + 1, i_3) - A_1(i_1, i_2 + 1, i_3) \right.$$
$$\left. - A_1(i_1 + 1, i_2, i_3) + A_1(i_1, i_2, i_3) \right|^2,$$

$$g_2(i_1, i_2, i_3) = \left| A_2(i_1, i_2 + 1, i_3 + 1) - A_2(i_1, i_2, i_3 + 1) \right.$$
$$\left. - A_2(i_1, i_2 + 1, i_3) + A_2(i_1, i_2, i_3) \right|^2,$$
$$g_3(i_1, i_2, i_3) = \left| A_3(i_1 + 1, i_2, i_3 + 1) - A_3(i_1 + 1, i_2, i_3) \right.$$
$$\left. - A_3(i_1, i_2, i_3 + 1) + A_3(i_1, i_2, i_3) \right|^2.$$

5.4 Conclusion

In this paper, we proposed a new evaluation criterion for a Self-Organizing Map, and showed its effectiveness by experimental results for artificial data and real-world data about cities and villages in Japan. By using the proposed criterion, we can compare trained SOMs according to the geometric naturalness. Moreover, we can compare SOM training algorithms.

Acknowledgment. We would like to thank National Statistics Center of Japan for our using data about cities and villages. http://www.nstac.go.jp/index.html

This research was partially supported by The Ministry of Education, Culture, Sports, Science and Technology in Japan, Grant-in-Aid for Scientific Research 23500172 and 25120013.

References

1. Adorno, M.C., Resta, M.: Reliability and Convergence on Kohonen Maps: An Empirical Study. In: Negoita, M.G., Howlett, R.J., Jain, L.C. (eds.) KES 2004. LNCS (LNAI), vol. 3213, pp. 426–433. Springer, Heidelberg (2004)
2. Barron, A.R.: Universal Approximation Bounds for Superpositions of a Sigmoidal Function. IEEE Transactions on Information Theory 39(3) (1993)
3. de Bodt, E., Cottrell, M., Verleysen, M.: Statistical tools to assess the reliability of self-organizing maps. Neural Networks 15, 967–978 (2002)
4. Fukui, K.-I., Numao, M.: Topographic Measure Based on External Criteria for Self-Organizing Map. In: Laaksonen, J., Honkela, T. (eds.) WSOM 2011. LNCS, vol. 6731, pp. 131–140. Springer, Heidelberg (2011)
5. Kohonen, T.: Self-Organizing maps. Springer (1995)
6. Ohara, S., Watanabe, S.: Correspondence of Different Self-Organizing Maps and its Application to Learning Evaluation. IEICE Technical Report, NC2012-149, 89–94 (2013)
7. Rumelhart, D.E., McClelland, J.L.: PDP Research Group. Parallel Distributed Processing. MIT Press (1986)
8. Saitoh, F., Watanabe, S.: On Generalization Error of Self-Organizing Map. In: Wong, K.W., Mendis, B.S.U., Bouzerdoum, A. (eds.) ICONIP 2010, Part II. LNCS, vol. 6444, pp. 399–406. Springer, Heidelberg (2010)
9. Watanabe, S.: Algebraic Geometry and Statistical Learning Theory. Cambridge University Press (2009)

AOF-Based Algorithm for Dynamic Multi-Objective Distributed Constraint Optimization

Tenda Okimoto[1,3], Maxime Clement[2], and Katsumi Inoue[3]

[1] Transdisciplinary Research Integration Center, Tokyo 1018430, Japan
[2] Pierre and Marie Curie University (Paris 6), Paris 75005, France
[3] National Institute of Informatics, Tokyo 1018430, Japan
{tenda,inoue}@nii.ac.jp, maxime.clement@etu.upmc.fr

Abstract. Many real world problems involve multiple criteria that should be considered separately and optimized simultaneously. A Multi-Objective Distributed Constraint Optimization Problem (MO-DCOP) is the extension of a mono-objective Distributed Constraint Optimization Problem (DCOP). A DCOP is a fundamental problem that can formalize various applications related to multi-agent cooperation. This problem consists of a set of agents, each of which needs to decide the value assignment of its variables so that the sum of the resulting rewards is maximized. An MO-DCOP is a DCOP which involves multiple criteria. Most researches have focused on developing algorithms for solving static problems. However, many real world problems are dynamic. In this paper, we focus on a change of criteria/objectives and model a Dynamic MO-DCOP (DMO-DCOP) which is defined by a sequence of static MO-DCOPs. Furthermore, we develop a novel algorithm for DMO-DCOPs. The characteristics of this algorithm are as follows: (i) it is a reused algorithm which finds Pareto optimal solutions for all MO-DCOPs in a sequence using the information of previous solutions, (ii) it utilizes the Aggregate Objective Function (AOF) technique which is the widely used classical method to find Pareto optimal solutions, and (iii) the complexity of this algorithm is determined by the induced width of problem instances.

1 Introduction

Many real world optimization problems involve multiple criteria that should be considered separately and optimized simultaneously. A *Multi-Objective Distributed Constraint Optimization Problem* (MO-DCOP) [6, 12] is the extension of a mono-objective Distributed Constraint Optimization Problem (DCOP) [14, 17] which is a fundamental problem that can formalize various applications related to multi-agent cooperation, e.g., distributed resource allocation problems including sensor networks [8], meeting scheduling [10], and the synchronization of traffic lights [7]. A DCOP consists of a set of agents, each of which needs to decide the value assignment of its variables so that the sum of the resulting rewards is maximized. An MO-DCOP is a DCOP which involves multiple criteria/objectives.

S. Ramanna et al. (Eds.): MIWAI 2013, LNCS 8271, pp. 175–186, 2013.

In MO-DCOPs, since trade-offs exist among objectives, there does not generally exist an ideal assignment, which maximizes all objectives simultaneously. Thus, the optimal solution of an MO-DCOP is characterized by using the concept of *Pareto optimality*. Solving an MO-DCOP is to find the Pareto front. The Pareto front is a set of reward vectors which are Pareto optimal solutions. An assignment is a Pareto optimal solution, if there does not exist another assignment that improves all of the criteria. An MO-DCOP can be represented using a constraint graph, in which a node/an edge represents an agent/a constraint.

Solving an MO-DCOP is intractable for large-scale problem instances. The size of the Pareto front of an MO-DCOP is exponential in the number of variables, i.e, in an MO-DCOP, all assignments are Pareto optimal solutions in the worst case. Since finding all Pareto optimal solutions becomes easily intractable, it is important to consider the following two approaches. The first approach is to find a subset of Pareto front instead of finding all Pareto optimal solutions. The second is to develop a fast but incomplete algorithm. In this paper, we consider the first approach. For the second approach, there exists an incomplete algorithm called the Bounded Multi-Objective Max-Sum algorithm (B-MOMS) [6] which is the extension of the bounded max-sum algorithm for DCOPs. This is the first and only existing incomplete MO-DCOP algorithm.

Most of previous works have focused on developing algorithms for solving static DCOPs and MO-DCOPs. However, many real world problems are dynamic, i.e., problems change at runtime. There exists some works for Dynamic DCOPs (DDCOPs) [1, 19]. Compared to previous works, there exists no work on considering multiple criteria in a dynamic environment, as far as the authors are aware. Let us introduce an example that can be formalized as Dynamic MO-DCOPs. Consider a security problem among several corporations where there exists an agent who acts as a secretary for each corporation [15]. When solving this problem with several criteria like security, privacy and reward in such a corporation network, we can apply MO-DCOP techniques. Also, the number of criteria changes at runtime, e.g., this week each agent tries to optimize the security level and the cost, and next week they additionally consider the privacy level. When we consider such problems, we can apply Dynamic MO-DCOP.

An *Aggregate Objective Function* (AOF) [5, 13] is the simplest and the most widely used classical method to find the Pareto solutions of a *Multi-Objective Optimization Problem* (MOOP). This method scalarizes the set of objective functions into a weighted mono-objective function, and finds the corresponding optimal solution. It is well known that the optimal solution obtained by AOF is a Pareto optimal solution of the original MOOP [13]. It is also guaranteed that AOF can find all Pareto optimal solutions when the Pareto front is convex. Otherwise, it cannot find all of them.

In this paper, we introduce a Dynamic Multi-Objective Distributed Constraint Optimization Problem (DMO-DCOP) which is the extension of an MO-DCOP and a DDCOP. A DMO-DCOP is defined by a sequence of static MO-DCOPs. As an initial step forward developing an algorithm for solving a DMO-DCOP, we focus on the change of criteria/objectives, i.e., we assume that only the number of objectives changes at runtime. However, the change is unpredictable, i.e., it is not known in advance how many objectives will be added or removed in the next problem in a sequence. This assumption requires a reactive approach. Furthermore, we develop two novel algorithms called Dynamic Programming based on AOF-technique (DP-AOF) and Dynamic Programming based on Reused-technique (DPR). DP-AOF utilizes an AOF-technique and Dynamic Programming (DP) and finds a subset of Pareto front of an MO-DCOP. DPR is the first algorithm for solving a DMO-DCOP which is based on DP-AOF.

A distributed search method with bounded cost vectors [12] is a complete algorithm which can guarantee to find all Pareto optimal solutions of MO-DCOPs. This algorithm is a generalized ADOPT algorithm [14] that performs tree-search and partial dynamic programming. Compared to this algorithm, DP-AOF computes a subset of the Pareto front of MO-DCOPs, while it solves all Pareto optimal solutions. Furthermore, DPR is an algorithm for solving a DMO-DCOP.

In a *Multi-Objective Constraint Optimization Problem* (MO-COP), various algorithms have been developed [11, 16, 18]. Compared to these algorithms, DP-AOF is for MO-DCOPs where variables and constraints are distributed among agents. Also, DPR is for DMO-DCOPs. Compared to evolutionary algorithms [2, 3] for solving a *Multi-Objective Optimization Problem* (MOOP), the advantage of our algorithms is that they can guarantee to find a Pareto optimal solution.

2   Preliminaries

In this section, we briefly describe the formalization of Distributed Constraint Optimization Problems (DCOPs), Multi-Objective Distributed Constraint Optimization Problems (MO-DCOPs), and introduce an Aggregate Objective Function (AOF), which is a widely used method to find a Pareto optimal solution.

2.1   Mono-Objective DCOP

A *Distributed Constraint Optimization Problem* (DCOP) [14, 17] is a fundamental problem that can formalize various applications for multi-agent cooperation. Without loss of generality, we make the following assumptions for simplicity. Relaxing these assumptions to general cases is relatively straightforward:

– Each agent has exactly one variable.
– All constraints are binary.
– Each agent knows all constraints related to its variable.
– All reward values are non-negative.

A DCOP is defined with a set of agents S, a set of variables X, a set of constraint relations C, and a set of reward functions O. An agent i has its own variable x_i. A variable x_i takes its value from a finite, discrete domain D_i. A constraint relation (i, j) means there exists a constraint relation between x_i and x_j. For x_i and x_j, which have a constraint relation, the reward for an assignment

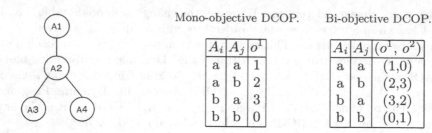

Fig. 1. Example of mono-objective and bi-objective DCOPs

$\{(x_i, d_i), (x_j, d_j)\}$ is defined by a reward function $r_{i,j}(d_i, d_j) : D_i \times D_j \to \mathbb{R}^+$. For a value assignment to all variables A, let us denote

$$R(A) = \sum_{(i,j)\in C, \{(x_i,d_i),(x_j,d_j)\}\subseteq A} r_{i,j}(d_i, d_j), \qquad (1)$$

where $d_i \in D_i$ and $d_j \in D_j$. Then, an optimal assignment A^* is given as $\arg\max_A R(A)$, i.e., A^* is an assignment that maximizes the sum of the value of all reward functions. A DCOP can be represented using a constraint graph, in which a node represents an agent/variable and an edge represents a constraint.

Example 1 (DCOP). Left table of Figure 1 shows a mono-objective DCOP with four variables x_1, x_2, x_3 and x_4. o^1 is an objective function where $i < j$. Each variable takes its value from a discrete domain $\{a, b\}$. The optimal solution of this problem is $\{(x_1, a), (x_2, b), (x_3, a), (x_4, a)\}$ and the optimal value is eight.

2.2   Multi-Objective DCOP

A *Multi-Objective Distributed Constraint Optimization Problem* (MO-DCOP) [6] is the extension of a mono-objective DCOP. An MO-DCOP is defined with a set of agents S, a set of variables X, multi-objective constraints $C=\{C^1, \ldots, C^m\}$, i.e., a set of sets of constraint relations, and multi-objective functions $O = \{O^1, \ldots, O^m\}$, i.e., a set of sets of objective functions. For an objective l ($1 \le l \le m$), a reward function $r^l_{i,j} : D_i \times D_j \to \mathbb{R}^+$, and a value assignment to all variables A, let us denote

$$R^l(A) = \sum_{(i,j)\in C^l, \{(x_i,d_i),(x_j,d_j)\}\subseteq A} r^l_{i,j}(d_i, d_j), \qquad (2)$$

where $d_i \in D_i$ and $d_j \in D_j$. Then, the sum of the values of all reward functions for m objectives is defined by a reward vector, denoted $R(A) = (R^1(A), \ldots, R^m(A))$. Finding an assignment that maximizes all objective simultaneously is ideal. However, in general, since trade-offs exist among objectives, there does not exist such an ideal assignment. Thus, the optimal solution of an MO-DCOP is characterized by using the concept of Pareto optimality. This problem can be also represented using a constraint graph.

Definition 1 (Dominance). For an MO-DCOP and two reward vectors $R(A)$ and $R(A')$ obtained by assignments A and A', we call that $R(A)$ *dominates* $R(A')$, denoted by $R(A') \prec R(A)$, iff $R(A')$ is partially less than $R(A)$, i.e., (i) it holds $R^l(A') \leq R^l(A)$ for all objectives l, and (ii) there exists at least one objective l', such that $R^{l'}(A') < R^{l'}(A)$.

Definition 2 (Pareto optimal solution). For an MO-DCOP and an assignment A, we call A is the *Pareto optimal solution*, iff there does not exist another assignment A', such that $R(A) \prec R(A')$.

Definition 3 (Pareto Front). For an MO-DCOP, we call the set of all Pareto optimal solutions to an MO-DCOP the *Pareto front. Solving an MO-DCOP is to find the Pareto front.*

Example 2 (MO-DCOP). Right table of Figure 1 shows a bi-objective DCOP. o^1 and o^2 represent objective functions. For example, when all agents take their value a, the obtained rewards for o^1 is 3, and 0 for o^2. The Pareto optimal solutions of this problem are $\{\{(x_1, b), (x_2, a), (x_3, b), (x_4, b)\}, \{(x_1, a), (x_2, b), (x_3, a), (x_4, a)\}\}$, and Pareto front obtained by these solutions is $\{(7, 8), (8, 7)\}$.

2.3 Aggregate Objective Function

An Aggregate Objective Function (AOF) [5, 13] is the simplest and the most widely used classical method to find the Pareto optimal solutions of an MOOP. This method scalarizes the set of objective functions into a weighted mono-objective function and find its optimal solution. For objective functions $o^1, \ldots, o^m$ of MOOPs, we define a weight denoted by $\alpha = (\alpha_1, \ldots, \alpha_m)$, where $\sum_{1 \leq i \leq m} \alpha_i = 1, \alpha_i > 0$. Next, we make a weighted mono-objective function $\alpha_1 o^1 + \ldots + \alpha_m o^m$, and finds an optimal solution for it. It is well known that this method can guarantee to find Pareto optimal solution [13]. In this paper, we use this method to find the Pareto optimal solutions of an MO-DCOP.

3 Dynamic MO-DCOP

In this section, we provide a model for a Dynamic MO-DCOP (DMO-DCOP). Furthermore, we develop a novel algorithm called Dynamic Programming based on AOF-technique (DP-AOF) for solving an MO-DCOP. Also, we introduce a novel algorithm called Dynamic Programming based on Reused-technique (DPR) for solving a DMO-DCOP. This algorithm utilizes DP-AOF and finds Pareto optimal solutions using the information of a previous MO-DCOP in a sequence. Finally, we provide the complexity of theses two algorithms respectively.

3.1 Model

A *Dynamic Multi-Objective Distributed Constraint Optimization Problem* (DMO-DCOP) is defined by a sequence of static MO-DCOPs.

$$< MO\text{-}DCOP_0, MO\text{-}DCOP_1, ..., MO\text{-}DCOP_{k-1} > . \tag{3}$$

Solving a DMO-DCOP is to find a sequence of Pareto front

$$< PF_0, PF_1, ..., PF_{k-1} >, \tag{4}$$

where PF_i $(0 \leq i \leq k-1)$ is the Pareto front of $MO\text{-}DCOP_i$. In this paper, we assume that only the number of objective functions changes, i.e., each MO-DCOP in a sequence has the same constraint graph structure. Since we do not know how many objective functions will be removed or added in the next problem, it requires to solve each MO-DCOP in a sequence one by one (*reactive*).

3.2   Algorithm

First, we propose a novel algorithm called Dynamic Programming based on AOF-technique (DP-AOF) for solving an MO-DCOP. This algorithm has the following two Phases:

- **Phase 1:** For each objective function, find an optimal solution.
- **Phase 2:** For a weighted mono-objective function, find its optimal solution.

Let us describe Phase 1. We use AOF-technique to find an optimal solution for each objective function, respectively. Specifically, for m objective functions of an MO-DCOP, we give the following m weights $(1, 0, ..., 0)$, $(0, 1, 0, ..., 0)$, $..., (0, ..., 0, 1)$ and make the m weighted objective functions $o^1, ..., o^m$. Then, we find an optimal solution for each weighted mono-objective function o^i $(1 \leq i \leq m)$, respectively, i.e., it is equivalent to solve m DCOP problems independently. In this paper, we denote the obtained m optimal values as $R_{max}^1, ..., R_{max}^m$.

In Phase 2, for m objective functions of an MO-DCOP, we provide the following weights [1] and create a weighted mono-objective function,

$$\alpha_1 = \frac{R_{max}^1}{R_{max}^1 + ... + R_{max}^m}, ..., \alpha_m = \frac{R_{max}^m}{R_{max}^1 + ... + R_{max}^m} \tag{5}$$

and find Pareto optimal solution for it. This algorithm find at most $(m+1)$ Pareto optimal solutions, if there exists and Pareto front of an MO-DCOP is convex, where m is the number of objectives. In this paper, we use a representative dynamic programming based DCOP algorithm called DPOP [17] to solve problems in Phase 1 and 2.

Example 3 (DP-AOF). We show the execution of DP-AOF using the bi-objective DCOP in example 2. In Phase 1, we solve the optimal solutions for objective o^1 and o^2, independently. The optimal solution for o^1 is $\{(x_1, a), (x_2, b), (x_3, a), (x_4, a)\}$ and the optimal value is 8 $(=R_{max}^1)$. For o^2, we obtain the optimal solution $\{(x_1, b), (x_2, a), (x_3, b), (x_4, b)\}$ and the optimal value 8 $(=R_{max}^2)$. In Phase 2, we create a mono-objective function using the following weights

$$\alpha_1 = \alpha_2 = \frac{R_{max}^1}{R_{max}^1 + R_{max}^2} = \frac{8}{8+8} = \frac{1}{2}$$

---

[1] Since we have no information about the weights in advance, we choose them where each objective has the same weight.

Algorithm 1. DPR

---

1 Given : <MO-DCOP$_0$,...,MO-DCOP$_{k-1}$ >, $\mathbb{PF}=\emptyset$
2 Solve initial MO-DCOP$_0$ by DP-AOF
3 $\mathbb{PF}=\{PF_0\}$ // Pareto front of MO-DCOP$_0$
4 **for each** $i = 1, \ldots, k-1$
5 **if** objectives are removed **then**
6   Remove the dominated solution from PF$_{i-1}$
7   Add PF$_i$ in $\mathbb{PF}$
8 **else** // new objectives are added
9   Solve MO-DCOP$_i$ by DP-AOF using PF$_{i-1}$
10   Add PF$_i$ in $\mathbb{PF}$
11 **end if**
12 **end for**

---

and compute the optimal solutions $\{\{(x_1, a), (x_2, b), (x_3, a), (x_4, a)\}, \{(x_1, b),$ $(x_2, a), (x_3, b), (x_4, b)\}\}$ which are Pareto optimal solutions of the bi-objective DCOP in example 2. The Pareto front obtained by DP-AOF is $\{(8, 7), (7, 8)\}$.

Next, we introduce a novel algorithm called Dynamic Programming based on Reused-technique (DPR) for solving a DMO-DCOP. This algorithm utilizes DP-AOF and can find Pareto optimal solutions using the information of a previous MO-DCOP in a sequence. Algorithm 1 represents the pseudo-code of DPR. We first solve an initial problem MO-DCOP$_0$ using DP-AOF (line 2) and adds the obtained Pareto front PF_0 in $\mathbb{PF}$ (line 3). We start to solve MO-DCOP$_i$ one by one in a sequence (line 4-12). For MO-DCOP$_1$, we check whether some objectives are removed (line 5) or added (line 8). In case some objectives are removed from MO-DCOP$_0$, for each reward vector in PF_0, we ignore the values of these removed objectives. Then, we remove the dominated reward vectors and add a set of the remaining reward vectors in $\mathbb{PF}$ as PF_1 (line 6 and 7). In case some objectives are added to MO-DCOP$_0$, we (i) find the optimal solutions for newly added objectives using DP-AOF (Phase 1), (ii) use the Pareto optimal solutions in PF_0 and compute the corresponding reward vectors in MO-DCOP$_1$, (iii) create a weighted mono-objective function using the information of (i) and (ii), i.e., compute the weights using the optimal values obtained in (i) and (ii), and find its optimal solution (Phase 2 of DP-AOF) (line 9). We add a set of Pareto optimal solutions obtained by (i)-(iii) in $\mathbb{PF}$ as PF$_1$ (line 10). It continues this process until $i = k-1$. Since DPR reuses the information of a previous problem, we can expect that it can efficiently solve a DMO-DCOP compared to a naive way (restart) which solves MO-DCOPs in a sequence independently.

For a sequence <MO-DCOP$_0$, MO-DCOP$_1$, ... , MO-DCOP$_{k-1}$ >, let m_{i-1} and m_i be the number of objectives of MO-DCOP$_{i-1}$ and MO-DCOP$_i$ $(1 \leq i \leq k-1)$. For simplicity, we assume $| m_i - m_{i-1} |= 1$, i.e., MO-DCOP$_i$ is a problem where an objective is removed from MO-DCOP$_{i-1}$, or an objective is added to MO-DCOP$_{i-1}$. Let us describe in the case $m_i < m_{i-1}$. Assume that MO-DCOP$_i$ is a problem where an objective o^h is removed from MO-DCOP$_{i-1}$. For an MO-DCOP$_{i-1}$ and its objective functions $o^1, .., o^h, ..., o^{m_{i-1}}$, let $PF_{i-1} = \{(R_{max}^1, 0, ...,$

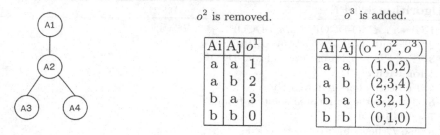

o^2 is removed. o^3 is added.

Ai	Aj	o^1
a	a	1
a	b	2
b	a	3
b	b	0

Ai	Aj	(o^1, o^2, o^3)
a	a	(1,0,2)
a	b	(2,3,4)
b	a	(3,2,1)
b	b	(0,1,0)

Fig. 2. Example of a DMO-DCOP

$0^h, .., 0), ..., (0, ..., 0, R_{max}^h, 0, ..., 0), ..., (0, ..., 0^h, ..., R_{max}^{m_{i-1}}), (r^1, ..., r^h, ..., r^{m_{i-1}})\}$ be the Pareto front of this problem, where $r^i > 0$ $(1 \leq i \leq m_{i-1})$, reward vectors including R_{max}^i $(1 \leq i \leq m_{i-1})$ are the reward vectors obtained by DP-AOF (Phase 1), 0^h is the reward (zero) for objective h, and $(r^1, ..., r^h, ..., r^{m_{i-1}})$ is the reward vector obtained by DP-AOF (Phase 2). Then, PF_i is a set of remaining reward vectors in which the values for objective h of each reward vector in PF_{i-1} are ignored/removed. Consider the reward vector $(0, ..., 0, R_{max}^h, 0, ..., 0)$ in PF_{i-1}. Then, this vector becomes a null vector and is dominated by another reward vectors. Thus, PF_i is $\{(R_{max}^1, 0, ..., 0^{h-1}, 0^{h+1}, ..., 0), ..., (0, ..., 0^{h-1}, 0^{h+1}, ..., R_{max}^{m_{i-1}}), (r^1, ..., r^{h-1}, r^{h+1}, ..., r^{m_{i-1}})\}$. Clearly, these reward vectors are not dominated by another vector. In DPR, we compute the Pareto front PF_i of MO-DCOP$_i$ removing the dominated reward vectors in PF_{i-1}.

Example 4 (Removal). Let the example 2 be an initial problem (MO-DCOP$_0$). The Pareto optimal solutions of this bi-objective DCOP are $\{\{(x_1, a), (x_2, b), (x_3, a), (x_4, a)\}, \{(x_1, b), (x_2, a), (x_3, b), (x_4, b)\}\}$ and Pareto front is $\{(8, 7), (7, 8)\}$. We assume that fig.2 table (left) represents a reward table for MO-DCOP$_1$, i.e., MO-DCOP$_1$ is a problem where an objective o^2 is removed from MO-DCOP$_0$. In DPR, for each reward vector in PF_0, we remove the value for an objective o^2 and check the dominance, and compute the Pareto front $PF_1 = \{8\}$ for MO-DCOP$_1$.

Let us describe in the case $m_i > m_{i-1}$. Assume that MO-DCOP$_i$ is a problem where an objective o^{m_i} is added to MO-DCOP$_{i-1}$. For an MO-DCOP$_{i-1}$, let $PF_{i-1} = \{(R_{max}^1, 0, ..., 0), ..., (0, ..., R_{max}^{m_{i-1}}), (r^1, ..., r^{m_{i-1}})\}$ be the Pareto front of this problem. Let $PF_i = \{(R_{max}^1, 0, .., 0, q_1^{m_i}), ..., (0, ..., 0, R_{max}^{m_{i-1}}, q_{m_{i-1}}^{m_i}), (r^1, ..., r^{m_{i-1}}, r^{m_i}), (0, ..., R_{max}^{m_i}), (s^1, ..., s^{m_i})\}$ be the Pareto front obtained by DP-AOF, where $s^i > 0$ $(1 \leq i \leq m_i)$, $(0, ..., R_{max}^{m_i})$ and $(s^1, ..., s^{m_i})$ represent reward vectors obtained by Phase 1 and 2 in DP-AOF, and the values $q_1^{m_i}, ..., q_{m_{i-1}}^{m_i}$ and r^{m_i} represent rewards for objective m_i obtained by Pareto optimal solutions of MO-DCOP$_{i-1}$. Clearly, reward vectors including one of $R_{max}^1, ..., R_{max}^{m_i}$ are not dominated by another reward vector. Also, $(r^1, ..., r^{m_{i-1}}, r^{m_i})$ and $(s^1, ..., s^{m_i})$ are not dominated by another vector excepting $(s^1, ..., s^{m_i})$ dominates $(r^1, ..., r^{m_{i-1}}, r^{m_i})$ vice versa. In DPR, we compute the Pareto front of MO-DCOP$_i$ utilizing the information of PF_{i-1}. In case $m_i = m_{i-1}$, MO-DCOP$_{i-1}$ and MO-DCOP$_i$ are same problems. Thus, PF_{i-1} and PF_i are same.

Example 5 (Addition). Let the example 2 be an initial problem (MO-DCOP$_0$). The Pareto front of this bi-objective DCOP is $PF_0 = \{(8,7),(7,8)\}$. We assume that fig.2 table (right) represents a reward table for MO-DCOP$_1$, i.e., MO-DCOP$_1$ is a problem where an objective o^3 is added to MO-DCOP$_0$. In DPR, we first find a Pareto optimal solution of objective function o^3 using DP-AOF (Phase 1) which is $\{\{(x_1,a),(x_2,a),(x_3,b),(x_4,b)\}$, and compute the reward vector $\{(5,6,10)\}$. Next, we compute the reward vectors $\{(8,7,6),(7,8,9)\}$ for MO-DCOP$_1$ using the Pareto optimal solutions of MO-DCOP$_0$, i.e., $\{\{(x_1,a),(x_2,b),(x_3,a),(x_4,a)\}, \{(x_1,b),(x_2,a),(x_3,b),(x_4,b)\}\}$. Finally, we find a Pareto optimal solution $\{\{(x_1,b),(x_2,a),(x_3,b),(x_4,b)\}$ and its reward vector $\{(7,8,9)\}$ of a weighted mono-objective function using DP-AOF (Phase 2). The Pareto optimal solutions of MO-DCOP$_1$ are $\{\{(x_1,a),(x_2,b),(x_3,a),(x_4,a)\}, \{(x_1,b),(x_2,a),(x_3,b),(x_4,b)\}, \{(x_1,a),(x_2,a),(x_3,b),(x_4,b)\}\}$ and the Pareto front PF_1 obtained by these solutions is $\{(8,7,6),(7,8,9),(5,6,10)\}$.

3.3 Complexity

We provide the complexity of DP-AOF and DPR, respectively. The complexity of DP-AOF is given by $O((m+1) \times |D_{max}|^{w*})$ where m is the number of objectives, $|D_{max}|$ is the maximal domain size, and $w*$ is the induced width [4] of a problem. Induced width is a parameter that determines the complexity of many DCOP algorithms. For example, if the induced width of a graph is one, it is a tree. Also, the induced width of a complete graph with n variables is $n-1$. In this paper, we utilize DPOP [17] to solve a mono-objective DCOP and the complexity of DPOP is given by $O(|D_{max}|^{w*})$. Since we need to solve $(m+1)$ mono-objective DCOPs in DP-AOF (Phase 1 and 2), the complexity is $O((m+1) \times |D_{max}|^{w*})$.

We provide the complexities of DPR for removal and additional cases. Let MO-DCOP$_{i-1}$ be the current problem and MO-DCOP$_i$ be the next problem in a sequence, and m_{i-1} and m_i be the number of objectives of MO-DCOP$_{i-1}$ and MO-DCOP$_i$. In case some objectives are removed from MO-DCOP$_{i-1}$, we first need to ignore/remove the values of removed objectives from the reward vectors of PF_{i-1}. Then, we check the dominance of remaining reward vectors of PF_{i-1}. Thus, the complexity of DPR to solve MO-DCOP$_i$ is given by $O(|PF_{i-1}| + |PF_{i-1}|^2)$, where $|PF_{i-1}|$ is the size of Pareto front of MO-DCOP$_{i-1}$.

In case some objectives are added to MO-DCOP$_i$, we first compute reward vectors for MO-DCOP$_i$ using the Pareto optimal solutions obtained in MO-DCOP$_{i-1}$. Then, we solve $(m_i - m_{i-1})$ mono-objective DCOPs by Phase 1 in DP-AOF, and create a weighted mono-objective DCOP and solve its optimal solution by Phase 2 in DP-AOF. Thus, the complexity of DPR to solve MO-DCOP$_i$ is given by $O(|PF_{i-1}| + ((m_i - m_{i-1}) + 1) \times |D_{max}|^{w*})$.

4 Evaluation

In this section, we evaluate the runtime of our algorithm for DMO-DCOPs and compare it with the naive method which solves each MO-DCOP in a sequence

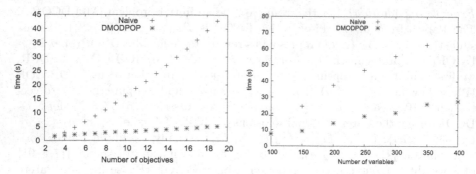

Fig. 3. Varying the number of objectives   **Fig. 4.** Varying the number of variables

independently. To evaluate our algorithm, we generate two different problem instances, where objectives are added (DMO-DCOP$_{add}$) and removed (DMO-DCOP$_{remove}$). In this paper, we show only the results for DMO-DCOP$_{add}$[2]. In DMO-DCOP$_{add}$, we add only one new objective at a time. We chose the reward values uniformly at random from the range $[0,\dots,100]$ for all objectives. The domain size of each variable is three. Each data point in a graph represents an average of 20 problem instances. We implemented our algorithms in Java and carried out all experiments on 6 core running at 2.6GHz with 12GB of RAM.

Figure 3 represents the results for problems with 20 nodes and the induced width is 6, varying the number of objectives. The results show the average of the total runtime for solving a sequence with the first MO-DCOP having two objectives and the last MO-DCOP having 20 objectives, i.e., each sequence contains 19 MO-DCOPs. We can see that the average runtime of our algorithm is almost constant, while the results of the naive method increases linearly. This is because our algorithm reuses the information of previous solutions, i.e., we only solve the added objective function, and not other objective functions which we have already solved. For example, for the last MO-DCOP with 20 objectives, our algorithm only solves the added objective function and a new created weighted mono-objective function using DP-AOF, but not another 19 objective function again, while the naive method solves 20 DCOPs independently as new problems.

Figure 4 shows the results for large-scale problem instances. We vary the number of variables from 100 to 400 [3]. The results show the average of the total runtime for solving a sequence of four MO-DCOPs, starting with 2 objectives and finishing with 5. We can see that both runtime increases slightly but with more variables, computing weight (solving a DCOP) becomes more difficult. Thus, since the naive method requires solving more DCOPs than our method, its runtime increases faster.

---

[2] Our algorithm is much faster in DMO-DCOP$_{remove}$ compared to that in DMO-DCOP$_{add}$, since this algorithm only checks the dominance among reward vectors in previous solutions PF_{i-1} for removal case which can be done in $O(|PF_{i-1}|^2)$.

[3] For an incomplete algorithm introduced in [6], the authors experimented problem instances with 100 nodes, and 25 nodes for a complete algorithm introduced in [12].

5   Discussion

In this section, we propose the extension of DP-AOF [4]. The advantage of our algorithm is that it can solve an MO-DCOP faster compared to an algorithm which finds all Pareto optimal solutions, since we solve a subset of Pareto front. However, one might consider that the size of Pareto front obtained by our algorithm is small, since it can find at most $(m+1)$ Pareto optimal solutions.

We introduce an extended version of our algorithm which utilizes DP-AOF iteratively. For simplicity, we explain using a bi-objective DCOP. In Phase 2 of the original DP-AOF, we provide the following weights and create a weighted mono-objective function π and find an optimal solution for it.

$$\alpha_1 = \frac{R_{max}^1}{R_{max}^1 + R_{max}^2}, \alpha_2 = \frac{R_{max}^2}{R_{max}^1 + R_{max}^2} \tag{6}$$

Let p_1 and p_2 be the solution points obtained by Phase 1 in DP-AOF, where p_1 contains R_{max}^1 for objective o^1 and p_2 contains R_{max}^2 for objective o^2. Also, let p_3 be the solution point obtained by Phase 2. In our extended algorithm, we create additional weighted mono-objective functions utilizing the obtained solution points. For example, we generate new weighted mono-objective functions π_1 and π_2, where the weights for π_1 is determined by the reward value of p_1 and p_3, and the weights for π_2 using p_2 and p_3. Intuitively, for each pair of solution points, this algorithm finds a new solution point between them, if the Pareto front is convex and there exists such solution points on Pareto surface. We can continue this process and obtain at most 2^t additional Pareto optimal solutions, where t is the number of iterations. If we want all Pareto optimal solutions of an MO-DCOP, we execute this process until it cannot find a new solution point. For tri-objective DCOP, we create a two-dimensional weighted plane (which we consider for further work). In case Pareto front is non-convex, it requires another scalarization technique, e.g., L_p-norm method [5]. We will develop an algorithm which can find solution points of the non-convex part on the Pareto surface.

6   Conclusion

In this paper, we introduced a Dynamic Multi-Objective Distributed Constraint Optimization Problem (DMO-DCOP). Furthermore, we developed two novel algorithms called DP-AOF and DPR. DP-AOF utilizes a AOF-technique and Dynamic Programming and finds a subset of Pareto front of an MO-DCOP. DPR is the first algorithm for solving a DMO-DCOP which is based on DP-AOF. In the experiments, we evaluated the runtime of our algorithm for DMO-DCOPs and showed that our algorithm outperforms the naive method. Also, we showed that it can efficiently solve large-scale problem instances. Finally, we proposed the extension of DP-AOF (and also DPR) which provides several Pareto optimal solutions for bi-objective MO-DCOPs. Our plans for future work include developing algorithms based on other scalarization methods, e.g., ϵ-constraint method [5] and weighted metrics methods [9] for solving an MOOP.

---

[4] Since DPR is based on DP-AOF, the extended version can be used in DPR.

References

[1] Billiau, G., Chang, C.F., Ghose, A.: Sbdo: A new robust approach to dynamic distributed constraint optimisation. In: Desai, N., Liu, A., Winikoff, M. (eds.) PRIMA 2010. LNCS, vol. 7057, pp. 11–26. Springer, Heidelberg (2012)

[2] Bringmann, K., Friedrich, T., Neumann, F., Wagner, M.: Approximation-guided evolutionary multi-objective optimization. In: IJCAI, pp. 1198–1203 (2011)

[3] Deb, K., Agrawal, S., Pratap, A., Meyarivan, T.: A fast and elitist multiobjective genetic algorithm: Nsga-ii. IEEE Trans. Evolutionary Computation 6(2), 182–197 (2002)

[4] Dechter, R.: Constraint Processing. Morgan Kaufmann Publishers (2003)

[5] Ehrgott, M.: Multicriteria Optimization. Springer (2005)

[6] Fave, F.M.D., Stranders, R., Rogers, A., Jennings, N.R.: Bounded decentralised coordination over multiple objectives. In: AAMAS, pp. 371–378 (2011)

[7] Junges, R., Bazzan, A.L.C.: Evaluating the performance of DCOP algorithms in a real world, dynamic problem. In: AAMAS, pp. 599–606 (2008)

[8] Lesser, V., Ortiz, C., Tambe, M. (eds.): Distributed Sensor Networks: A Multiagent Perspective, vol. 9. Kluwer Academic Publishers (2003)

[9] Lin, J.G.: On min-norm and min-max methods of multi-objective optimization. Mathematical Programming 103(1), 1–33 (2005)

[10] Maheswaran, R.T., Tambe, M., Bowring, E., Pearce, J.P., Varakantham, P.: Taking DCOP to the real world: Efficient complete solutions for distributed multi-event scheduling. In: AAMAS, pp. 310–317 (2004)

[11] Marinescu, R.: Best-first vs. depth-first and/or search for multi-objective constraint optimization. In: ICTAI, pp. 439–446 (2010)

[12] Matsui, T., Silaghi, M., Hirayama, K., Yokoo, M., Matsuo, H.: Distributed search method with bounded cost vectors on multiple objective dcops. In: Rahwan, I., Wobcke, W., Sen, S., Sugawara, T. (eds.) PRIMA 2012. LNCS, vol. 7455, pp. 137–152. Springer, Heidelberg (2012)

[13] Miettinen, K.: Nonlinear Multiobjective Optimization. Kluwer Academic Publishers, Boston (1999)

[14] Modi, P., Shen, W., Tambe, M., Yokoo, M.: Adopt: asynchronous distributed constraint optimization with quality guarantees. Artificial Intelligence 161(1-2), 149–180 (2005)

[15] Okimoto, T., Ikegai, N., Ribeiro, T., Inoue, K., Okada, H., Maruyama, H.: Cyber security problem based on multi-objective distributed constraint optimization technique. In: DSN Workshop (to appear, 2013)

[16] Perny, P., Spanjaard, O.: Near admissible algorithms for multiobjective search. In: ECAI, pp. 490–494 (2008)

[17] Petcu, A., Faltings, B.: A scalable method for multiagent constraint optimization. In: IJCAI, pp. 266–271 (2005)

[18] Rollon, E., Larrosa, J.: Bucket elimination for multiobjective optimization problems. Journal of Heuristics 12(4-5), 307–328 (2006)

[19] Yeoh, W., Varakantham, P., Sun, X., Koenig, S.: Incremental dcop search algorithms for solving dynamic dcops. In: AAMAS, pp. 1069–1070 (2011)

Relational Change Pattern Mining Based on Modularity Difference

Yoshiaki Okubo[1], Makoto Haraguchi[1], and Etsuji Tomita[2]

[1] Graduate School of Information Science and Technology
Hokkaido University
N-14 W-9, Sapporo 060-0814, Japan
{mh,yoshiaki}@ist.hokudai.ac.jp
[2] The Advanced Algorithms Research Laboratory
The University of Electro-Communications
Chofugaoka 1-5-1, Chofu, Tokyo 182-8585, Japan
tomita@ice.uec.ac.jp

Abstract. This paper is concerned with a problem of detecting *relational changes*. Many kinds of graph data including social networks are increasing nowadays. In such a graph, the relationships among vertices are changing day by day. Therefore, it would be worth investigating a data mining method for detecting significant patterns informing us about what changes. We present in this paper a general framework for detecting relational changes over two graphs to be contrasted. Our target pattern with relational change is defined as a set of vertices common in both graphs in which the vertices are almost disconnected in one graph, while densely connected in the other. We formalize such a target pattern based on the notions of *modularity* and *k-plex*. A depth-first algorithm for the mining task is designed as an extension of k-plex enumerators with some pruning mechanisms. Our experimental results show usefulness of the proposed method for two pairs of graphs representing actual reply-communications among Twitter users and word co-occurrence relations in Japanese news articles.

1 Introduction

In order to find interesting things that are hard to be realized when observing only a single domain or a single data set, we can take a strategy of *"compare and contrast"* as one of general and reasonable heuristics. This kind of strategy has been also applied in the field of data mining. For instance, when the target patterns are evaluated by frequency, probability and correlation, then the *degree of change* can be defined as their *ratio* or *difference*. The definitions and their corresponding methods have been investigated in the literature about *Emerging Pattern* [4], *Contrast Set* [5] and *Correlation Change* [3], all of which are targeting itemsets or variable sets whose evaluations increase from one transaction database to another one. On the other hand, we discuss in this paper a general method that can solve a problem of finding *relational changes between two (undirected) graphs*. For example, communication-links (relationships) among Twitter users at different time points might be represented as such a pair of graphs, as

S. Ramanna et al. (Eds.): MIWAI 2013, LNCS 8271, pp. 187–198, 2013.
© Springer-Verlag Berlin Heidelberg 2013

will be discussed later. It is noted here that we do not assume any particular restriction on graphs to be contrasted. Our framework can be applied to many cases, e.g., relational changes at different time points, in different domains, in different areas, and so forth.

We are given a pair of undirected graphs, $G^{(0)}$ and $G^{(1)}$. For the graphs, we can observe four types of relational changes among the vertices common in both graphs: a pair of vertices is

- 01-type: disconnected in $G^{(0)}$, while connected in $G^{(1)}$,
- 10-type: connected in $G^{(0)}$, while disconnected in $G^{(1)}$,
- 00-type: disconnected in $G^{(0)}$ and still disconnected in $G^{(1)}$, or
- 11-type: connected in $G^{(0)}$ and still connected in $G^{(1)}$.

In what follows, a vertex pair of its corresponding type is simply referred to as a 01, 10, 00 or 11-pair, respectively.

In our case of relational change, as a target, we try to detect a set of vertices common in both graphs that are *densely connected* in $G^{(1)}$, while *almost disconnected* in $G^{(0)}$. That is, our target would be a vertex set X in which most of the vertex pairs are of 01-type. A typical example of our target is shown in Figure 1.

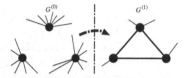

Fig. 1. From diverging independent set to nearly-isolated clique

As a simple and intuitive way to find such a vertex set, one might suggest that a *visualization-based approach* would be helpful. By visualizing a given graph [9], we can easily recognize its structure. Therefore, comparing visualization results of a given pair of graphs, we can identify our target vertex sets without any difficulty. This simple approach will surely work well for small graphs. However, if we try to detect our targets from larger graphs with several hundreds vertices and edges, such an approach would be no longer useful for the task.

In order to extract our target vertex set, we utilize in this paper the notion of *modularity* introduced in [6]. The notion of modularity has been commonly used to evaluate a community (densely connected group of vertices) in graph clustering/partitioning. On the other hand, we propose to use the same measure to evaluate a relational change. More concretely, a vertex set X is evaluated in each of $G^{(0)}$ and $G^{(1)}$, respectively, with the modularity measure, and then we evaluate the relational change by the *difference of their modularity values*. Particularly, to obtain a vertex set with denser connections in $G^{(1)}$, we evaluate a vertex set X by subtraction of X's modularity in $G^{(0)}$ from that in $G^{(1)}$ and prefer X with higher evaluation values. It should be noted here that from the characteristics of modularity, our evaluation value of X becomes higher if X forms an *anti-community* in $G^{(0)}$, while a community in $G^{(1)}$, where an anti-community consists of almost-isolated vertices with larger degrees as has been discussed in [6]. As a typical example, therefore, we can detect a vertex set shown in Figure 1 as our target.

In [10], a formulation of a change detection problem in terms of modularity difference has been discussed. Technically speaking, the modularity difference can be divided into two factors, *positive* and *negative* ones. Then, Li et al. have designed in [10] a constrained miner for finding vertex sets satisfying constraints for positive and negative factors so that the positive factors of the solution vertex sets exceed a given lower bound and the negative ones are less than a given upper bound. However, due to the huge number of very small modularity difference values, their algorithm have not worked well for severe parameter settings. The problem is now simply solved in this paper by introducing more relaxed constraint that a limited number of exceptional values are allowed.

We first divide the whole vertex set into several connected components of vertices showing higher positive modularity changes, and then apply the following constrained search for each component. The revised constraint is based on k-plexes [7] that are also used for *Structure Change Detection* [11] to find candidate vertex sets, while we make use of them to allow at most $(k-1)$ exceptional difference values. The solution set is then defined as a k-plex with larger positive factors with a limited number of exceptions determined by the value of k. We present an extension of standard k-plex enumerator enjoying branch-and-bound pruning techniques specialized to the evaluation of positive difference in addition to the standard pruning technique as in CLIQUES [1] or in parallel maximal k-Plex generator [8]. We show usefulness of the algorithm for two pairs of graphs representing actual reply-communications among Twitter users and word co-occurrence relations in Japanese news articles.

2 Modularity and Its Change Problem

Given a vertex set and a number of edges to be spanned, if the trial of drawing edges is statistically independent, and if the probability is uniform, then the resulting random graph will be too much far from a real graph we actually have. For this reason, in the notion of modularity [6], for each vertex, the sum of expectation of edges spanned around the vertex is required to be equal to the actual degree. As a result, the expectation of two vertices v_i and v_j to be linked is derived as $P_{ij} = \dfrac{k_i k_j}{2m}$, where k_i is the degree of v_i in the graph G and m is the number of edges in G. Then the modularity for two vertices is defined as the difference $B_{ij} = w_{ij} - P_{ij}$, where $w_{ij} = 1$ if v_i and v_j are adjacent, and $w_{ij} = 0$ otherwise. Although the notion of modularity is normally used to find a graph clustering/partitioning, we here use the measure to express the degree of change for vertex sets, as we have discussed in Section 1.

Using an index $\ell = 0, 1$, graphs at different time points are denoted as $G^{(0)}$ and $G^{(1)}$, respectively. Similarly, any expression $exp^{(\ell)}$ denotes the expression exp at $G^{(\ell)}$. Under the notation, the measure for *modularity change* for common vertices in both $G^{(0)}$ and $G^{(1)}$ is defined by subtracting modularity in $G^{(0)}$ from modularity in $G^{(1)}$ as follows:

$$D_{ij} = B_{ij}^{(1)} - B_{ij}^{(0)} = \left(w_{ij}^{(1)} - P_{ij}^{(1)}\right) - \left(w_{ij}^{(0)} - P_{ij}^{(0)}\right), \quad eval(X) = \sum_{v_i, v_j \in X} D_{ij}.$$

Then, a vertex set X with greater $eval(X)$ is preferred.

Applying the standard matrix decomposition, some approximation method to find near optimal X would be possible. For instance, given a graph with edges assigned positive and negative weights, as in the case of modularity difference problem, a linear programming approach is proposed in [2]. Although such an approach is adequate for finding a few near optimal solutions, we rather like to enumerate possible solutions under some constraints. This is simply because, by changing graphs, there would be various directions of changes some of which may attract one's attention while another one does not.

3   Observations on Modularity Difference

In order to reasonably formalize our problem of modularity change, we here discuss characteristics of modularity observed in real data on Twitter users.

As we will see in Section 6, we have constructed a pair of undirected graphs, $G^{(0)}$ and $G^{(1)}$, each of which represents actual communication links among Twitter users in some time period. The number of vertices (users) common in both graphs is 872. From characteristics of the communications, for each vertex $v_i^{(*)}$, the degree of $v_i^{(*)}$, $k_i^{(*)}$, is sufficiently less than the total number of edges in $G^{(*)}$, $m^{(*)}$, that is, $k_i^{(*)} \ll m^{(*)}$. Therefore, if a pair of vertices, $v_i^{(*)}$ and $v_j^{(*)}$, are connected, then the modularity for the vertices, $B_{ij}^{(*)}$, tends to be a positive value around 1.0. Conversely, if they are disconnected, then $B_{ij}^{(*)}$ tends to be negative values around zero.

Subtracting $B_{ij}^{(0)}$ from $B_{ij}^{(1)}$ for each pair of common vertices v_i and v_j, we can observe the distribution of modularity difference shown in Figure 2. We can see three peaks at around -1.0, 0.0 and 1.0. In the distribution, 01-pairs give D-values around 1.0, and those of 10 around -1.0. The other 0-0 or 1-1-pairs are almost distributed around 0.0. It should be emphasized that a huge number of vertex pairs, about $372K$ of $379K$ in this case, intensively have D-values around zero, implying that most of the ver-

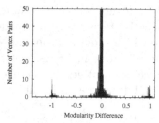

Fig. 2. Distributions of Modularity Difference

tex connections are unchanged. How to cope with this phenomenon, called CZN (Concentration to Zero Neighborhood), is the key for the combinatorial approach, as we will discuss later.

Since we are interested in detecting change patterns typically shown in Figure 1, our attention must be paid mainly to 01-pairs of vertices. That is, we basically try to extract a set of vertices X in which most of the vertex pairs are those of 01. From the distribution of modularity differences, since those pairs have large D-values around 1.0, $eval(X)$ is expected to be a large positive. In order to have such an X with larger $eval(X)$, it would be reasonable to include some vertex pairs with D-value near to zero as *exceptions*.

4   Problem of Relational Change Pattern Mining

As has been mentioned, a method for finding some class of change patterns basically based on the idea just above has been proposed in [10]. Although it has been verified that some interesting patterns can be extracted based on the method, examining combinations of vertices with D-values in the exceptional range makes its computational efficiency worse due to CZN. In order to cope with the issue, this paper presents a revised formulation of change pattern mining.

As is similar to the formulation in [10], for a threshold δ (> 0.0), we assume an *admissible range*, (δ, ∞), on D-values. Then, for a threshold ϵ ($0.0 < \epsilon \leq \delta$), an *exceptional range* is defined as $[-\epsilon, \epsilon]$. In addition, we regard the range $(-\infty, -\epsilon)$ as a *prohibitive* one. We call the remaining $(\epsilon, \delta]$ a *do-not-care range*, because it is not certain whether medium positive values are meaningful.

Here, let V be the set of common vertices in both $G^{(0)}$ and $G^{(1)}$. We consider a graph $G_\delta = (V, E)$, where $(x_i, x_j) \in E$ iff $D_{ij} > \delta$. In order to cope with a huge number of combinations of vertices with D-values in the exceptional range, we require our target pattern X to be a k-plex in G_δ.

The notion of k-plex has been first introduced in [7] as a relaxation model of clique. A set of vertices X is called a k-plex if for each vertex $v \in X$, v is not adjacent to *at most* k vertices in X. Thus, 1-plex is equivalent to a clique.

By requiring our target to be a k-plex in G_δ, we can obtain a vertex set X in which most of the vertex pairs in X are expected to be those of 01. In such an X, some limited number of pairs can have D-values in the exceptional, prohibitive, or do-not-care range. Since vertex pairs in the prohibitive range are probably those of 01 undesirable for us, we allow X to include vertex pairs with D-values only in the exceptional or do-not-care range, where it is required the sum of their D-values in the exceptional range must be less than ϵ. Note that any vertex pair with D-value in the do-not-care range is not taken into account for our evaluation of X. In addition to the constraints, we actually try to extract a pattern X in which the sum of D-values in the admissible range is greater than a threshold ζ. Our mining problem is now summarized as follows:

Definition 1. Let δ and ϵ be range parameters for modularity difference values and ζ a threshold such that $0.0 < \epsilon \leq \delta < \zeta$. For an integer k and a pair of graphs $G^{(0)}$ and $G^{(1)}$, the problem of mining relational change patterns is to enumerate every maximal set of vertices shared in both graphs, X, satisfying the following constraints:

(1) X is a k-plex in G_δ,

(2) $-\epsilon \leq D_{ij}$ for any pair of vertices v_i and v_j in X such that $i \neq j$,

(3) $f(X) = \left(\sum_{v_i, v_j \in X, i \neq j, D_{ij} > \delta} D_{ij} \right) > \zeta$,   and,

(4) $g(X) = \left(\sum_{v_i, v_j \in X, i \neq j, 0 \leq D_{ij} \leq \epsilon} D_{ij} \right) - \left(\sum_{v_i, v_j \in X, i \neq j, -\epsilon \leq D_{ij} < 0} D_{ij} \right) < \epsilon$.   ∎

5   Extracting Relational Change Patterns

In this section, we present an algorithm for extracting our change patterns which is an extension of a k-plex enumerator [8,1]. It is basically a depth-first tree expansion procedure with some pruning mechanisms. Each search node of the tree is a set of vertices X common in both $G^{(0)}$ and $G^{(1)}$ and forms a k-plex in G_δ satisfying the constraints (2) and (4) in the problem definition. With the root node of $\emptyset$, a path from the root to a leaf is a sequence $\emptyset = X_0, \ldots, X_n$ such that $X_{i+1} = X_i \cup \{x\}$, where x is called a *candidate* for X_i and required to satisfy the constraints (2) and (4) for $X \cup \{x\}$. That is, the set of candidates for X is defined as

$$Cand(X) = \{\text{ a vertex } x \notin X \text{ common in } G^{(0)} \text{ and } G^{(1)} |$$
$$Xx \text{ is a } k\text{-plex in } G_\delta \text{ satisfying the constraints (2) and (4) } \},$$

where Xx is an abbreviation of $X \cup \{x\}$. Similarly, for sets X and Y, XY denotes $X \cup Y$. Each tentative X on a path and each $x \in Cand(X)$ keep the values $f(X)$, $g(X)$, $f(Xx)$ and $g(Xx)$, respectively, as their internal information.

5.1   Branch-and-Bound Control

At each search node X, $g(X)$ and $f(X)$ are computed. If $Cand(X) = \emptyset$, there exists no vertex to be added. This means that X is a maximal vertex set (k-plex) satisfying the constraints (2) and (4). When $f(X)$ for such a maximal X is less than or equal to the bound ζ, X is never our solution and the path from the root to X is meaningless. A branch-and-bound control tries to detect such useless paths as earlier as possible.

For this aim, we compute $\tilde{f}(X) = f(X \cup Cand(X))$. It is noted that $X \subseteq Y \subseteq (X \cup Cand(X))$ holds for any maximal k-plex Y satisfying the two constraints such that $X \subseteq Y$. Moreover, we always have $f(X) \le f(X')$ for any X and X' such that $X \subseteq X'$. Therefore, we can safely cut off X and the subtree rooted with X whenever $\tilde{f}(X) \le \zeta$, because we can never obtain our target satisfying the constraint (1) on f-value by expanding X.

For the computation of $\tilde{f}(X)$, we use the following equation:

$$\tilde{f}(X) = f(X) + \sum_{x \in Cand(X)} (f(Xx) - f(X)) + f(Cand(X)).$$

Except $f(Cand(X))$, all the necessary values, $f(X)$ and $f(Xx)$, are stored at X and x, respectively. On the other hand, $f(Cand(X))$ can be computed just as the sum of D_{ij} for any vertex pair, v_i and v_j ($i \neq j$), in the subgraph of G_δ induced by $Cand(X)$.

5.2   Right Candidate Control

For a search node X, each candidate $x \in Cand(X)$ gives a potential branch for generating the next node of search tree. However, when our task is to generate every maximal solution, some of the branches guiding to some maximal ones can be replaced with another branch in the sense that those maximal solutions can be also obtained via another branch. In the case of maximal clique enumeration [1], such an observation can be stated as follows.

For a clique X and its candidate u meaning that Xu is also a clique, consider the set R of X's candidates adjacent to u. For any subset $Q \subseteq R$, if XQ is a clique, then XQ is never maximal as XuQ is a clique. In other words, any maximal clique including Xx for $x \in R$ can be obtained as an expansion of Xu or Xw such that w is another candidate of X not adjacent to u. As a result, we do not need to examine any branch with $x \in R$ and can safely prune all of those branches without loosing our solutions. Although our current targets are not defined as cliques, we can introduce a similar pruning mechanism as follows.

Assume we are at a search node X and have a candidate $u \in Cand(X)$. We consider the set of vertices R defined as $R = N_{G_\delta}(u) \cap Cand(X)$, where $N_{G_\delta}(u)$ is the set of vertices adjacent to u in G_δ. For any subset $Q \subseteq R$, if XQ is a k-plex, then XuQ is also a k-plex in G_δ, that is, XQ can never be maximal.

Furthermore, we can identify another set of vertices, R', we do not need to examine. A vertex $x \in Cand(X)$ not adjacent to u is included in R' iff x is not adjacent to any vertex in X. For any *connected* maximal k-plex Q such that $Xx \subset Q$, there is a vertex $v \in Cand(X)$ which is adjacent to some vertex in X and there exists a path from x to v. Therefore, Q can be obtained through the search branch with v and we do not need to examine the search branch with x.

We call the union $R \cup R'$ a right candidate set (RCS). Without violating the completeness of generating every connected maximal k-plex, any candidate in RCS has no need to be tried.

6   Experimental Results

This section presents our experimental results to verify usefulness of our method[1]. We have tried to extract relational change patterns observed in Twitter user relations and word co-occurrence relations in news articles.

Twitter User Relations: We have collected tweets including names of political parties in Japan with Twitter search API available at https://dev.twitter.com. From the collected tweets, we can construct an undirected graph representing reply-communications among users, where each vertex corresponds to a user and each edge between x and y means that x sent a reply tweet to y or vice versa.

More concretely speaking, we have prepared a pair of Tweet collections, $\mathcal{T}^{(0)}$ and $\mathcal{T}^{(1)}$, where $\mathcal{T}^{(0)}$ consists of tweets in the period of 4th to 6th April in 2012 and $\mathcal{T}^{(1)}$ those in the period of 13th to 15th in the same month. For each collection $\mathcal{T}^{(i)}$, we have created an undirected graph $G_{\mathcal{T}}^{(i)}$ representing reply-communications among users in the corresponding period, where the numbers of vertices in $G_{\mathcal{T}}^{(0)}$ and $G_{\mathcal{T}}^{(1)}$ are 4150 and 3755, and those of edges 3380 (density:0.0004) and 3755 (density:0.0005), respectively.

---

[1] Our system has been implemented in JAVA and executed on a PC with Intel® Core^TM-i3 M380 (2.53GHz) CPU and 8GB main memory.

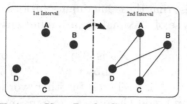
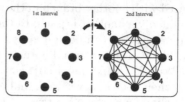

| (a) Twitter User Reply-Communications | (b) Word Co-Occurences |

Fig. 3. Relational Change Patterns

Word Co-occurrence Relations: We have also prepared a pair of collections of news articles, $\mathcal{M}^{(0)}$ and $\mathcal{M}^{(1)}$, appeared in a Japanese newspaper "*Mainichi*". $\mathcal{M}^{(0)}$ is a collection of articles in the category "*Politics*" in 1997 and $\mathcal{M}^{(1)}$ consists of those in 1998.

After applying a morphological analysis and removing too rare and too frequent words, we have selected remaining nouns as feature terms. Then, for each $\mathcal{M}^{(i)}$, we have created an undirected graph $G_{\mathcal{M}}^{(i)}$ with feature terms as vertices, where for a threshold α, a pair of terms x and y are connected by an edge if they co-occur in

Table 1. Characteristics of $G_{\mathcal{M}}^{(i)}$

		$G_{\mathcal{M}}^{(0)}$	$G_{\mathcal{M}}^{(1)}$
Number of vertices		3149	2648
Number of edges	$\alpha = 0.5\%$	38144	45314
	$\alpha = 0.8\%$	15527	19949
	$\alpha = 1.0\%$	9631	12596
Density	$\alpha = 0.5\%$	0.0077	0.0129
	$\alpha = 0.8\%$	0.0031	0.0057
	$\alpha = 1.0\%$	0.0019	0.0036

at least α % articles in $\mathcal{M}^{(i)}$. Since density of graphs directly depends on values of α, we have created our graphs under several α values to observe computational performance of our system affected by graph density. Table 1 summarizes characteristics of the graphs.

6.1   Examples of Extracted Relational Change Patterns

Relational Change Patterns for Twitter User Reply-Communications
Under the parameter setting [2], $\delta = 0.7$, $\epsilon = 0.1$, and $\zeta = 2.1$, we have tried to extract solution sets showing relational changes of twitter user-communications observed in the pair of graphs, $G_{\mathcal{T}}^{(0)}$ and $G_{\mathcal{T}}^{(1)}$. In the graphs, 872 users are commonly appeared and actually used for finding solution patterns of users.

An example of extracted pattern is shown in Figure 3 (a). The pattern consists of four twitter users, A, B, C and D. Their profiles tell us that A and B are members of a Japanese major party, C and D are (ordinary) active users who are interested in politics. In the former period, these four users are not communicated each other. However, in the latter period, each ordinary user sends reply tweets to the both members. This change can be explained as follows.

As a fact in Japan, the party X has supported Postal Privatization before. However, on 12th April, a legislation for reversing steps towards Postal Privatization has been approved in the House of Representatives. Particularly, most of

---
[2] The setting has been based on our preliminary experimentation.

the X's members of the House have actually agreed to reverse the progressing steps. Therefore, many people had distrust of the party X. Thus, the ordinary users sent some messages directly to the party members, e.g., *"I expect you to oppose the legislation in the House of Councilors."*

Relational Change Patterns for Word Co-occurrences
Under the parameter setting, $\delta = 0.7$, $\epsilon = 0.1$, $\zeta = 20.0$ and $k = 3$, we have tried to extract solution sets showing relational changes of word co-occurrences observed in the pair of graphs, $G_{\mathcal{M}}^{(0)}$ and $G_{\mathcal{M}}^{(1)}$. The number of common words (vertices) appeared in both graphs is 2343. Some of them forms a solution pattern representing a relational change in the graphs as is shown in Figure 3 (b).

An example of solution patterns actually extracted consists of eight words, 1:*Public*, 2:*Management*, 3:*Ministry of Finance*, 4:*Disposal*, 5:*Bank*, 6:*Capital*, 7:*Failure* and 8:*Revitalization*

For articles in 1997, we can observe there exists no pair of them co-occurred in the same articles (with the minimum co-occurrence threshold $\alpha = 1.0\%$). That is, the pattern is appeared as an independent set in $G_{\mathcal{M}}^{(0)}$. On the contrary, for articles in 1998, most of the word pairs co-occur in the same articles under the threshold, that is, the pattern forms a nearly (pseudo) clique in $G_{\mathcal{M}}^{(1)}$. Based on the relational change pattern, we can guess something noticeable has happened to the financial world in 1998.

After the collapse of the bubble economy in Japan, people in Japan were hit by a severe long-term economic recession. Every kind of business was greatly depressed and particularly in 1997 to 1998, several big financial institutions went bankrupt. In March 1998, especially, Long-Term Credit Bank of Japan (LTCB for short) was reported prominently because they got public funds of 176 billion JPY by hiding many of their bad debts. The bank, however, ultimately had a failure in October 1998 and then was temporarily nationalized. We can actually observe the words of the relational change pattern in original news articles concerned with this matter.

6.2   Computational Performance

Comparison with Structural Change Pattern Miner
Our system is compared with the k-plex-based structural change pattern (SCP) miner proposed in [11]. Particularly, under various k-value settings, we observe their performance for two pairs of graphs. One is the pair of the graphs, $G_{\mathcal{T}}^{(0)}$ and $G_{\mathcal{T}}^{(1)}$, with 872-common vertices already used to detect relational change patterns for Twitter user reply-communications. The other is a pair of larger graphs similarly constructed from a more number of Twitter users, where they have 2109-common vertices.

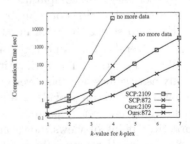

Fig. 4. Computation Time

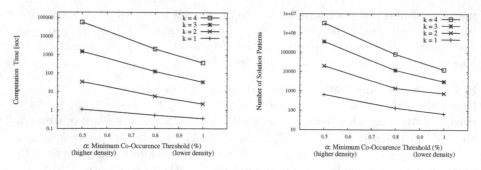

Fig. 5. Indirect Influence of Graph Density in *"Politics"* under α-Values

By adjusting parameter settings adequately, for each pair of graphs, both systems have tried to examine k-plexes in the same G_δ. The main difference is that k-plexes by SCP can include vertex-pairs with any type, whereas those by ours can never have vertex-pairs of 10-type as prohibitive ones.

Figure 4 shows their computational behavior. "SCP:872" and "SCP:2109" are performance curves of SCP for the graph pairs with 872 and 2109-common vertices, respectively. "Ours:872" and "Ours:2109" are those of our system. As we can see, as k-values become larger, computation time grows exponentially. Particularly, the growth for SCP is quite fast and SCP fails to detect their targets even for $k = 5$. On the other hand, our system can succeed to extract our targets at least for $k = 7$. This computational advantage comes from the restriction that our targets can never include any 10-pairs. Moreover, the results also show that our branch-and-bound control can work effectively in our search.

Influence of Graph Density

We then try to observe influence of graph density on our computational performance. The performance is affected by density of graphs in two senses. Since the primary computation in our framework is to extract k-plexes in δ-graph constructed from a given pair of graphs, the performance is *directly* affected by the parameter δ. Moreover, it might also be *indirectly* affected by density of each given graph itself. In case of word co-occurrence relations, density of given graphs can be easily controlled by α, a minimum co-occurrence threshold. Therefore, we present here our computational performance for word co-occurrence relations under several settings of α and δ.

Figure 5 shows the total numbers of solutions as well as computation times under several α-values for the graph pair, "$G_\mathcal{M}^{(0)}$ and $G_\mathcal{M}^{(1)}$".

We can see the computation time and the number of solutions grow exponentially as α-values become lower (that is, the given graphs become denser). Since a lower α-value can provide us a more number of solutions, our chance of finding potentially interesting patterns would be enhanced under lower α-values. From practical point of views, however, it seems adequate to take α-values around 0.8 or more. If we strongly require to have lower α-values, we may reasonably assume k-values of *three* or less for our practical computation.

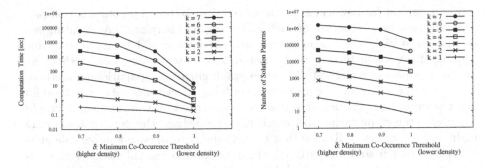

Fig. 6. Direct Influence of Graph Density in *"Politics"* under δ-Values

We now observe the total numbers of solutions as well as computation times under several δ-values in order to observe "direct" influence of graph density. Figure 6 shows the result for the graph pair, "$G_{\mathcal{M}}^{(0)}$ and $G_{\mathcal{M}}^{(1)}$".

As is similar to the previous figure, the computation time and the number of solutions grow rapidly as δ-values become lower (that is, the δ-graphs become denser). Although a lower δ-value can provide us a more number of solutions, they often involve undesirable patterns because many edges with such lower D-values might not always be 01-type we prefer. In this sense, it would be better to take higher δ-values around 0.9 or more. Such δ settings with higher values are also desirable from the viewpoint of practical computation. If we have a higher δ, we can extract our target patterns with practical time even for $k = 7$. Otherwise, k-values would be bounded by 5 or less. Thus, higher δ-values are adequate for both quality of target patterns and practical computation.

As has been mentioned above, the parameter δ controls density of graphs by which our computation is affected directly. In this sense, adjusting δ-values would be easier than adjusting α-values to control graph density.

7   Concluding Remarks

Based on the characteristics about the distribution of modularity difference, we have presented a method for finding relational change patterns observed in given two graphs. Our target pattern is defined as a maximal set of vertices which is a k-plex in G_δ satisfying some constraints on modularity difference. A branch-and-bound algorithm with right candidate control has been designed as an extension of a k-plex enumerator. In our experimentations with Twitter data and news article data, we have verified that our method can efficiently detect interesting relational change patterns including those never obtained by the framework of structural change patterns [11].

Our method is not restricted to graphs at different time points. Since it is a general framework, we can assume graphs constructed for different domains, areas and so forth, as long as we have a set of vertices common in both graphs.

In our current framework, all of the constraints imposed on our solution patterns are based on structure of each given graph. Therefore, investigating additional constraints from semantical point of views would be interesting and valuable. In order to further verify usefulness of our k-plex-based method, we need to compare our method with some graph clustering/partitioning methods for the graph G_δ. It would be also worth verifying effectiveness of our method for larger scale graphs. Particularly, if we are concerned with much denser graphs and prefer target patterns with much higher k-values, the current framework might not work very well. In such a case, we need to impose another adequate constraints on our solution patterns. Moreover, another social media, e.g., Facebook, would be interesting candidates to which our method can be applied.

Acknowledgments. The authors would like to thank the anonymous reviewers for their very valuable suggestions and helpful comments.

References

1. Tomita, E., Tanaka, A., Takahashi, H.: The Worst-Case Time Complexity for Generating All Maximal Cliques and Computational Experiments. Theoretical Computer Science 363(1), 28–42 (2006)
2. Dilkina, B., Gomes, C.P.: Solving Connected Subgraph Problems in Wildlife Conservation. In: Lodi, A., Milano, M., Toth, P. (eds.) CPAIOR 2010. LNCS, vol. 6140, pp. 102–116. Springer, Heidelberg (2010)
3. Li, A., Haraguchi, M., Okubo, Y.: Contrasting Correlations by an Efficient Double-Clique Condition. In: Perner, P. (ed.) MLDM 2011. LNCS (LNAI), vol. 6871, pp. 469–483. Springer, Heidelberg (2011)
4. Terlecki, P., Walczak, K.: Efficient Discovery of Top-K Minimal Jumping Emerging Patterns. In: Chan, C.-C., Grzymala-Busse, J.W., Ziarko, W.P. (eds.) RSCTC 2008. LNCS (LNAI), vol. 5306, pp. 438–447. Springer, Heidelberg (2008)
5. Bay, S.D., Pazzani, M.J.: Detecting Group Differences: Mining Contrast Sets. Data Mining and Knowledge Discovery 5(3), 213–246 (2001)
6. Newman, M.E.J.: Finding Community Structure in Networks Using the Eigenvectors of Matrices. Physical Review E 74(3), 36104 (2006)
7. Seidman, S.B., Foster, B.L.: A Graph Theoretic Generalization of the Clique Concept. Journal of Mathematical Sociology 6(1), 139–154 (1978)
8. Wu, B., Pei, X.: A Parallel Algorithm for Enumerating All the Maximal k-Plexes. In: Washio, T., Zhou, Z.-H., Huang, J.Z., Hu, X., Li, J., Xie, C., He, J., Zou, D., Li, K.-C., Freire, M.M. (eds.) PAKDD 2007. LNCS (LNAI), vol. 4819, pp. 476–483. Springer, Heidelberg (2007)
9. Kaufmann, M., Wagner, D. (eds.): Drawing Graphs. LNCS, vol. 2025. Springer, Heidelberg (2001)
10. Li, A., Haraguchi, M., Okubo, Y., Tomita, E.: Finding What Changes for Two Graphs Constructed from Different Time Intervals. In: The 2012 IIAI International Conference on Advanced Applied Informatics, pp. 48–53. CPS, California (2012)
11. Okubo, Y., Haraguchi, M., Tomita, E.: Structural Change Pattern Mining Based on Constrained Maximal k-Plex Search. In: Ganascia, J.-G., Lenca, P., Petit, J.-M. (eds.) DS 2012. LNCS, vol. 7569, pp. 284–298. Springer, Heidelberg (2012)

Reasoning with Near Set-Based Digital Image Flow Graphs*

James F. Peters[1], Doungrat Chitcharoen[2], and Sheela Ramanna[3]

[1] Computational Intelligence Laboratory, ECE Department,
University of Manitoba Winnipeg, MB R3T 5V6
james.peters3@ad.umanitoba.ca
[2] Department of Mathematics, King Mongkut's Institute of Technology,
Ladkrabang, Bangkok, Thailand
doungratmath@gmail.com
[3] Department of Applied Computer Science, University of Winnipeg,
Winnipeg, Manitoba R3B 2E9 Canada
s.ramanna@uwinnipeg.ca

Abstract. This paper introduces sufficiently near points in pairs of digital image flow graphs (DIFGs). This work is an extension of earlier work on a framework for layered perceptual flow graphs, where analysis of such graphs was performed in terms of flow graph nodes, branches, and paths using near set theory. A description-based method for determining nearness between flow graphs is given in terms of a practical application to digital image analysis.

Keywords: digital image flow graphs, near sets, perceptual system, probe functions, sufficiently near.

Introduction

This paper continues the work which was first reported in [1] where perceptual flowgraphs were introduced to represent and reason about sufficiently near visual points in images. In [2], a framework for extended layered perceptual flow graphs was established, where analysis of such graphs was performed using near set theory [3, 4]. In this paper, the framework has been extended to include set of points between pairs of digital image flow graphs (DIGF). In a perceptual flow graph induced by a perceptual system, a node in the graph is an object in a perceptual system with normalized flows σ derived from probe functions. A perceptual system is a specialised form of information system consisting of a set of objects equipped with a family of probe functions. The probe functions give rise to a number of perceptual relations between objects of a perceptual system [5]. This approach is useful when decisions on nearness are made in the context of a perceptual systems. This is especially important in image retrieval

---

* This research has been supported by the Natural Sciences and Engineering Research Council of Canada (NSERC) Discovery grants 185986 and 194376.

S. Ramanna et al. (Eds.): MIWAI 2013, LNCS 8271, pp. 199–210, 2013.

and applications such as robotics. Decision Tables have been represented as flow graphs using Fuzzy Sets [6] as well as Rough Sets [7]. Rough set-based flow graphs (also referred to as Pawlak flow graphs in this paper) represent a model of information flow from a given data set. The branches of a Pawlak flow graph can be represented as decision rules, while the entire flow graph can be viewed as a learning structure. Recent work includes theoretical and algorithmic aspects of Pawlak flow graphs [8–10]. A more thorough literature survey and an in-depth discussion of mathematical aspects of flow graph approaches to data analysis and connections to association rules can be also be found in [11]. This paper is organized as follows: Basic defintions concerning near sets and ordered perceptual system are given in Sec. 1. Details concerning Pawlak flow graphs are given in Sec. 2 followed by concluding remarks.

1   Preliminaries

Underlying perceptual flow graphs, is the notion of an ordered perceptual system. In this section, we give the basic definitions of near set theory, Pawlak flow graphs and extended layered flow graphs.

1.1   Perceptual System

The basic structure which underlies near set theory is a perceptual system [12].

Definition 1. *A perceptual system is a pair $\langle O, F \rangle$, where O is a nonempty set of perceptual objects and F is a countable set of probe functions $\phi_i : O \to \mathbb{R}$.*

An object description is defined by means of a tuple of probe function values $\Phi(x)$ associated with an object $x \in X$, where $X \subseteq O$ and $\Phi(x) = (\phi_1(x), \phi_2(x) \ldots, \phi_n(x))$ [3] and where $\phi_i : O \to \mathbb{R}$ is a probe function of a single feature. In this paper, probe functions are defined in terms of features of pixels in digital images (e.g., colour, texture, edges, ridges, junctions, and corners). The standard distance $\Delta\Phi_{x,y}$ is the taxicab(cityblock) distance between feature vectors $\Phi(x), \Phi(y), x, y \in X$, *i.e.*,

$$\Delta\Phi_{x,y} = \sum_{i=1}^{n} |\phi_i(x) - \phi_i(y)|.$$

Definition 2. *Let $P \subseteq F$ be a given set of probe functions representing features of sample objects O where $|P| = |\Phi| = n$. Objects $x, x' \in O$ are minimally near each other if, and only if there exists $\phi_i \in P$, such that $\Delta\phi_i = 0$.*

Definition 3. *Let $X, X' \subseteq O$, $P \subseteq F$. Set X is near X' if, and only if, there exists $x \in X, x' \in X'$, such that $\Delta\Phi_{x,x'} = 0$.*

1.2 Pawlak Flow Graphs

In this section, recent work on flow graphs is considered (see, e.g., [13, 9]).

Definition 4 ([7]). *Let $G = (N, B, \varphi)$ be a directed, acyclic, finite graph, where N is a set of nodes, $B \subseteq N \times N$ a set of directed branches, $\varphi : B \to R^+$ a flow function and R^+ denotes a set of non-negative real numbers.*

Let G be a flow graph (introduced in Def.4). If $(x, y) \in B$ then x is an *input* of node y denoted by $I(y)$ and y is an *output* of node x denoted by $O(x)$. Next, *input* and *output* of a flow graph G are defined respectively by $I(G) = \{x \in N : I(x) = \emptyset\}$ and $O(G) = \{x \in N : O(x) = \emptyset\}$. These inputs and outputs of G are called *external nodes* of G whereas other nodes are called *internal nodes* of G. If $(x, y) \in B$ then we call (x, y) a *throughflow* from x to y. We will assume in what follows that $\varphi(x, y) \neq 0$ for every $(x, y) \in B$.

Definition 5. *Let $\mathcal{G} = (N, B, \varphi, \sigma)$ be a normalized flow graph, where N is a set of nodes, $B \subseteq N \times N$ a set of directed branches, $\varphi : B \to R^+$ and $\sigma : B \to [0, 1]$ a normalized flow between nodes.*

We now elaborate on normalized flow graphs (introduced in Def. 5) where the normalized flow or *strength* of (x, y) is defined as $\sigma(x, y) = \frac{\varphi(x,y)}{\varphi(G)}$ where $0 \leq \sigma(x, y) \leq 1$. With every node x of a normalized flow graph $\mathcal{G}$, the associated *normalized inflows* and *outflows* are defined as: $\sigma_+(x) = \frac{\varphi_+(x)}{\varphi(G)} = \sum_{y \in I(x)} \sigma(y, x)$ and $\sigma_-(x) = \frac{\varphi_-(x)}{\varphi(G)} = \sum_{y \in O(x)} \sigma(x, y)$. For any internal node x, it holds that $\sigma_+(x) = \sigma_-(x) = \sigma(x)$, where $\sigma(x)$ is a *normalized throughflow* of x. Similarly, *normalized inflows* and *outflows* for the flow graph $\mathcal{G}$ are defined as: $\sigma_+(G) = \frac{\varphi_+(G)}{\varphi(G)} = \sum_{x \in I(\mathcal{G})} \sigma_-(x)$ and $\sigma_-(\mathcal{G}) = \frac{\varphi_-(G)}{\varphi(G)} = \sum_{x \in O(G)} \sigma_+(x)$. It also holds that $\sigma_+(\mathcal{G}) = \sigma_-(\mathcal{G}) = \sigma(\mathcal{G}) = 1$. With every branch (x, y) of a normalized flow graph $\mathcal{G}$, the *certainty* and the *coverage* of (x, y) are defined respectively as: $cer(x, y) = \frac{\sigma(x,y)}{\sigma(x)}$, and $cov(x, y) = \frac{\sigma(x,y)}{\sigma(y)}$, where $\sigma(x), \sigma(y) \neq 0$.

In a normalized flow graph $\mathcal{G}$, if we concentrate on relationships of a sequence of nodes in a flow graph, we can find these relationships using the concept of paths. A (directed) *path* from x to y ($x \neq y$), denoted by $[x \ldots y]$, is a sequence of nodes $x_1, \ldots, x_n$ such that $x_1 = x$ and $x_n = y$ and $(x_i, x_{i+1}) \in B$ for every i, $1 \leq i \leq n-1$. The *certainty*, *coverage* and *strength* of the path $[x_1 \ldots x_n]$ are defined as: $cer[x_1 \ldots x_n] = \prod_{i=1}^{n-1} cer(x_i, x_{i+1})$, $cov[x_1 \ldots x_n] = \prod_{i=1}^{n-1} cov(x_i, x_{i+1})$, and $\sigma[x_1 \ldots x_n] = \sigma(x_1) cer[x_1 \ldots x_n] = \sigma(x_n) cov[x_1 \ldots x_n]$. If $[x \ldots y]$ is a path such that x and y are input and output of the graph $\mathcal{G}$, respectively, then $[x \ldots y]$ will be referred to as a *complete path*.

2 Digital Image Flow Graphs

The basics of digital image flow graphs are briefly presented in this section. A *perceptual object*, in this case, is a digital image viewed a set of picture elements (pixels) that carry information for flowing between and into flow graph nodes.

Definition 6 ([2]). *A digital image flow graph (denoted DIFG) that represents a perceptual information system is a directed, acyclic, finite graph* $G_\mathcal{N} = (O, \mathcal{N}, \mathcal{B}, \varphi, \psi)$, *where*

$O = \{x_1, x_2, x_3, \ldots, x_n\}$ *is a non-empty set of objects in the graph,*

$\mathcal{N} = \{n \mid n \text{ is a node in } G_\mathcal{N}\}$ *is a non-empty set of nodes,*

$\mathcal{B} = \{b_{n_i,n_j} \mid b_{n_i,n_j} = (n_i, n_j) \text{ and } (n_i, n_j) \in \mathcal{N} \times \mathcal{N}\}$ *is a non-empty set of directed branches,*

$\varphi : \mathcal{B} \to \mathcal{P}(O)$ *is a flow function of branch to a collection of objects,*

$\psi : \mathcal{N} \to \mathcal{P}(O)$ *is a throughflow function of node to a collection of objects and* $\mathcal{P}(O)$ *is a collection of all subsets of objects.*

In combination, the functions representing object features provide a basis for an object description $\Phi(x) = (\Phi_1(x), \Phi_2(x) \ldots, \Phi_n(x))$, and equivalence classes of feature $\Phi_i(x)$ are represented by nodes in i^{th} layer. In other words, nodes with the same features (e.g., pixel, colour and gradient) form a layer. In this paper, nodes of the DIFG are depicted by circles labeled by $n_i^{j_i}$ such that node $n_i^{j_i}$ is a j_i^{th} node in the ith layer, where $1 \leq i \leq n$, $1 \leq j_i \leq k_i$. $\psi(n_i^{j_i})$ is an k_i equivalence classes relative to one feature (feature $\Phi_i(x)$). We say that the set of equivalence classes of ψ form a partition of O such that every pair of distinct equivalence classes has empty intersection (i.e., $\bigcup_{l_i=1}^{k_i} \psi(n_i^{l_i}) = O$ and $\psi(n_i^{l_i}) \cap \psi(n_j^{l_j}) = \emptyset$, $i \neq j$). Since a DIFG is arranged in several layers and in each layer, there is no branch between nodes in that layer.

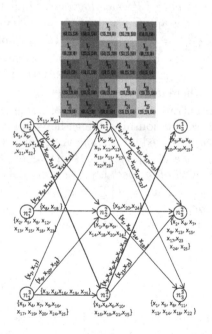

Fig. 1. A DIFG $G_\mathcal{N}$

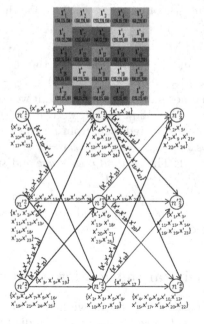

Fig. 2. A DIFG $G_{\mathcal{N}'}$

Example. We illustrate the above definitions by means of a simple example. Let $\langle O, F \rangle$ be a perceptual system and $X, X' \subseteq O$ where $X = \{x_1, x_2, x_3, \ldots, x_{25}\}$, $X' = \{x'_1, x'_2, x'_3, \ldots, x'_{25}\}$ and the values of probe functions $P = \{R(red), G(green), B(blue)\}$, where $P \subseteq F$.

A perceptual DIFG $G_\mathcal{N}$ ($X \subseteq O$) results in indiscernibility relations such as $\sim_R, \sim_G, \sim_B$ leading to the following partitions from the perceptual system shown in Figure 1.

$$X_{/\sim_R} = \{\{x_1, x_6, x_{10}, x_{11}, x_{14}, x_{21}, x_{22}\}, \{x_2, x_5, x_8, x_{12}, x_{13}, x_{15}, x_{18}, x_{23}\},$$
$$\{x_3, x_4, x_7, x_9, x_{16}, x_{17}, x_{19}, x_{20}, x_{24}, x_{25}\}\},$$
$$X_{/\sim_G} = \{\{x_2, x_5, x_7, x_{11}, x_{12}, x_{13}, x_{15}, x_{17}, x_{22}, x_{23}\}, \{x_1, x_8, x_9, x_{14}, x_{18}, x_{20},$$
$$x_{24}\}, \{x_3, x_4, x_{16}, x_{10}, x_{16}, x_{21}, x_{25}\}\},$$
$$X_{/\sim_B} = \{\{x_3, x_4, x_6, x_{10}, x_{16}, x_{19}\}, \{x_2, x_4, x_7, x_9, x_{13}, x_{15}, x_{17}, x_{23}\},$$
$$\{x_1, x_5, x_8, x_{11}, x_{12}, x_{14}, x_{18}, x_{22}\}\}.$$

There are three layers (one for each probe function) in a DIFG $G_\mathcal{N}$ as shown in Figure 1. The first layer consists of nodes: n_1^1, n_1^2 and n_1^3, where objects at node n_1^1, n_1^2 and n_1^3 form an equivalence classes relative to intensity value of red is 60, 150 and 235, respectively. The second layer consists of nodes n_2^1, n_2^2 and n_2^3, where objects at node n_2^1, n_2^2 and n_2^3 form an equivalence classes relative to intensity value of green is 35, 125 and 220, respectively. The third layer consists of nodes n_3^1, n_3^2 and n_3^3, where objects at node n_3^1, n_3^2 and n_3^3 form an equivalence classes relative to intensity value of blue is 80, 150 and 230, respectively.

Similarly, a perceptual DIFG $G_{\mathcal{N}'}$ ($X' \subseteq O$) results in indiscernibility relations such as $\sim_R, \sim_G, \sim_B$ leading to the following partitions from the perceptual system shown in Figure 2 where intensity values for red are: 60, 150 and 235, intensity values for green are: 35, 125 and 220 and intensity values for blue are: 80, 150 and 230.

$$X'_{/\sim_R} = \{\{x'_5, x'_8, x'_{10}, x'_{15}, x'_{17}, x'_{22}\}, \{x'_1, x'_2, x'_6, x'_{11}, x'_{12}, x'_{13}, x'_{14}, x'_{18}, x'_{20}, x'_{23}\},$$
$$\{x'_3, x'_4, x'_7, x'_9, x'_{16}, x'_{19}, x'_{21}, x'_{24}, x_{25}\}\},$$
$$X'_{/\sim_G} = \{\{x'_4, x'_7, x'_8, x'_{11}, x'_{12}, x'_{14}, x'_{16}, x'_{22}, x'_{24}\}, \{x'_1, x'_6, x'_{13}, x'_{18}, x'_{20}, x'_{21}, x'_{23}$$
$$x'_{25}\}, \{x'_2, x'_3, x'_5, x'_9, x'_{10}, x'_{17}, x'_{19}\}\},$$
$$X'_{/\sim_B} = \{\{x'_2, x'_5, x'_7, x'_9, x'_{21}, x'_{22}, x'_{24}\}, \{x'_1, x'_3, x'_{11}, x'_{13}, x'_{14}, x'_{16}, x'_{19}, x'_{23}\},$$
$$\{x'_4, x'_6, x'_8, x'_{10}, x'_{12}, x'_{15}, x'_{17}, x'_{18}, x'_{20}, x'_{22}\}\}.$$

Next, if we consider relationship between node in i^{th} and $(i+1)^{th}$ layers in a DIFG (relationship of equivalence class between i^{th} features and $(i+1)^{th}$ features e.g., between an equivalence classes relative to intensity value of red and an equivalence classes relative to intensity value of green), then we create a directed branch between that node. A non-empty set of objects which flow through directed branches $b_{n_i^{l_i}, n_{i+1}^{l_{i+1}}}$ that starts with node $n_i^{l_i}$ and ends with node $n_{i+1}^{l_{i+1}}$, is denoted by $\varphi(b_{n_i^{l_i}, n_{i+1}^{l_{i+1}}})$, and is equal to the intersection between set of objects that flow through nodes of that directed branch $b_{n_i^{l_i}, n_{i+1}^{l_{i+1}}}$. Then throughflow of directed branches $b_{n_i^{l_i}, n_{i+1}^{l_{i+1}}}$ is defined as: $\varphi(b_{n_i^{l_i}, n_{i+1}^{l_{i+1}}}) = \psi(n_i^{l_i}) \cap \psi(n_{i+1}^{l_{i+1}})$.

Example. To illustrate how to find the throughflow of directed branches of the DIFG $G_{\mathcal{N}}$ (in Fig. 1), we present some examples: the set of objects that flow through nodes n_1^1 and n_2^1 can be determined by $\psi(n_1^1) = \{x_1, x_6, x_{10}, x_{11}, x_{14}, x_{21}, x_{22}\}$ and $\psi(n_2^1) = \{x_2, x_5, x_7, x_{11}, x_{12}, x_{13}, x_{15}, x_{17}, x_{22}, x_{23}\}$, respectively, then the set of objects that flow through directed branches b_{r_1,g_1} is given by $\psi(n_1^1) \cap \psi(n_2^1) = \{x_1, x_6, x_{10}, x_{11}, x_{14}, x_{21}, x_{22}\} \cap \{x_2, x_5, x_7, x_{11}, x_{12}, x_{13}, x_{15}, x_{17}, x_{22}, x_{23}\} = \{x_{11}, x_{22}\}$. ■

In a DIFG, if we consider a sequence of nodes $n_1^{l_1}$ to $n_n^{l_n}$ in $G_{\mathcal{N}}$, then we can use the concept of a directed path. A directed path from $n_1^{l_1}$ to $n_n^{l_n}$ in $G_{\mathcal{N}}$, denoted by $[n_1^{l_1} \ldots n_n^{l_n}]$, is a sequence of nodes $n_1^{l_1}, \ldots, n_n^{l_n}$ where $b_{n_i^{l_i}, n_{i+1}^{l_{i+1}}} \in \mathcal{B}$ and $\bigcap_{i=1}^{n-1} \varphi(b_{n_i^{l_i}, n_{i+1}^{l_{i+1}}}) \neq \emptyset, 1 \leq i \leq n-1$.

Example. To illustrate how to find the throughflow of objects in a directed path of the DIFG $G_{\mathcal{N}}$ (in Fig. 1), we present some examples: the set of objects that flow through directed path $[n_1^1 n_2^1 n_3^3]$ can be determined by $\psi(n_1^1) = \{x_1, x_6, x_{10}, x_{11}, x_{14}, x_{21}, x_{22}\}$, $\psi(n_2^1) = \{x_2, x_5, x_7, x_{11}, x_{12}, x_{13}, x_{15}, x_{17}, x_{22}, x_{23}\}$ and $\psi(n_3^3) = \{x_1, x_5, x_8, x_{11}, x_{12}, x_{14}, x_{18}, x_{22}\}$, respectively, then the set of objects that flow through directed path $[n_1^1 n_2^1 n_3^3]$ is given by $\psi(n_1^1) \cap \psi(n_2^1) \cap \psi(n_3^3) = \{x_1, x_6, x_{10}, x_{11}, x_{14}, x_{21}, x_{22}\} \cap \{x_2, x_5, x_7, x_{11}, x_{12}, x_{13}, x_{15}, x_{17}, x_{22}, x_{23}\} \cap \{x_1, x_5, x_8, x_{11}, x_{12}, x_{14}, x_{18}, x_{22}\} = \{x_{11}, x_{22}\}$. ■

Lemma 1. *Let $G_{\mathcal{N}} = (\mathcal{N}, \mathcal{B}, \delta, \sigma)$ be a DIFG with n layers and the ith layer having k_i nodes where $1 \leq i \leq n$, $1 \leq j_i \leq k_i$: $n_i^1, n_i^2, \ldots, n_i^{j_i}, \ldots, n_i^{k_i}$. Let $G_{\mathcal{N}'} = (\mathcal{N}', \mathcal{B}', \delta', \sigma')$ be another DIFG with n' layers and the i'^{th} layer having k_i' nodes: $n_i'^1, n_i'^2, \ldots, n_i'^{j_i}, \ldots, n_i'^{k_i}$, the following properties hold:*

1. *The objects in node $n_i^{j_i} \in \mathcal{N}$ and the objects in node $n_i'^{j_i} \in \mathcal{N}'$ such that nodes $n_i^{j_i}$ and node $n_i'^{j_i}$ represent probe functions for the same feature are minimally near each other.*

2. *The objects flow through $b_{n_i^{j_i}, n_{i+1}^{j_{i+1}}} \in \mathcal{B}$ and the objects flow through $b_{n_i'^{j_i}, n_{i+1}'^{j_{i+1}}} \in \mathcal{B}'$ such that node $n_i^{j_i} \in \mathcal{N}$ and node $n_i'^{j_i} \in \mathcal{N}'$ and node $n_{i+1}^{j_{i+1}} \in \mathcal{N}$ and node $n_{i+1}'^{j_{i+1}} \in \mathcal{N}'$ represent probe functions for the same set of features are minimally near each other.*

3. *The objects flow through complete path $\left[n_1^{j_1} \ldots x_{j_1}^{j_n}\right]$ in $\mathcal{G}$ and the objects flow through complete path complete path $\left[n_1'^{j_1} \ldots n_{j_1}'^{j_n}\right]$ in $\mathcal{G}'$ such that node $n_i^{j_i} \in \mathcal{N}$ and node $n_i'^{j_i} \in \mathcal{N}', 1 \leq i \leq n$, represent probe functions for the same set of features are near objects.*

Proof. 1. Let x be a set of objects in node $n_i^{j_i}$ and x' be a set of objects in node $n_i'^{j_i}$ and since nodes $n_i^{j_i}$ and node $n_i'^{j_i}$ represent probe functions for the same feature, implies that there exists $\phi_i \in P$ such that $\phi_i(x) = \phi_i(x')$, then $\Delta\phi_i = 0$. By Definition 2, the objects stored in a node $n_i^{j_i}$ and node $n_i'^{j_i}$ are minimally near each other.

2. Let x be a set of objects flowing through $b_{n_i^{j_i}, n_{i+1}^{j_{i+1}}}$ and x' be a set of objects flowing through $b_{n_i'^{j_i}, n_{i+1}'^{j_{i+1}}}$ such that node $n_i^{j_i} \in \mathcal{N}$ and node $n_i'^{j_i} \in \mathcal{N}'$ and node $n_{i+1}^{j_{i+1}} \in \mathcal{N}$ and node $n_{i+1}'^{j_{i+1}} \in \mathcal{N}'$ represent probe functions for the same feature, implies that there exists at least for two features $\phi_i \in P$ such that $\phi_i(x) = \phi_i(x')$, then $\Delta \phi_i = 0$. By Definition 2, the objects flowing through $b_{n_i^{j_i}, n_{i+1}^{j_{i+1}}} \in \mathcal{B}$ and the objects flowing through $b_{n_i'^{j_i}, n_{i+1}'^{j_{i+1}}} \in \mathcal{B}'$ are minimally near each other.

3. Let x be set of objects flowing through complete path $\left[n_1^{j_1} \ldots x_{j_1}^{j_n} \right]$ in $\mathcal{G}$ and x' be a set of objects flowing through a complete path $\left[n_1'^{j_1} \ldots n_{j_1}'^{j_n} \right]$ in $\mathcal{G}'$ such that node $n_i^{j_i} \in \mathcal{N}$ and node $n_i'^{j_i} \in \mathcal{N}'$, $1 \le i \le n$, represent probe functions for the same feature, then $\Phi(x) = \Phi(y)$. By definition 2, the objects flowing through a complete path $\left[n_1^{j_1} \ldots x_{j_1}^{j_n} \right]$ in $\mathcal{G}$ and the objects flowing through a complete path $\left[n_1'^{j_1} \ldots n_{j_1}'^{j_n} \right]$ in $\mathcal{G}'$ are near objects.

Example. We illustrate the above lemma by means of a simple example.

- An object x_1 flowing through node n_1^1 in $G_{\mathcal{N}}$ and an object x_5' flowing through node $n_1'^1$ in $G_{\mathcal{N}'}$. Since node n_1^1 and node $n_1'^1$ represent same feature, then by Lemma 1, we have x_1 and x_5' minimally near each other.
- An object x_{11} flowing through $d_{n_1^1, n_2^1}$ in $G_{\mathcal{N}}$ and an object x_8' flowing through $d_{n_1'^1, n_2'^1}$ in $G_{\mathcal{N}'}$. Since node n_1^1 and node $n_1'^1$ and n_2^1 and node $n_2'^1$ represent same set of features, then by Lemma 1, we have x_{11} and x_8' minimally near each other.
- An object x_{22} flowing through $[n_1^1 n_1^1 n_3^1]$ in $G_{\mathcal{N}}$ and an object x_{22}' flowing through $[n_1'^1 n_1'^1 n_3'^1]$ in $G_{\mathcal{N}'}$. Since node n_1^1 and node $n_1'^1$, node n_2^1 and node $n_2'^1$ and node n_3^1 and node $n_3'^3$ represent same set of features, then by Lemma 1, x_{22} and x_{22}' are near objects. ∎

Definition 7 ([2]). *A normalized digital image of flow graph(denoted DINFG) is a directed, acyclic, finite graph* $\mathcal{G}_{\mathcal{N}} = (O, \mathcal{N}, \mathcal{B}, \delta, \sigma)$, *where*

$O = \{x_1, x_2, x_3, \ldots, x_n\}$ *is a non-empty set of objects flowing in the graph,*

$\mathcal{N} = \{n \mid n \text{ is a node in } \mathcal{G}_{\mathcal{N}}\}$ *is a non-empty set of nodes,*

$\mathcal{B} = \{ b_{n_i, n_j} \mid b_{n_i, n_j} = (n_i, n_j) \text{ and } (n_i, n_j) \in \mathcal{N} \times \mathcal{N} \}$ *is a non-empty set of directed branches,*

$\delta : \mathcal{N} \to [0, 1]$ *is normalized throughflow function of node* $n_i^{l_i}$, *where* $\delta(n_i^{l_i}) = \frac{\psi(n_i^{l_i})}{\psi(G_{\mathcal{N}})}$ *and*

$\sigma : \mathcal{B} \to [0, 1]$ *is normalized flow function of* $b_{n_i^{l_i}, n_{i+1}^{l_{i+1}}}$, *where* $\sigma(b_{n_i^{l_i}, n_{i+1}^{l_{i+1}}}) = \frac{\left| \varphi(b_{n_i^{l_i}, n_{i+1}^{l_{i+1}}}) \right|}{|\psi(G_{\mathcal{N}})|}$.

With every node $n_i^{l_i}$ of the DINFG, the associated *normalized inflow* and *outflow* are defined as: $\sigma_+(n_i^{l_i}) = \frac{\varphi_+(n_i^{l_i})}{\varphi(G)} = \sum_{n_{i-1}^{l_{i-1}} \in I(n_i^{l_i})} \sigma(d_{n_{i-1}^{l_{i-1}}, n_i^{l_i}})$ and $\sigma_-(n_i^{l_i}) =$

$\frac{\varphi_-(n_i^{l_i})}{\varphi(G)} = \sum_{n_{i+1}^{l_{i+1}} \in O(n_i^{l_i})} \sigma(d_{n_i^{l_i}, n_{i+1}^{l_{i+1}}})$. For any internal node $n_i^{l_i}$, it holds that $\sigma_+(n_i^{l_i}) = \sigma_-(n_i^{l_i}) = \sigma(n_i^{l_i})$, where $\sigma(n_i^{l_i})$ is a *normalized throughflow* of Φ. Similarly, *normalized inflow* and *outflow* for the flow graph G are defined as: $\sigma_+(G) = \frac{\varphi_+(G)}{\varphi(G)} = \sum_{n_i^{l_i} \in I(G)} \sigma_-(n_{i-1}^{l_{i-1}})$ and $\sigma_-(G) = \frac{\varphi_-(G)}{\varphi(G)} = \sum_{n_i^{l_i} \in O(\Phi)} \sigma_+(n_{i+1}^{l_{i+1}})$. It also holds that $\sigma_+(G) = \sigma_-(G) = \sigma(G) = 1$.

Definition 8. *Let G be a DINFG, certainty and coverage of $b_{n_i^{l_i}, n_{i+1}^{l_{i+1}}}$ define respectively as,* $cer(b_{n_i^{l_i}, n_{i+1}^{l_{i+1}}}) = \frac{|\varphi(b_{n_i^{l_i}, n_{i+1}^{l_{i+1}}})|}{|\psi(n_i^{l_i})|}$ *and* $cov(b_{n_i^{l_i}, n_{i+1}^{l_{i+1}}}) = \frac{|\varphi(b_{n_i^{l_i}, n_{i+1}^{l_{i+1}}})|}{|\varphi(n_{i+1}^{l_{i+1}})|}$.

The normalized flow graph of Figures. 1 and Figure. 2 is shown in Figure. 3 and Figure. 4 respectively. The strength coefficient is calculated using definition 7 and certainity and coveragecoefficient are calculated using definition 8.

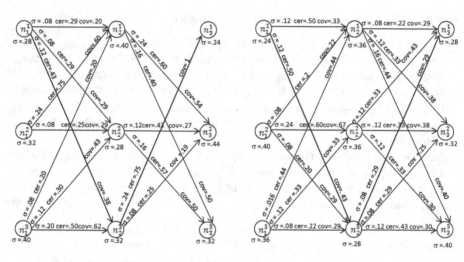

Fig. 3. A DINFG G_N **Fig. 4.** A DINFG $G_{N'}$

Definition 9. *Let $\mathcal{G}_N = (\mathcal{N}, \mathcal{B}, \delta, \sigma)$ be a DINFG with n layers and the ith layer having k_i nodes where $1 \leq i \leq n$, $1 \leq j_i \leq k_i$: $n_i^1, n_i^2, \ldots, n_i^{j_i}, \ldots, n_i^{k_i}$. Let $\mathcal{G'}_N = (N', B', \varphi', \sigma')$ be a DINFG with n' layers and the i'^{th} layer having k_i' nodes: $n_i'^1, n_i'^2, \ldots, n_i'^{j_i}, \ldots, n_i'^{k_i}$.*

The distance between node $n_i^{j_i}$ in N and corresponding node $x_i'^{j_i}$ in N' such that nodes $n_i^{j_i}$ and $n_i'^{j_i}$ represent probe functions for the same feature (maximum distance being 1), is defined as

$$\rho_n\left(n_j^{l_j}, n_j'^{l_j}\right) = \frac{\left|\sigma(n_i^{l_j}) - \sigma(n_j'^{l_j})\right|}{max\left\{\sigma(n_j^{l_j}), \sigma(n_j'^{l_j})\right\}},$$

The distance between branches $b_{n_i^{j_i},n_{i+1}^{j_{i+1}}}$ *in* B *and the corresponding branches* $b_{n_i^{\prime j_i},n_{i+1}^{\prime j_{i+1}}}$ *in* B' *such that node* $n_i^{j_i}$ *in* $\mathcal{G}$ *and node* $n_i^{\prime j_i}$ *in* $\mathcal{G}'$ *and* $n_{i+1}^{j_{i+1}}$ *in* $\mathcal{G}$ *and node* $n_{i+1}^{\prime j_{i+1}}$ *in* $\mathcal{G}'$ *represent probe functions for the same set of features (maximum distance being 1), is defined as*

$$\rho_b\left(b_{n_j^{l_j},n_{j+1}^{l_{j+1}}}, b_{n_j^{\prime l_j},n_{j+1}^{\prime l_{j+1}}}\right) = \frac{\left|\sigma(b_{n_j^{l_j},n_{j+1}^{l_{j+1}}}) - \sigma(b_{n_j^{\prime l_j},n_{j+1}^{\prime l_{j+1}}})\right|}{max\left\{\sigma(b^{n_j^{l_j},n_{j+1}^{l_{j+1}}}), \sigma(b_{n_j^{\prime l_j},n_{j+1}^{\prime l_{j+1}}})\right\}},$$

The distance between path $\left[n_1^{j_1}\ldots n_{j_1}^{j_n}\right]$ *in* $\mathcal{G}_N$ *and path* $\left[n_1^{\prime j_1}\ldots n_{j_1}^{\prime j_n}\right]$ *in* $\mathcal{G}'_N$ *such that node* $n_i^{j_i}$ *in* $\mathcal{G}_N$ *and node* $n_i^{\prime j_i}$ *in* $\mathcal{G}'_N$, $1 \le i \le n$, *represent probe functions for the same set of features ((maximum distance being 1), is defined as*

$$\rho_p\left(\left[n_1^{l_1}\ldots n_n^{l_n}\right], \left[n_1^{\prime l_1}\ldots n_n^{\prime l_n}\right]\right) = \frac{\left|\sigma\left[n_1^{l_1}\ldots n_n^{l_n}\right] - \sigma\left[n_1^{\prime l_1}\ldots n_n^{\prime l_n}\right]\right|}{max\left\{\sigma\left[n_1^{l_1}\ldots n_n^{l_n}\right], \sigma\left[n_1^{\prime l_1}\ldots n_n^{\prime l_n}\right]\right\}},$$

Lemma 2. *Let* $\mathcal{G}_N = (\mathcal{N}, \mathcal{B}, \delta, \sigma)$ *be a DIFG with* n *layers and the* i^{th} *layer having* k_i *nodes where* $1 \le i \le n$, $1 \le j_i \le k_i$: $n_i^1, n_i^2, \ldots, n_i^{j_i}, \ldots, n_i^{k_i}$. *Let* $\mathcal{G}'_{N'} = (\mathcal{N}', \mathcal{B}', \delta', \sigma')$ *be a DIFG with* n' *layers and the* i'^{th} *layer having* k_i' *nodes:* $n_i^{\prime 1}, n_i^{\prime 2}, \ldots, n_i^{\prime j_i}, \ldots, n_i^{\prime k_i}$, *we can calculate distance between two nodes, two branches and two paths of* G_N *and* G'_N, *respectively as:*

$$\rho_b\left(n_j^{l_j}, n_j^{\prime l_j}\right) = \frac{\left||\varphi(G')|\cdot|\varphi(n_j^{l_j})| - |\varphi(G)|\cdot|\varphi(n_j^{\prime l_j})|\right|}{max\left\{|\varphi(G')|\cdot|\varphi(n_j^{l_j})|, |\varphi(G)|\cdot|\varphi(n_j^{\prime l_j})|\right\}},$$

$$\rho_b\left(b_{n_j^{l_j},n_{j+1}^{l_{j+1}}}, b_{n_j^{\prime l_j},n_{j+1}^{\prime l_{j+1}}}\right) = \frac{\left||\varphi(G')|\cdot|\varphi(n_j^{l_j})\cap\varphi(n_{j+1}^{l_{j+1}})| - |\varphi(G)|\cdot|\varphi(n_j^{\prime l_j})\cap\varphi(n_{j+1}^{\prime l_{j+1}})|\right|}{max\left\{|\varphi(G')|\cdot|\varphi(n_j^{l_j})\cap\varphi(n_{j+1}^{l_{j+1}})|, |\varphi(G)|\cdot|\varphi(n_j^{\prime l_j})\cap\varphi(n_{j+1}^{\prime l_{j+1}})|\right\}},$$

and

$$\rho_p\left(\left[n_1^{l_1}..n_n^{l_n}\right], \left[n_1^{\prime l_1}..n_n^{\prime l_n}\right]\right) = \frac{\left||\varphi(G')|\cdot|\varphi(n_1^{l_1})\cap\ldots\cap\varphi(n_n^{l_n})| - |\varphi(G)|\cdot|\varphi(n_1^{\prime l_1})\cap\ldots\cap\varphi(n_n^{\prime l_n})|\right|}{max\left\{|\varphi(G')|\cdot|\varphi(n_1^{l_1})\cap\ldots\cap\varphi(n_n^{l_n})|, |\varphi(G)|\cdot|\varphi(n_1^{\prime l_1})\cap\varphi(n_n^{\prime l_n})|\right\}}.$$

Proof. From Definitions 7 and 9, we have
$$\rho_n\left(n_j^{l_j}, n_j^{\prime l_j}\right)$$

$$= \frac{\left|\sigma(n_i^{l_j}) - \sigma(n_j^{\prime l_j})\right|}{max\left\{\sigma(n_i^{l_j}), \sigma(n_j^{\prime l_j})\right\}} = \frac{\left|\frac{|\varphi(n_i^{l_j})|}{|\psi(G)|} - \frac{|\varphi(n_j^{\prime l_j})|}{|\psi(G')|}\right|}{max\left\{\frac{|\varphi(n_i^{l_j})|}{|\psi(G)|}, \frac{|\varphi(n_j^{\prime l_j})|}{|\psi(G')|}\right\}} = \frac{\frac{|\psi(G')||\varphi(n_i^{l_j})| - |\psi(G)||\varphi(n_j^{\prime l_j})|}{|\varphi(G)||\varphi(G')|}}{max\left\{\frac{|\varphi(n_i^{l_j})|}{|\psi(G)|}, \frac{|\varphi(n_j^{\prime l_j})|}{|\psi(G')|}\right\}}$$

$$= \frac{\left||\psi(G')||\varphi(n_i^{l_j})| - |\psi(G)||\varphi(n_j^{\prime l_j})|\right|}{|\psi(G)||\psi(G')|max\left\{\frac{|\varphi(n_i^{l_j})|}{|\psi(G)|}, \frac{|\varphi(n_j^{\prime l_j})|}{|\psi(G')|}\right\}} = \frac{\left||\psi(G')|\cdot|\varphi(n_j^{l_j})| - |\psi(G)|\cdot|\varphi(n_j^{\prime l_j})|\right|}{max\left\{|\psi(G')|\cdot|\varphi(n_j^{l_j})|, |\psi(G)|\cdot|\varphi(n_j^{\prime l_j})|\right\}}.$$

$$\rho_b\left(b_{n_j^{l_j},n_{j+1}^{l_{j+1}}}, b_{n_j^{\prime l_j},n_{j+1}^{\prime l_{j+1}}}\right)$$

$$= \frac{\left|\sigma(b_{n_i^{l_j},n_{j+1}^{l_{j+1}}})-\sigma(b_{n_j^{l_j},n_{j+1}^{l_{j+1}}})\right|}{max\left\{\sigma(b^{n_j^{l_j},n_{j+1}^{l_{j+1}}}),\sigma(b_{n_j^{l_j},n_{j+1}^{l_{j+1}}})\right\}}$$

$$= \frac{\left|\frac{|\varphi(b_{n_i^{l_j},n_{j+1}^{l_{j+1}}})|}{|\psi(G)|}-\frac{|\varphi(b_{n_j^{l_j},n_{j+1}^{l_{j+1}}})|}{|\psi(G')|}\right|}{max\left\{\frac{|\varphi(b_{n_i^{l_j},n_{j+1}^{l_{j+1}}})|}{|\psi(G)|},\frac{|\varphi(b_{n_i^{l_j},n_{j+1}^{l_{j+1}}})|}{|\psi(G')|}\right\}}$$

$$= \frac{\left|\frac{|\varphi(n_i^{l_j})\cap\varphi(n_{j+1}^{l_{j+1}})|}{|\psi(G)|}-\frac{|\varphi(n_i^{l_j})\cap\varphi(n_{j+1}^{l_{j+1}})|}{|\psi(G')|}\right|}{max\left\{\frac{|\varphi(n_i^{l_j})\cap\varphi(n_{j+1}^{l_{j+1}})|}{|\psi(G)|},\frac{|\varphi(n_i^{l_j})\cap\varphi(n_{j+1}^{l_{j+1}})|}{|\psi(G')|}\right\}}$$

$$= \frac{\left||\psi(G')||\varphi(n_i^{l_j})\cap\varphi(n_{j+1}^{l_{j+1}})|-|\psi(G)||\varphi(n_i^{l_j})\cap\varphi(n_{j+1}^{l_{j+1}})|\right|}{|\psi(G)||\psi(G')|max\left\{\frac{|\varphi(n_i^{l_j})\cap\varphi(n_{j+1}^{l_{j+1}})|}{|\psi(G)|},\frac{|\varphi(n_i^{l_j})\cap\varphi(n_{j+1}^{l_{j+1}})|}{|\psi(G')|}\right\}}$$

$$= \frac{\left||\varphi(G')|\cdot\left|\varphi(n_j^{l_j})\cap\varphi(n_{j+1}^{l_{j+1}})\right|-|\varphi(G)|\cdot\left|\varphi(n_j^{l_j})\cap\varphi(n_{j+1}^{l_{j+1}})\right|\right|}{max\left\{|\varphi(G')|\cdot\left|\varphi(n_j^{l_j})\cap\varphi(n_{j+1}^{l_{j+1}})\right|,|\varphi(G)|\cdot\left|\varphi(n_j^{l_j})\cap\varphi(n_{j+1}^{l_{j+1}})\right|\right\}}.$$

Similary, $\rho_p\left(\left[n_1^{l_1}...n_n^{l_n}\right],\left[n_1^{l_1}...n_n^{l_n}\right]\right)$

$$= \frac{\left||\varphi(G')|\cdot\left|\varphi(n_1^{l_1})\cap...\cap\varphi(n_n^{l_n})\right|-|\varphi(G)|\cdot\left|\varphi(n_1^{l_1})\cap...\cap\varphi(n_n^{l_n})\right|\right|}{max\left\{|\varphi(G')|\cdot\left|\varphi(n_1^{l_1})\cap...\cap\varphi(n_n^{l_n})\right|,|\varphi(G)|\cdot\left|\varphi(n_1^{l_1})\cap\varphi(n_n^{l_n})\right|\right\}}.$$

Definition 10. *Let* $\mathcal{G}_N = (\mathcal{N},\mathcal{B},\delta,\sigma)$ *be a DIFG with n layers and the ith layer having k_i nodes where $1 \le i \le n$, $1 \le j_i \le k_i$: $n_i^1, n_i^2, \ldots, n_i^{j_i}, \ldots, n_i^{k_i}$. Let $\mathcal{G}'_{N'}$ $= (\mathcal{N}',\mathcal{B}',\delta',\sigma')$ be a DIFG with n' layers and the i'^{th} layer having k_i' nodes: $n_i'^1, n_i'^2, \ldots, n_i'^{j_i}, \ldots, n_i'^{k_i}$,*
The distance between set of nodes of two DINFGs is defined by

$$D_{\rho_b}(\mathcal{G}_N,\mathcal{G}_{N'}) = \begin{cases} \inf\left\{\rho_b\left(n_{j_l}^l,n_{j_l}'^l\right) : n_{j_l}^l \in \mathcal{N}, n_{j_l}'^l \in \mathcal{N}'\right\} & \text{if } \mathcal{N} \ne \emptyset \text{ and } \mathcal{N}' \ne \emptyset, \\ \infty & ; \text{if } \mathcal{N} = \emptyset \text{ or } \mathcal{N}' = \emptyset, \end{cases}$$

where $\mathcal{N}$ and $\mathcal{N}'$ denote set of all nodes in $\mathcal{G}_N$ and $\mathcal{G}_{N'}$, respectively. The distance between set of branches of two DINFGs is defined by $D_{\rho_b}(\mathcal{G}_N,\mathcal{G}_{N'}) =$

$$\begin{cases} \inf\left\{\rho_b\left(b_{n_{j_l}^l,n_{jm}^m}, b_{n_{j_l}'^l,n_{jm}'^m}\right) : b_{n_{j_l}^l,n_{jm}^m} \in \mathcal{B}, b_{n_{j_l}'^l,n_{jm}'^m} \in \mathcal{B}'\right\}; \\ \qquad\qquad\qquad\qquad\qquad\qquad\qquad \text{if } \mathcal{B} \ne \emptyset \text{ and } \mathcal{B}' \ne \emptyset, \\ \infty; \qquad\qquad\qquad\qquad\qquad\qquad \text{if } \mathcal{B} = \emptyset \text{ or } \mathcal{B}' = \emptyset, \end{cases}$$

where $\mathcal{B}_{\mathcal{G}_N}$ and $\mathcal{B}_{\mathcal{G}_{N'}}$ denote set of all branches in $\mathcal{G}_N$ and $\mathcal{G}_{N'}$, respectively.
The distance between set of paths of two DINFGs is defined by $D_{\rho_p}(\mathcal{G}_N,\mathcal{G}_{N'}) =$

$$\begin{cases} \inf\left\{\rho_p\left(\left[n_1^{l_1}...n_n^{l_n}\right],\left[n_1'^{l_1}...n_{n'}'^{l_{n'}}\right]\right) : \left[n_1^{l_1}...n_L^{l_L}\right] \in \mathcal{P}_{\mathcal{G}_N},\left[n_1'^{l_1}...n_{n'}'^{l_{n'}}\right] \in \mathcal{P}_{\mathcal{G}_{N'}}\right\}; \\ \qquad\qquad\qquad\qquad\qquad\qquad\qquad \text{if } \mathcal{P}_{\mathcal{G}_N} = \emptyset \text{ or } \mathcal{P}_{\mathcal{G}_{N'}} = \emptyset, \\ \infty; \qquad\qquad\qquad\qquad\qquad\qquad \text{if } \mathcal{P}_{\mathcal{G}_N} = \emptyset \text{ or } \mathcal{P}_{\mathcal{G}_{N'}} = \emptyset, \end{cases}$$

where $\mathcal{P}_G$ and $\mathcal{P}_{G'}$ denote set of all paths in $\mathcal{G}_N$ and $\mathcal{G}_{N'}$, respectively.

Definition 11. *A DINFG $\mathcal{G}_N$ and $\mathcal{G}_{N'}$ are minimally near each other if and only if $D_{\rho_n}(\mathcal{G}_N,\mathcal{G}_{N'}) = 0$ or $D_{\rho_b}(\mathcal{G}_N,\mathcal{G}_{N'}) = 0$.*

Lemma 3. *DINFGs $\mathcal{G}_N$ and $\mathcal{G}_{N'}$ are minimally near each other if and only if there is $n_{j_l}^l \in \mathcal{N}$ and $n_{j_l}^{'l} \in \mathcal{N}'$ that have same strength coefficient or $b_{n_{j_l}^l, n_{j_m}^m} \in \mathcal{B}_G$ and $b_{n_{j_l}^{'l}, n_{j_m}^{'m}} \in \mathcal{B}_{G'}$ have same strength coefficient.*

Proof. From Definitions 9, 10 and 11.

Definition 12. *DINFGs $\mathcal{G}_N$ and $\mathcal{G}_{N'}$ are near each other if and only if $D_{\rho_p}(\mathcal{G}_N, \mathcal{G}_{N'}) = 0$.*

Lemma 4. *A DINFG $\mathcal{G}_N$ and $\mathcal{G}_{N'}$ are near each other if and only if there is $\left[n_1^{l_1} \ldots n_n^{l_n}\right] \in \mathcal{P}_{\mathcal{G}_N}$ and $\left[n_1^{'l_1} \ldots n_{n'}^{'l_{n'}}\right] \in \mathcal{P}_{\mathcal{G}_{N'}}$ have same strength coefficient.*

Proof. From Definition 9 and 12.

Definition 13. *A DINFG $\mathcal{G}_N$ and $\mathcal{G}_{N'}$ are sufficiently near each other if and only if $D_\rho(\mathcal{G}_N, \mathcal{G}_{N'}) \leq \epsilon$ and far each other if and only if $D_\rho(\mathcal{G}_N, \mathcal{G}_{N'}) > \epsilon$, where $\epsilon \in [0, \infty]$.*

Example. We illustrate the above ideas by means of a simple example, from the example in Figures 3 and 4,

if we consider the distance between nodes, by definition 9, we have

$$\rho_n\left(n_1^1, n_1^{'1}\right) = \frac{|\sigma(n_1^1) - \sigma(n_1^{'1})|}{max\{\sigma(n_1^1), \sigma(n_1^{'1})\}} = \frac{|0.28 - 0.24|}{max\{0.28, 0.24\}} = 0.1429,$$

$$\vdots$$

$$\rho_n\left(n_3^3, n_3^{'3}\right) = \frac{|\sigma(n_3^3) - \sigma(n_3^{'3})|}{max\{\sigma(n_3^3), \sigma(n_3^{'3})\}} = \frac{|0.32 - 0.40|}{max\{0.32, 0.40\}} = 0.2000,$$

then by using Definition 10, distance between set of nodes $\mathcal{N}$ and $\mathcal{N}'$,

$$D_{\rho_n}(\mathcal{G}_N, \mathcal{G}_{N'}) = inf\left\{\rho\left(n_1^1, n_1^{'1}\right), \rho\left(n_1^2, n_1^{'2}\right), \ldots, \rho\left(n_3^3, n_3^{'3}\right)\right\}$$
$$= inf\{0.1429, 0.2000, 0.1000, 0, 0.1000, 0.2222, 0.1250,$$
$$0.2857, 0.1429, 0.2727, 0.2000\} = 0,$$

if we consider the distance between branches, by definition 9, we have

$$\rho_b\left(d_{n_1^1, n_2^1}, d_{n_1^{'1}, n_2^{'1}}\right) = \frac{|\sigma(d_{n_1^1, n_2^1}) - \sigma(d_{n_1^{'1}, n_2^{'1}})|}{max\{\sigma(d_{n_1^1, n_2^1}), \sigma(d_{n_1^{'1}, n_2^{'1}})\}} = \frac{|0.08 - 0.12|}{max\{0.08, 0.12\}} = 0.3333,$$

$$\vdots$$

$$\rho_b\left(d_{n_2^3, n_3^2}, d_{n_2^{'3}, n_3^{'2}}\right) = \frac{|\sigma(d_{n_2^3, n_3^2}) - \sigma(d_{n_2^{'3}, n_3^{'2}})|}{max\{\sigma(d_{n_2^3, n_3^2}), \sigma(d_{n_2^{'3}, n_3^{'2}})\}} = \frac{|0.08 - 0.08|}{max\{0.08, 0.08\}} = 1.000,$$

then by using Definition 10, distance between set of branches $\mathcal{B}$ and $\mathcal{B}'$,

$$D_{\rho_b}(\mathcal{G}_N, \mathcal{G}_{N'}) = inf\left\{\rho\left(d_{n_1^1, n_2^1}, d_{n_1^{'1}, n_2^{'1}}\right), \ldots, \rho\left(d_{n_2^2, n_2^2}, d_{n_2^{'2}, n_2^{'2}}\right)\right\}$$
$$= inf\{0.333, 0.0000, 0.500, 0.6667, \ldots, 0.0000\} = 0.0000,$$

if we consider the distance directed path, by definition 9, we have

$$\rho_p\left([n_1^1 n_2^1 n_3^3], [n_1^{'1} n_2^{'1} n_3^{'3}]\right) = \frac{|\sigma([n_1^1 n_2^1 n_3^3]) - \sigma([n_1^{'1} n_2^{'1} n_3^{'3}])|}{max\{\sigma([n_1^1 n_2^1 n_3^3]), \sigma([n_1^{'1} n_2^{'1} n_3^{'3}])\}} = 0.0278,$$

$$\vdots$$

$$\rho_p\left([n_1^3 n_2^3 n_3^2], [n_1^{'3} n_2^{'3} n_3^{'2}]\right) = \frac{|\sigma([n_1^3 n_2^3 n_3^2]) - \sigma([n_1^{'3} n_2^{'3} n_3^{'2}])|}{max\{\sigma([n_1^3 n_2^3 n_3^2]), \sigma([n_1^{'3} n_2^{'3} n_3^{'2}])\}} = 0.0278,$$

then by using Definition 10, distance between set of directed path $\mathcal{P}_G$ and $\mathcal{P}_{G'}$,

$$D_{\rho_b}(\mathcal{G}_N, \mathcal{G}_{N'}) = inf\left\{\rho\left(d_{n_1^1, n_2^1}, d_{n_1^{'1}, n_2^{'1}}\right), \ldots, \rho\left(d_{n_2^2, n_2^2}, d_{n_2^{'2}, n_2^{'2}}\right)\right\}$$

$$= inf\{0.0278, 0.0278, 0.0556\ldots, 0.0278\} = 0.0278.$$

Since $D_{\rho_n}(\mathcal{G}_N, \mathcal{G}_{N'}) = 0$ and $D_{\rho_b}(\mathcal{G}_N, \mathcal{G}_{N'}) = 0$, then by Definition 3, DINFG $\mathcal{G}_N$ and $\mathcal{G}_{N'}$ are minimally near each other. Recall that the determination of near and far depends on the value of ϵ. If we assume $\epsilon > 0.0278$ (e.g., $\epsilon = 0.028$), then graphs $\mathcal{G}$ and $\mathcal{G}'$ are sufficiently near (by definition 13), but if we assume that the value of $\epsilon < 0.0278$ (e.g., $\epsilon = 0.027$), then graphs $\mathcal{G}$ and $\mathcal{G}'$ are far.   ∎

3   Conclusion

The contribution of this paper is the introduction of a formal framework for *sufficient nearness* of pairs of flowgraphs in general, and of digital image flow graphs in particular. Future work includes a proximal view of flow graphs in which the sets of points in a flow graph is endowed with a proximity relation.

References

1. Peters, J.F., Chitcharoen, D.: Sufficiently near neighbourhoods of points in flow graphs. A Near Set Approach. Fundamenta Informaticae 124(1), 175–196 (2013)
2. Ramanna, S., Chitcharoen, D.: Flow graphs. analysis with near sets. Mathematics in Computer Science 7(1), 11–29 (2013)
3. Peters, J.: Near sets. General theory about nearness of objects. Applied Math. Sci. 1(53), 2609–2629 (2007)
4. Peters, J.F., Naimpally, S.: Applications of near sets. Amer. Math. Soc. Notices 59(4), 536–542 (2012)
5. Wolski, M.: Perception and classification. A Note on Near sets and Rough sets. Fund. Inform. 101, 143–155 (2010)
6. Mieszkowicz-Rolka, A., Rolka, L.: Flow graphs and decision tables with fuzzy attributes. In: Rutkowski, L., Tadeusiewicz, R., Zadeh, L.A., Żurada, J.M. (eds.) ICAISC 2006. LNCS (LNAI), vol. 4029, pp. 268–277. Springer, Heidelberg (2006)
7. Pawlak, Z.: Decision algorithms, bayes theorem and flow graphs. AISC, pp. 18–24. Physica-Verlag, Heidelberg (2003)
8. Kostek, B., Czyzewski, A.: A processing of musical metadata employing Pawlak's flow graphs. In: Peters, J.F., Skowron, A., Grzymała-Busse, J.W., Kostek, B., Swiniarski, R.W., Szczuka, M.S. (eds.) Transactions on Rough Sets I. LNCS, vol. 3100, pp. 279–298. Springer, Heidelberg (2004)
9. Butz, C., Yan, W., Yang, B.: An Efficient Algorithm for Inference in Rough Set Flow Graphs. In: Peters, J.F., Skowron, A. (eds.) Transactions on Rough Sets V. LNCS, vol. 4100, pp. 102–122. Springer, Heidelberg (2006)
10. Sun, J., Liu, H., Zhang, H.: An extension of pawlak's flow graphs. In: Wang, G.-Y., Peters, J.F., Skowron, A., Yao, Y. (eds.) RSKT 2006. LNCS (LNAI), vol. 4062, pp. 191–199. Springer, Heidelberg (2006)
11. Chitcharoen, D.: Mathematical Aspects of Flow Graph Approaches to Data Analysis. PhD thesis, Department of Applied Mathematics, King Mongkut Institute of Technology, Ladkrabang (2010)
12. Peters, J., Wasilewski, P.: Foundations of Near Sets. Inf. Sci. 179(18), 3091–3109 (2009)
13. Pawlak, Z.: Decision trees and flow graphs. In: Greco, S., Hata, Y., Hirano, S., Inuiguchi, M., Miyamoto, S., Nguyen, H.S., Słowiński, R. (eds.) RSCTC 2006. LNCS (LNAI), vol. 4259, pp. 1–11. Springer, Heidelberg (2006)

An Efficient Interval-Based Approach to Mining Frequent Patterns in a Time Series Database

Phan Thi Bao Tran, Vo Thi Ngoc Chau, and Duong Tuan Anh

Faculty of Computer Science & Engineering, HCMC Uni. of Technology, Vietnam
phanthibaotran@gmail.com, {chauvtn,dtanh}@cse.hcmut.edu.vn

Abstract. In this paper, we introduce an interval-based approach to mining frequent patterns in a time series database. As compared to frequent patterns in the existing approaches, frequent patterns in our approach are more informative with explicit time gaps automatically discovered along with the temporal relationships between the components in each pattern. In addition, our interval-based frequent pattern mining algorithm on time series databases, called IFPATS, is more efficient with a single database scan and a looking-ahead mechanism for a reduction in non-potential candidates for frequent patterns. Experimental results have been conducted and have confirmed that our IFPATS algorithm outperforms both the existing interval-based algorithm on sequential databases and the straightforward approach with post processing for explicit time gaps in the temporal relationships of the resulting patterns. Especially as a time series database gets larger and time series get longer in a higher dimensional space, our approach is much more efficient.

Keywords: Time Series, Frequent Pattern, Time Series Mining, Interval-based Temporal Data Mining, ARMADA, Support.

1 Introduction

Frequent pattern mining is one of the famous mining tasks to discover interesting patterns based on how frequent they are in the database. They are knowledge useful for users to get more understanding about the objects and phenomena of interest. In 1990s, some works in [1, 6, 9] were defined for frequent pattern mining on point-based sequential datasets. Several works in [10, 13, 14] were later proposed on interval-based sequential datasets. Recently, frequent pattern mining on time series databases has been investigated in [3–5, 7, 8, 11, 12].

In this paper, our work is dedicated to mining frequent patterns in a database of multiple time series as time series mining is one of the most challenging problems in data mining research pointed out in [15]. Among the related works on interval-based datasets, we use an existing interval-based approach in [13] because of its good idea of prefix spanning and index sets to generate frequent patterns at a higher level from frequent patterns at a lower level. Besides, it is stressed that [13] used memory-based indexing to record the first positions of a pattern in the database. However, [13] was not originally proposed on time series

S. Ramanna et al. (Eds.): MIWAI 2013, LNCS 8271, pp. 211–222, 2013.
© Springer-Verlag Berlin Heidelberg 2013

databases. Therefore, we had to adapt such an idea to mining frequent patterns in a time series database. In comparison with the approach in [13], our resulting approach is an efficient interval-based approach with a single database scan and a looking-ahead mechanism to early abandon non-potential candidates for frequent patterns. As a result and contribution of our work, an efficient interval-based frequent pattern mining algorithm on a time series database, called IF-PATS algorithm, is proposed for discovering frequent patterns with explicit time gaps in the temporal relationships between their elements. The reason for our approach is that comparison between a proposed work with its previous related works is present rarely in frequent pattern mining in time series databases. Indeed, there was no experimental comparison with the related works in [13]. This situation is understandable because there were different datasets and algorithms with different intended purposes in the related works and thus, data and implementation bias could exist in experimental comparison. Differently, in our work, the proposed IFPATS algorithm is designed with theoretical improvements and examined with experimental results for confirmation of its efficiency. Above all, our resulting frequent patterns are informative and valuable knowledge to enable us get fascinating insights into the relationships and properties of the objects of interest in many various domains rich of time series such as finance, medicine, geology, meteorology, and so on.

2   Related Works

In the most basic form, motifs can be considered as primitive frequent patterns in time series mining. There exist many approaches to finding motifs in time series named a few as [8, 11, 12]. Our work is different from those because the scope of our algorithm does not include the phase of finding primitive patterns with a motif discovery algorithm. We suppose that those primitive patterns are available to our work in terms of a *list of subsequences of interest*.

As for more complex patterns from a time series database, we examine several related works in [3–5, 7]. First, [4] has introduced a notion of perception-based pattern in time series mining with a so-called methodology of computing with words and perceptions. Also towards perception-based time series mining, [5] presented a duration-based linguistic trend summarization of time series using a few features such as the slope of the line, the fairness of the approximation of the original data points by line segments and the length of a period of time comprising the trend. Differently, we concentrate on discovering relationships among primitive patterns. It is worth noting that primitive patterns are given to our algorithm. Moreover, [3] has recently focused on discovering recent temporal patterns from interval-based sequences of temporal abstractions with two temporal relationships: before and co-occur. Mining recent temporal patterns in [3] is one step in learning a classification model for event detection problems. Different from [3], our work discovers each frequent pattern along with explicit time gaps in the temporal relationships between its elements. Besides, our work belongs to the time series rule mining task. For more applications, such patterns can be used in other time series mining tasks such as clustering, classification, and prediction in time series. Based on the temporal concepts of duration, coincidence, and partial order in interval time series, [7] defined pattern types from multivariate time series as Tone, Chord, and Phrase. Tones representing durations are labeled time intervals, which are basic primitives. Chords representing coincidence are formed by simultaneously occurring Tones. Phrases are formed by several Chords connected with a partial order which is actually the temporal relationship "before" in Allen's terms. As compared to [7], our work supports more informative frequent patterns with explicit time automatically discovered with the temporal relationships.

To the best of our knowledge, frequent patterns with explicit time gaps in the temporal relationships between their elements defined in our work have not yet been received much consideration from the existing works in such a way that these time gaps are all automatically discovered from the database. In this paper, we propose an interval-based approach and the IFPATS algorithm which is efficiently designed and outperforms the existing ARMADA algorithm.

3   Problem Definition

3.1   Fundamental Concepts

Definition 1. (*time series database (DB)*) Given a time series database $DB = \{ts_1, \ldots, ts_n\}$, each element ts_i is a times series composed of a sequence of values recorded regularly over the time and all time series have the same size for i = 1...n. $|DB|$ is the number of time series in the database DB.

Definition 2. (*subsequence (s)*) Let S be a set of all subsequences in DB, $s \in S$ denotes a subsequence in S. For simplicity, s in this work is predefined by 3 characters which are from {D, I, U} where D means *decrease*, I means *increase*, and U means *unchanged*. We further use [ts_i, *start-time*, *s*, *end-time*] to represent a subsequence s in time series ts_i where the start time of s is *start-time* and its end time is *end-time*.

Example 1: For the rest of the paper, a time series DB is given as a running example composed of three time series ts_0, ts_1, and ts_2 with the same length 17. These time series are encoded with letters D, I, and U as follows: $ts_0 = $ (I,D,D,U,D,U,U,D,U,I,U,U,I,I,U,I,U), $ts_1 = $ (D,D,D,D,U,I,D,U,D,I,D,D,U, D, U, U, I), and $ts_2 = $ (D,U,I,I,U,D,D,U,I,I,U,D,U,D,U,U,I). A subsequence s_1 of these time series is DUI whose shape includes *decrease*, then *unchanged*, and then *increase* while a subsequence s_2 which is DDU has a *decrease-decrease-unchanged* shape. Further, s_1 from point 3 to 5 in ts_1 is represented as [1, 3, DUI, 5].

Definition 3. (*k-pattern (p)*) A pattern p is defined as a collection of several subsequences which are sequentially connected to each other by an Allen's temporal relationship in [2] with a specific time gap. For simplicity, our current work considers only the "before" relationship. A *k-pattern* is a pattern that consists of k subsequences for $k \geq 1$ which do not overlap each other in a single time series ts. Especially, each single subsequence in S is a *1-pattern* denoted as (s).

To represent a *k-pattern* for $k > 1$, we use a $k \times k$ matrix M whose elements $M[i,j]$ for $i = 1 \ldots k$ and $j = 1 \ldots k$ are "before" relationships with explicit time gaps between subsequence s_i and subsequence s_j included in the pattern. Each pattern is associated with its own interestingness measured by a well-known measure *support* to show how often the pattern appears in the database DB as regards how many time series in DB contain the pattern. *Support* of a pattern p is defined as $\sigma(p) = \frac{|TS_p|}{|DB|}$, where $|TS_p|$ is the number of time series that support, i.e. contain, the pattern p, and $|DB|$ is the number of time series in DB.

Example 2: Given a 3-pattern $p = \dfrac{\begin{pmatrix} DDU & UDU & UUI \\ = & b2 & b6 \\ * & = & b1 \\ * & * & = \end{pmatrix}}{\sigma = 0.6}$, let us explain the content of this 3-pattern. The pattern p contains 3 subsequences DDU, UDU, and UUI which are connected to each other via "before" temporal relationships such that with b2, DDU occurs before UDU with a time gap 2; with b1, UDU before UUI with a time gap 1; and with b6, DDU before UUI with a time gap 6. *Support* of p is $\sigma(p) = 0.6$. In the representation of a *k-pattern*, we use "=" to denote a relationship between a subsequence and itself and "*" to denote a relationship between two subsequences that has been listed in the matrix.

Definition 4. (*subpattern*) A pattern p is a subpattern of pattern p' if p can be obtained by removing some subsequences from pattern p'.

Definition 5. (*instance of pattern*) An instance of pattern p is an appearance of subsequences in the pattern with respect to their relationships denoted in p.

Definition 6. (*frequent pattern*) Given a minimum support *minsup*, a pattern p is a frequent pattern if its support σ is greater than or equal to *minsup*. In other words, a pattern is frequent if $|TS_p|$, the number of time series in the database DB that contain or support p, is greater than or equal to $|DB| \times minsup$.

Definition 7. (*index set (p-idx)*) Based on a similar notion in [13], an index set *p-idx* defined for a pattern p is a set that records every instance of p as entries (or rows) in itself. An entry in an index set *p-idx* includes four fields: field *ts* that records the index of time series to which subsequences belong; field *list of subsequences* that records all subsequences in the instance of pattern p with theirs locations [*start-time, s, end-time*] with respect to their relationships in p; field *pos* that records the end time of the last subsequence in p; and field *maxgap* that records the possibly maximum time gap between the last subsequences in p with other subsequences in the same time series. Table 1 is an example of an index set with respect to the given database DB.

Definition 8. (*stem*) A stem is a subsequence which is added to a frequent *k-pattern* p to form a frequent *[k+1]-pattern* p'. That is, *[k+1]-pattern* $p' = k$-pattern p + *stem*.

For details with the given database DB, in ts_0, subsequence [ts_0, 0, IDD, 2] occurs one time unit before subsequence [ts_0, 4, DUU, 6]. We use <IDD>b1<DUU> to show the relationship between these two subsequences. Similarly, in ts_1, subsequence $[ts_1, 10, DDU, 12]$ occurs one time unit before subsequence $[ts_1, 14, UUI, 16]$ and that relationship is denoted as $<DDU>b1<UUI>$. Note that the relationship between subsequence $[ts_1, 2, DDU, 4]$ and subsequence $[ts_1, 14, UUI, 16]$ is not valid as their combination for a *2-pattern* overlaps an instance of a *1-pattern* (DDU) at 10.

3.2 A Problem Definition of Mining Frequent Patterns in a Time Series Database

In this work, we aim at mining frequent patterns in a time series database with an efficient interval-based approach. Given a time series database *DB* and a specific minimum support *minsup*, the purpose is to discover all frequent patterns whose *supports* are greater than or equal to *minsup*. As compared to the problem of [13] in the context of interval-based data mining, our problem defined in a time series mining area is more challenging to automatically determine explicit time gaps inherently associated with the "before" temporal relationships between subsequences in discovered frequent patterns. The input and output of our problem are elaborated as follows:

- **Input** includes a time series database whose time series have the same length and a minimum support *minsup*. For demonstration, the input also includes a set of subsequences of interest on which our segementation of time series is based to obtain a set of interval-based subsequences for each time series.
- **Output** is a set of all frequent patterns with explicit time gaps in "before" temporal relationships. These explicit time gaps are automatically discovered along with before temporal relationships from a time series database.

With the time series database *DB* in *Example 1*, a user-specified minimum support *minsup* = 0.5, and a list *lstInterestSS* of all interesting subsequences which have the same length 3, all the discovered frequent patterns with explicit time gaps in the "before" temporal relationships are shown in Fig. 1. The arrows display the connections between a frequent pattern and its next-level patterns.

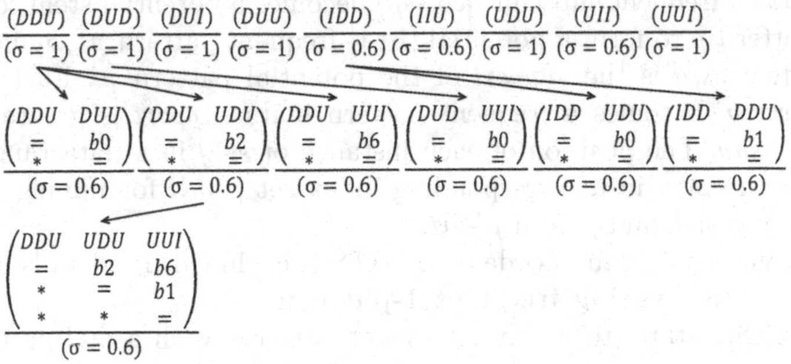

Fig. 1. Frequent patterns with explicit time gaps in the "before" temporal relationships

4 An Efficient Interval-Based Approach to Mining Frequent Patterns in a Time Series Database

In order to resolve the problem described above, we propose an efficient interval-based approach to mining frequent patterns in a time series database by means of an interval-based frequent pattern mining algorithm on a time series database, called IFPATS, taking the input and producing the output defined in section 3.

4.1 An Interval-Based Frequent Pattern Mining Algorithm on a Time Series Database - IFPATS

IFPATS is based on the main idea using index sets that record the positions of instances of a pattern to continue mining next-level patterns. This idea was employed from ARMADA in [13] for time series instead of interval-based temporal data. It is emphasized that ARMADA is unable to discover frequent patterns defined in **Definitions 3** and **6**. However, there is an interesting question "Is it possible for us to use the output of ARMADA on a time series database and then further process the original frequent patterns in that output to add more time information for each temporal relationship in each frequent pattern?" We believe that it is possible for us to do that; unfortunately, algorithm design of ARMADA is not originally dedicated to time series and might lead to an inefficient solution. Thus, the problem in this paper is challenging and needs an efficient approach for time series databases.

In our approach, the proposed IFPATS algorithm has two phases as follows:

Phase 1 aims to identify the frequent 1-patterns. Each interesting subsequence $ss[i]$ becomes a potential 1-pattern $<ss[i]>$. *Support* of $ss[i]$ is the *support* of the potential pattern $<ss[i]>$. Potential 1-pattern $<ss[i]>$ becomes a frequent 1-pattern if its *support* is greater than or equal to *minsup*. All positions of appearance of $ss[i]$ in the database will be recorded as entries in a corresponding index set $<ss[i]>$-*idx* for discovering next-level patterns.

Phase 2 is responsible for mining $[k+1]$-frequent patterns p' from k-patterns p and p-*idx*. Each frequent subsequence $ss[i]$ becomes a potential stem to combine with a k-pattern p to form a potential $[k+1]$-frequent pattern p'. *Support* of the potential stem $ss[i]$ is the *support* of the potential pattern p'. Each potential $[k+1]$-pattern p' becomes a frequent pattern if its *support* is greater than or equal to *minsup*. The position of each instance of $ss[i]$ in a data range will be recorded as an entry in a corresponding index set p'-*idx* for mining next-level patterns from just-found p' and p'-*idx*.

For space saving, the pseudo code of IFPATS is excluded. Its details are below:

(a) Phase 1-Discovering frequent 1-patterns.

The algorithm starts to discover frequent patterns with a null pattern $< >$-pattern. Each subsequence $ss[i]$ becomes a potential stem of $< >$-pattern to form a potential 1-pattern $<ss[i]>$-*idx*.

First, the algorithm scans each time series by using a sliding window to detect all appearances of each subsequence $ss[i]$ and updates the number of time series that contain $ss[i]$. Each $ss[i]$ becomes a potential 1-pattern $<ss[i]>$. Each position of appearance of the subsequence $ss[i]$ is recorded as an entry or a row in the index set, respectively. As presented in **Definition 7**, each index set entry contains 4 fields: ts, $list$ of $subsequences$, pos, and $maxgap$.

For example of (DDU)-idx in Table 1, the algorithm found the subsequence DDU in three time series ts_0, ts_1, and ts_2. In ts_0, DDU is found, at location [1, DDU, 3]; therefore, $ts = ts_0$, $list$ of $sequences$ = [1, DDU, 3], the end-$time$ of the last subsequence [1, DDU, 3] is 3, thus $pos = 3$, the maximum time gap between this subsequence [1, DDU, 3] with other subsequence equals the gap between [1, DDU, 3] and [14, UIU, 16] and equals 14-3-1 = 10; thus, $maxgap$ = 10. In ts1, DDU is found at location [2, DDU, 4]; therefore, $ts = ts_1$, $list$ of $sequences$ = [2, DDU, 4], the end-$time$ of the last subsequence [2, DDU, 4] is 4; thus, $pos = 4$, the maximum time gap between this subsequence [2, DDU, 4] with other subsequence equals the gap between [2, DDU, 4] and [7, UDI, 9] and equals 7-4-1 = 2; so, $maxgap = 2$. In ts_1, DUU is found again at location [10, DDU, 12]; therefore, $ts = ts_1$, $list$ of $sequences$ = [10, DDU, 12], the end-$time$ of the last subsequence [10, DDU, 12] is 12; thus, $pos = 12$, the maximum time gap between [10, DDU, 12] with other subsequence equals the gap between [10, DDU, 12] and [14, UUI, 16] and equals 14-12-1 = 1; thus, $maxgap$ = 1. It is similarly done for DUU in ts_2.

Table 1. Example of (DDU)-idx

ts	List of subsequences	pos	maxgap
ts_0	[1, DDU, 3]	3	10
ts_1	[2, DDU, 4]	4	2
ts_1	[10, DDU, 12]	12	1
ts_2	[5, DDU, 7]	7	6

Table 2. Example of an index set

$$\begin{pmatrix} DDU\ UDU \\ = \quad b2 \\ * \quad = \end{pmatrix} \text{-}idx$$

ts	List of subsequences	pos	maxgap
ts_0	[1, DDU, 3], [6, UDU, 8]	8	5
ts_2	[5, DDU, 7], [11, UDU, 12]	12	1

Whenever finishing scanning a time series in DB to update the number $|TS_p|$ of time series that contain subsequence $ss[i]$, IFPATS always looks ahead in the following way: let $|LeftTS|$ denote the number of the remaining time series which have not yet been scanned for the subsequence $ss[i]$, suppose that all $LeftTS$ remaining time series contain the subsequence $ss[i]$, if the total sum of $|TS_p|$ and $|LeftTS|$ does not satisfy minsup, then we remove $ss[i]$, remove its corresponding index set, and stop searching for $ss[i]$. This action helps reducing the computational cost of the algorithm. A condition of $minsup$ for looking ahead is formulated as: $(|TS_p| + |LeftTS|)/|DB| < minsup$.

After scanning all time series in DB, the algorithm obtains only the frequent 1-patterns p_i and the corresponding index sets p_i-idx. Both p_i and p_i-idx are used to find next-level frequent patterns by invoking a procedure $Mining(p_i, p_i$-$idx)$.

From the database in $Example$ 1, the algorithm finds frequent 1-patterns as follows: (DDU) ($\sigma = 0.67$), (DII) ($\sigma = 0.67$), (DUD) ($\sigma = 1$), (DUI) ($\sigma = 0.67$),

(DUU) ($\sigma = 1$), (IIU) ($\sigma = 1$), (IUI) ($\sigma = 0.67$), (UDU) ($\sigma = 1$), (UII) ($\sigma = 1$), (UIU) ($\sigma = 0.67$), (UUD) ($\sigma = 1$). Each frequent 1-pattern will become prefix of next-level patterns.

(b) Phase 2-Mining [k+1]-patterns from a k-pattern and its index set.

The purpose of this phase is to continue discovering all frequent [k+1]-patterns p' from a frequent k-pattern p and its corresponding index set p-idx. The algorithm considers any frequent subsequence whose *start-time* is greater than the value of *pos* as a potential stem with respect to p in order to form a potential [k+1]-pattern p'. The *support* of a potential stem is also the *support* of a potential frequent pattern. The pseudo code of the procedure $Mining(p, p$-$idx)$ for mining all next-level frequent patterns p' from a frequent pattern p and its index set p-idx is omitted for space limitation.

First, the procedure defines the list of potential stems *lstPtnStem* corresponding to the pattern p. *lstPtnStem* should include all items in the list of interesting subsequence *lstInterestSS* but not in the pattern, i.e. *lstPtnStem* = *lstInterestSS* *list of subsequences* in the pattern. Second, it defines the maximum time gap *maxgap* of each entry in the index set and then defines the maximum time gap *MaxGap* of the whole index set. The maximum time gap *MaxGap* of the index set is the maximum of all *maxgaps*. For example of $Mining(\text{<DDU>}, \text{<DDU>-}idx)$ in Table 1, *maxgaps* of entries in the index set are 10, 2, 1, 6. Therefore, *MaxGap* of the index set is 10. After that, the procedure checks each time gap *gap[j]* present in the index set from 0 to *MaxGap*. For each *gap[j]*, the procedure finds out all stems to form all patterns p' with an explicit time equal to *gap[j]*. Each entry in p-idx has its own *maxgap* and the procedure only needs to deal with the entries whose *maxgaps* are greater than or equal to *gap[j]*. By each suitable entry, the procedure checks the appearance of *ss[i]* at position *pos* + 1 + *gap[j]* in the time series in field *ts* and update the *support* of *ss[i]*.

Continuing with the example of mining an index set (DDU)-idx in Table 1 by calling the procedure $Mining((DDU), (DDU)$-$idx)$, each step is explained as follows. *MaxGap* of the index set <DDU>-idx is 10. So, the procedure examines all time gaps *gap[j]* for $j=0\ldots10$. For instance, it checks *gap[j]* = 2 and the first entry p-$idx[0]$. We have a corresponding time series ts_0, *pos* = 3, *gap[j]* = 2, *maxgap* = 10, and a potential stem <UDU> present at [ts_0, 6, UDU, 8]. So, the procedure increases $|TS_p|$ of <UDU> by 1 and inserts a new entry into the index set p'-idx with the following content: $ts = ts_0$, *list of subsequences* updated with an insertion of [6, UDU, 8], *pos* = 8, *maxgap* = 5. If $|LeftTS| = 2$ and these two time series contain <UDU>, and if $(ss[i].|TS_p| + |LeftTS|)/|\text{DB}| = (1+2)/3 = 1 \geq minsup = 0.5$, then <UDU> might be a stem. Otherwise, the procedure does not need to examine the remaining time series for *gap[j]* = 2 with the potential stem <UDU> anymore. The resulting index set from $Mining(\text{<DDU>}, \text{<DDU>-}idx)$ with an explicit time gap = 2 is shown in Table 2.

4.2   Improvements in the Design of the IFPATS Algorithm

In order to efficiently discover frequent patterns from a time series database with explicit time gaps in the relationships of their components, we adopted and adapted the ARMADA algorithm in [13] for time series instead of interval-based temporal data. Our resulting IFPATS algorithm is designed with two key improvements to deal with a time series database in a higher dimensional space: reducing the number of database scans and reducing the number of candidates for frequent patterns. These improvements make our approach more efficient.

In [13], ARMADA needs two database scans to determine a set of frequent 1-patterns and their corresponding index sets. It is realized that we can use just one database scan by computing *support* of each interesting subsequence and recording their occurrences in each time series for its index set at the same time. Later, if any interesting subsequence has *support* greater or equal to *minsup*, this subsequence becomes a frequent 1-pattern. Otherwise, we remove it and destroy its index set. This approach helps us save processing time for database scans but asks us for more memory. Thus, we define the second improvement.

As for the second one, while examing each time series in a database scan to compute *support* of an interesting subsequence, our algorithm looks ahead and see if *support* of the subsequence is able to meet the condition of *minsup*. Suppose all the remaining time series which have not yet been examined contain the subsequence, if the total number of those time series and the time series examined so far which contain the subsequence is not large enough for the subsequence to have a *support* satisfying *minsup*, it is obvious that the subsequence is not a frequent 1-pattern. So, our algorithm stops checking the subsequence and then stops building its corresponding index set. By looking ahead in that manner, our algorithm reduces the number of candidates for frequent patterns. Consequently, we can save both time and memory.

5 Experiments and Discussion

5.1 Experiment Setting

To evaluate the effect of the proposed algorithm IFPATS on a time series database, we conducted three sets of experiments with varying the values of a minimum support, the numbers of time series, and time series lengths on the synthetic database. The synthetic database was generated randomly from three letters D, I and U. Generation parameters are the number $|DB|$ of time series and the length $|TS|$ of time series. The rationale behind the use of synthetic datasets in our work is that there is neither standardized definition of the frequent pattern mining problem on time series nor data benchmarking for this problem. In each experiment, we recorded the processing time in millisecond and the number of resulting frequent patterns with explicit time gaps in their "before" relationships. All programs were programmed using Visual C# on a 1.66 GHz Intel Core 2 PC with 2GB RAM.

In this paper, we compared our proposed IFPATS algorithm with another straightforward approach based on the original interval-based ARMADA algorithm with post processing for explicit time gaps in the "before" temporal relationships. We also examined our improvements on the original interval-based ARMADA algorithm whose results are associated with no explicit time gaps. All the experiments were conducted on the above synthetic time series database.

- **Time1:** The processing time in millisecond of IFPATS for explicit time gaps.
- **Time2:** The processing time in millisecond of the ARMADA-based algorithm with no explicit time gaps. Its output is not the same as the one from our algorithm and from the straightforward ARMADA-based approach with post processing. Actually, its output is a preliminary set which needs to be further processed to derive explicit time gaps in the resulting frequent patterns. Nevertheless, our work has a significant improvement on the adopted ARMADA algorithm so that time series can be its input instead of interval-based temporal data. For more clarity, after mining on a time series database in *Example 1* with the ARMADA-based algorithm, we got a frequent pattern $\begin{pmatrix} DDU\ UDU \\ =\quad b \\ *\quad = \end{pmatrix}$, but after post processing, we obtained $\begin{pmatrix} DDU\ UDU \\ =\quad b2 \\ *\quad = \end{pmatrix}$, a frequent pattern with an explicit 2-point time gap in the "before" temporal relationship between DDU and UDU.
- **Time3:** The processing time in millisecond of the ARMADA-based algorithm with post processing for explicit time gaps. Its output is exactly the same as the one from our algorithm. The pseudo code of the post processing procedure is omitted for space limitation. In our resulting reports, Time3 is the sum of Time2 and the processsing time of the post processing phase.

5.2 Experimental Results

In each experiment, we reported the number #FPattern of the resulting frequent patterns from our proposed IFPATS algorithm. Ratio T2/T1 of Time2 to Time1 and ratio T3/T1 of Time3 to Time1 are also recorded.

With the improvements made in subsection 4.3, in almost all the cases, IFPATS performs better than the straightforward ARMADA-based approach with post processing. Even without post processing, IFPATS is more efficient than the interval-based ARMADA algorithm on a time series database. As database size $|DB|$ is small, e.g. 20, time series length $|TS|$ is small, e.g. 20, a larger minimum support *minsup*, e.g. 0.2 to 0.3, the two approaches are quite equivalent. This can be explained in such a way that the larger *minsup*, the smaller number of candidates, and thus, the smaller number of frequent patterns. At that moment, a little computation is required and how well each algorithm performs is hard to determine. However, as database size $|DB|$ is large, e.g. 70, time series length $|TS|$ is large, e.g. 50, a smaller minimum support, e.g. 0.1, IFPATS is about 112 times better than the interval-based ARMADA algorithm and 198 times more efficient than the ARMADA-based algorithm with post processing for frequent patterns with explicit time gaps in their temporal relationships.

Particularly in Table 3.(a), the processing time of IFPATS keeps increasing slowly as a minimum support gets smaller while the others increase faster to process a larger number of candidates for frequent patterns with two database scans. In Table 3.(b), it is very clear to realize that IFPATS outperforms the others very much as the database gets larger. Previously reasoned, IFPATS has reduced the number of database scans as compared to the adopted ARMADA algorithm on a time series database so that the number of database scans in our work is minimum, one. This is a significant im-provement when we examine how frequent a pattern is in a very large time series database. Similarly, in Table 3.(c), as time series in the database get longer, i.e. the time series will be in a higher dimensional space, IFPATS is very efficient to examine the relationships between subsequences in frequent patterns and automatically discover their associated time gaps. In contrast, the others are a few times up to about two hundred times less efficient. Such results confirm the efficiency of IFPATS and how appropriate our approach is for minning frequent patterns in a time series database.

Table 3. Epxerimental results by varying (a).*minsup*, (b).|*DB*|, and (c).|*TS*|

Varying	|DB|	|TS|	minsup	#FPattern	Time1	Time2	Time3	T2/T1	T3/T1
(a). minsup	20	20	0.1	177	57	178	287	3.1	5
	20	20	0.15	46	40	69	101	1.7	2.5
	20	20	0.2	30	37	42	58	1.1	1.6
	20	20	0.25	27	37	31	40	0.8	1.1
	20	20	0.3	26	35	13	21	0.4	0.6
(b).|DB|	30	20	0.1	69	62	145	221	2.3	3.6
	40	20	0.1	43	78	175	259	2.2	3.3
	50	20	0.1	33	95	175	255	1.8	2.7
	60	20	0.1	28	110	184	260	1.7	2.4
	70	20	0.1	27	130	221	321	1.7	2.5
(c).|TS|	70	30	0.1	30	330	1828	2918	5.5	8.8
	70	35	0.1	79	533	7478	12227	14	22.9
	70	40	0.1	130	832	17981	30080	21.6	36.2
	70	45	0.1	193	1085	62544	109839	57.6	101.2
	70	50	0.1	262	1449	162604	286970	112.2	198

6   Conclusion

In this paper, we introduced an effective interval-based approach to mining frequent patterns in a database of multiple time series. Our main contribution is an interval-based frequent pattern mining algorithm on time series databases, IFPATS, for more informative frequent patterns with automatically discovered time gaps in the "before" temporal relationships between their subsequences. Through algorithm design and experiments, IFPATS can discover frequent patterns with explicit time gaps in the "before" temporal relationships more efficiently than the ARMADA algorithm on time series databases with no explicit time gaps and also more efficiently than the straightforward ARMADA-based algorithm with post processing for explicit time gaps. This is because IFPATS performed only one database scan and looked ahead to early remove non-potential candidates for frequent patterns. Such an approach enables our work to mine a larger time series database in a higher dimensional space.

Nonetheless, we plan to extend IFPATS in the future to support the other Allen's temporal relationships. Besides, the algorithm will be enabled to find frequent cross-section patterns whose subsequences can belong to different time series. Also, the scalability of IFPATS needs checking with more experiments.

References

1. Agrawal, R., Srikant, R.: Mining Sequential Patterns. In: Proc. ICDE (1995)
2. Allen, J.F.: Maintaining Knowledge about Temporal Intervals. Communications of the ACM 26, 832–843 (1983)
3. Batal, I., Fradkin, D., Harrison, J., Mörchen, F., Hauskrecht, M.: Mining Recent Temporal Patterns for Event Detection in Multivariate Time Series Data. In: KDD, pp. 280–288 (2012)
4. Batyrshin, I., Sheremetov, L., Herrera-Avelar, R.: Perception Based Patterns in Time Series Data Mining. In: Batyrshin, I., Kacprzyk, J., Sheremetov, L., Zadeh, L.A. (eds.) Perception-based Data Mining and Decision Making in Economics and Finance. SCI, vol. 36, pp. 85–118. Springer, Heidelberg (2007)
5. Kacprzyk, J., Wilbik, A., Zadrożny, S.: On Linguistic Summarization of Numerical Time Series Using Fuzzy Logic with Linguistic Quantiers. In: Chountas, P., Petrounias, I., Kacprzyk, J. (eds.) Intelligent Techniques and Tools for Novel System Architectures. SCI, vol. 109, pp. 169–184. Springer, Heidelberg (2008)
6. Lin, M., Lee, S.: Fast Discovery of Sequential Patterns through Memory Indexing and Database Partitioning. Journal of Information Science and Engineering 21, 109–128 (2005)
7. Mörchen, F., Ultsch, A.: Efficient Mining of Understandable Patterns from Multivariate Interval Time Series. Data Min. Knowl. Disc. 15, 181–215 (2007)
8. Mueen, A., Keogh, E., Zhu, Q., Cash, S.S., Westover, M.B., BigdelyShamlo, N.: A Disk-Aware Algorithm for Time Series Motif Discovery. Data Min. Knowl. Disc. 22, 73–105 (2011)
9. Pei, J., Han, J., Mortazavi-Asl, B., Wang, J., Pinto, H., Chen, Q., Dayal, U., Hsu, M.-C.: Prexspan: Mining Sequential Patterns Efficiently by Prex–projected Pattern Growth. In: Proc. ICDE (2001)
10. Papapetrou, P., Kollios, G., Sclaroff, S.: Discovering Frequent Arrangements of Temporal Intervals. In: Proc. ICDM (2005)
11. Tanaka, Y., Iwamoto, K., Uehara, K.: Discovery of Time Series Motif from Multidimensional Data Based on MDL Principle. Machine Learning 58, 269–300 (2005)
12. Tang, H., Liao, S.S.: Discovering Original Motifs with Different Lengths from Time Series. Knowledge-Based Systems 21, 666–671 (2008)
13. Winarko, E., Roddick, J.F.: ARMADA - An Algorithm for Discovering Richer Relative Temporal Association Rules from Interval-based Data. Data & Knowledge Engineering 63, 76–90 (2007)
14. Wu, S.-Y., Chen, Y.-L.: Mining Nonambiguous Temporal Patterns for Interval-based Events. IEEE Tran. on Knowledge and Data Engineering 19, 742–758 (2007)
15. Yang, Q., Wu, X.: 10 Challenging Problems in Data Mining Research. International Journal of Information Technology & Decision Making 5, 597–604 (2006)

A Secure and Privacy-Aware Cloud-Based Architecture for Online Social Networks

Kasun Senevirathna and Pradeep K. Atrey*

Department of Applied Computer Science, The University of Winnipeg,
515 Portage Avenue, Winnipeg MB R3B 2E9, Canada
senevirathna-k@webmail.uwinnipeg.ca, p.atrey@uwinnipeg.ca

Abstract. The use of social networks has grown exponentially in recent years, and these social networks continue to have an ever-increasing impact on human lives. There are many concerns regarding the privacy of users in these environments, such as how trustworthy the social network operators are, in addition to the external adversaries. In this paper we propose a new architecture for online social networking, based on distributed cloud-based datacenters and using secret sharing as the method of encrypting user profile data, for enhanced privacy and availability. This proposed architecture is theoretically analyzed for its security and performance along with some experimental analysis. We show that the proposed architecture is highly secure at an acceptable level of time complexity overhead in comparison to existing online social networks, as well as the models proposed in previous studies targeting the same research problem.

Keywords: Online Social Networks, Security, Privacy, Secret Sharing, Cloud Computing.

1 Introduction

One important fact that online social network (OSN) users overlook is the potential of the social network operator (SNO) itself becoming an adversary to their privacy. It is observed that most OSN users place absolute faith in SNOs. SNOs are mostly profit-oriented entities. As a profit-creating mechanism, these SNOs might share user information with third parties. Furthermore, it is a known fact that SNOs share user information with third party application developers to build applications within their OSNs. Even the SNO might not be able to guarantee the privacy of users when such information is shared with third parties. Privacy is a subjective measure and it can be difficult to be defined, particularly in an OSN environment. But at a minimum we can assume that the users expect their data being observed by intended parties only. There have also been certain concerns that SNOs may keep user data even after users delete them from their

* This research was supported by the Natural Sciences and Engineering Research Council of Canada (through Discovery Grant No. 408206) and the University of Winnipeg (through Major Research Grant No. 607362).

S. Ramanna et al. (Eds.): MIWAI 2013, LNCS 8271, pp. 223–234, 2013.

accounts (or delete the accounts themselves), and the SNO may provide a false assurance to the user about the deletion of the posted data.

To overcome this challenge of securing the data and maintaining the privacy of users of an OSN from SNOs and other adversaries while providing high availability, we propose a decentralized architecture for online social networking. The core idea behind our method is using cryptographic secret sharing schemes for multimedia, along with private user data storage in cloud based commercial datacenters. We use secret sharing in the proposed architecture mainly due to the following two reasons: first, the flexibility of secret sharing in supporting dynamic networks, and second, the high level of security provided by secret sharing.

The above justifications will be discussed in detail in Sections 3 and 4. We also provide a comprehensive security analysis of the proposed architecture followed by performance feasibility. Section 2 discusses other related works on online social networking privacy. Section 3 describes the proposed method. Section 4 presents the security analysis of the proposed solution and section 5 presents some experimental analyses of the proposed architecture. Sections 6 and 7 provide limitations and conclude the paper with a discussion of future works.

2   Related Works and Preliminaries

Lucas and Borisov [1] have come up with an approach for protecting information published on Facebook through encryption. The authors have built a prototype Facebook application named "flyByNight" to implement the proposed idea through proxy cryptography. However, the implemented prototype application was unable to achieve the encryption of images that can be highly sensitive when it comes to user privacy. In another work, Anderson et al. [3] proposed an architecture for OSNs that builds an OSN out of smart clients and an untrusted central server. The server resides as a hub. It is not trustworthy, and it provides only the availability of data. Alternatively, the clients must be smart and are responsible for their own confidentiality and integrity. There is still the vulnerability of carrying out traffic analysis at the central server. Similarly, Shakimov et al. [4] presented Vis-'a-Vis, a decentralized framework for online social networks based on a virtual individual server (VIS), a personal virtual machine running in a paid compute utility. In Vis-'a-Vis, a person stores his/her data on his/her own VIS, which arbitrates access to that data by others. Still, user data is visible to the vendor who is providing the VIS service. Baden et al. [5] presented "Persona", a decentralized OSN that hides personal information from aggregators and hides personal information from colleagues as is appropriate. It gives more control to users over their data.Users can decide where to store their data. Rather than trusting the data storage provider it relies on cryptographic techniques to protect user privacy. Cutillo et al. [6] proposed an OSN that builds upon peer-to-peer networks in order to remove the need for a potentially untrusted central entity. In this work, the challenge is to maintain continuous availability via distributed peers.

Recently, Atrey [2] proposed a secret-sharing-based key management scheme for encrypting user data. Secret-sharing-based key management is argued to be unconditionally secure, and also to be computationally efficient. The transfer of all messages and key shares takes place through the SNO in the encrypted form. Both of these models have the advantage of integrating with existing OSNs. However, they might face the challenge of gaining the required support from today's SNOs.

After the inception of the concept of secret sharing [7], the concept has been used in many studies on protecting different multimedia data types [8], [9]. It involves creating multiple shares (n) of the data to be kept secret, where of n shares, at least k ($k \leq n$) shares are required to reconstruct the original secret. A number of shares less than k (even k-1 shares) would not reveal any information about the original secret [7].

To the best of our knowledge, this is the first study to propose and assess the feasibility of using secret sharing in the online social networking domain to benefit from the high level of security provided by secret sharing, and to ensure security and privacy against network infrastructure providers and eavesdroppers.

3 Proposed Work

The proposed architecture attempts to achieve security and privacy along with high availability at an affordable cost to users. It requires storing user data in highly available commercial cloud-based datacenters of their choice. The approach presented in this paper is an alternative to the existing freely available centralized OSN architectures, but with higher security and privacy.

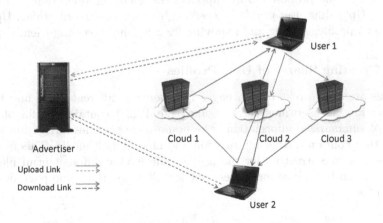

Fig. 1. An abstract view of the proposed architecture

This approach proposes storing data in the encrypted form at the datacenters using secret sharing. This creates the requirement to have multiple (at least two) cloud-based datacenters to hold the data of a particular user. Figure 1 shows an abstract view of this architecture. The next challenge is building a friend network for a particular user. To address this we introduce a central server (or multiple servers, theoretically) to maintain a database (DB) of users who have subscribed to the OSN, named the "Advertiser". This is depicted in Figure 1.

A user must present his/her name and email address to register with the Advertiser. Other information such as a profile photo, date of birth, current city, education related information, etc., which can be used as search filters, are optional. The email address is not publicly displayed to other users. It is only used for communication purposes between the Advertiser and the user, in scenarios such as adding friends, which is explained later.

Before going into further detail, it is important to establish some terminology to avoid possible ambiguities.

- The **Advertiser** is an online server, as explained above, that contains users' public profiles that can be searched and viewed by other users.
- A **user profile** refers to all the information and images that a user intends to share with his/her friends.
- The **public profile** has the information (e.g. name, profile picture, etc.) that a user intends to share publicly.
- A **user** refers to the owner of a particular user profile, while a **friend** refers to another user who can access profile information of the user.
- A pre-assigned computer is named the **access computer**.
- The client application is termed the **OSN application** in the rest of the discussion.

This architecture assumes a secret sharing scheme of (k, n). While k is a design parameter, users are free to decide the n value ($n \geq 2$).

Presently, the proposed OSN supports the following operations: User registration, Uploading user profiles, Accessing/reading one's own profile, Updating one's own profile, Adding and removing friends, and Accessing friends' profiles.

3.1 Creating Shares of User Profiles

Profiles are a hierarchy of web pages representing different levels and types of information. All users must have a home page. It may contain a profile photo and a list of summarized information. The home page provides navigation to other pages that contain further information and shared multimedia. This essentially results in a tree structure for the user profile having different html files as its nodes. It can be unique for a user, but for all users, the home page lies at the root of the tree.

$$f(x) = (a_0 + a_1x + a_2x^2 + \ldots + a_{k-1}x^{k-1}) \bmod p \tag{1}$$

The encryption of a user profile means encrypting a sequence of web pages. Secret sharing of a web page can be implemented by secret sharing the text content of the html file [9]. All of the embedded images are separately secret shared [8] and stored in the same directories as referenced by the original html file. In a (k, n) secret sharing scheme, shares are created for different values of x as per (1). The first coefficient a_0 is taken to be the secret number that needs to be protected. The remaining coefficients, i.e. a_1, a_2, a_3, ..., a_{k-1}, are chosen randomly. In (1), p is a prime number higher than the secret number. The secret can be reconstructed by any k shares using Lagrange interpolation for the k number of pairs of points as per (2).

$$g(x) = [\sum_{j=1}^{k}(y_j \times \prod_{i=1,i\neq j}^{k} \frac{x - x_i}{x_j - x_i})] \bmod p \qquad (2)$$

3.2   Accessing One's Own Profile

In the **user registration phase** with the datacenters, users must receive necessary access information from the datacenters, i.e. a publicly accessible IP address, and a unique authentication code generated by the datacenter. Only the user has "write" accessibility for his data. A database that can be accessed by the OSN application is maintained within the access computer.

3.3   Updating One's Own Profile

Profile data shares stored at multiple cloud datacenters need to be updated only when a user decides to update his/her profile. Updating a profile involves one of the following operations: adding new profile information, modifying existing profile information, or deleting existing profile information.

A profile update can involve any combination of text, images, and video. The OSN application ensures that only the html files affected by the update get updated at the cloud datacenters. This may involve changes in multiple leaf nodes, intermediary nodes and even in the root of the user profile tree. Even though a share update at cloud datacenters is required only when a user updates his/her profile, as a security measure users may decide to periodically update the shares at cloud datacenters without a profile update.

3.4   Accessing Friends' Profiles

A separate database is maintained at the access computer, keeping the required information to access friends' profiles. Let D_{ij} be the jth friend's i^{th} datacenter, the primary key, which consists of the friend's email address and the sub-index of the datacenter. The authentication codes in this database give only "read" access.

3.5   Adding and Removing Friends

The addition of friends to the OSN of a user is done via the Advertiser. It involves a public-key cryptographic algorithm, RSA. The OSN application creates a pair consisting of a public key and a private key. The user then registers with

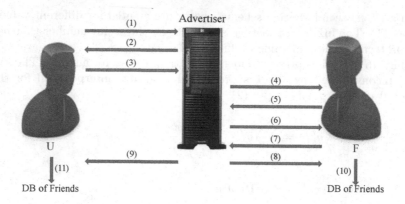

Fig. 2. The handshake protocol of adding a friend

the Advertiser by providing his/her name, email address and the public key. The user (U) accesses the Advertiser using the OSN application and searches for a particular friend (F). Once F is found, U sends a friend request via the Advertiser. This involves a two-way handshake, as shown in Figure 2. The steps are as follows: 1) U communicates his/her intention of adding F as his/her friend to the Advertiser. 2) The Advertiser sends the public key of F (pub_key_F). 3) The user header file (H_U) is encrypted using the pub_key_F and is sent to the Advertiser. It is kept at the Advertiser for the time being. 4) The Advertiser informs F about the friend request from U. 5) F accepts the friend request and communicates his/her intention to the Advertiser. 6) The Advertiser sends the public key of U (pub_key_U) to F. 7) F's header file (H_F) is encrypted using the pub_key_U and is sent to the Advertiser. 8) and 9) The Advertiser exchanges the two encrypted header files between U and F. 10) and 11) The OSN applications of U and F read the respective header files and update the database of friends.

The authentication code generated by a datacenter for the new friend is unique from others. This facilitates the removal of a friend from a user's OSN. It simply involves blocking access to user profile shares at all the cloud datacenters with particular authentication codes unique to the friend who needs to be removed. This is one of the major strengths of the proposed architecture when compared to the previous works. It is also one of the main advantages of adopting secret sharing over traditional key-based cryptographic schemes like AES, DES, etc. Key-based encryption schemes involve the additional overhead of changing the encryption key, notifying this change and distributing the new key to other existing friends.

4 Analysis for Security and Privacy

Below is a vulnerability assessment of the proposed architecture, considering different possible threat agents, along with its availability.

4.1   Unrelated Attackers and Non-friend OSN Users

In the OSN, there is no way for the third party attackers to know the contact list of a user. Even a friend does not have access to that information. A user will only have the information of friends up to one tier, i.e., his own friends. In addition, a third party lacks any information with regard to the datacenters where the user keeps his shares, other required information to reveal the secret, and the authentication codes to retrieve shares from the cloud datacenters. Furthermore, knowing the above information for a single datacenter is not sufficient to reveal the secrets and find a user's profile information. Therefore, the user profile can be considered to be immune against third party attacks, both in privacy and in integrity.

4.2   The Advertiser

A risk associated with the Advertiser is the ability to carry a man-in-the-middle attack. This is a significant risk since this architecture assumes the Advertiser to be untrustworthy. To combat this attack it is necessary to decouple the Advertiser and the public key storage. This requires another party (or possible multiple parties) to store users' public keys. This gives the extra flexibility of using the same public key for multiple applications, not just for the OSN. Exchanging header messages and/or public keys out-of-band also gives the extra level of protection from man-in-the-middle attacks by the Advertiser.

4.3   Cloud Datacenters

The cloud datacenters will have only the shares of user profiles that do not reveal any information about the user profiles. In addition, the cloud datacenter has no information about other datacenters where the user has stored his data. Therefore, it is safe to assume that user privacy is immune to any attacks from the cloud datacenter.

4.4   Network Infrastructure Provider and Eavesdroppers

When revealing secrets, in addition to the share, an index associated with that share is required for all shares. This corresponds to the x value used to create that share (from a polynomial view of secret sharing). This x value is also stored with each share and must be communicated to the friend to successfully reveal the secret.

So rather than encrypting all the shares resulting from secret sharing, it is efficient to pass only the x values in encrypted form to all the datacenters. This way, any eavesdropper between the user and the datacenters who can get access to all the shares of a secret, is still unable to reveal the secret successfully. Also, rather than relying on a symmetric key, the user and the datacenter can adopt the Diffie-Hellman key exchange protocol for each session of communication. It will remove the extra burden at key management from the user. Figure 3

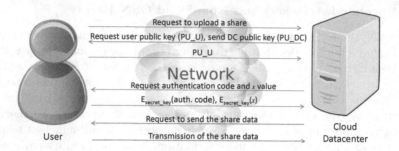

Fig. 3. Share uploading protocol

depicts this share uploading protocol. The steps involved in this protocol are as follows: 1) An authenticated user communicates his intention to upload a share of a secret. 2) The datacenter requests the user's public key (PU_U) while sending the datacenter's public key (PU_DC). 3) The user sends PU_U to the datacenter. The user and the datacenter can now derive the session key to be used, using the exchanged public keys. 4) The datacenter requests that the user send his/her unique authentication code (assigned during the initial registration) and the x value corresponding to the share to be uploaded. 5) The user encrypts both the authentication code and the x value using the derived session key (E_{secret_key}(auth code) and $E_{secret\_key}(x)$, respectively) and sends them to the datacenter. 6) The datacenter stores the x value and requests that the user send data in context to the share to be uploaded. 7) The user uploads share data to the datacenter as it is.

Table 1. Trust relationships in different OSN architectures

Architecture	External Adversaries	Other OSN Users (Non-Friends)	SNO	Network Infrastructure Provider and Eavesdroppers	Data Storage Provider	Friends
Conventional OSNs	Yes and No	Yes and No	Yes	Yes	Yes	Yes
Anderson et al. [3]	No	No	No	Yes	No	Yes
Shakimov et al. - Vis-'a-Vis [4]	No	No	Not Applicable	Yes	Yes	Yes
Baden et al. - Persona [5]	No	No	Not Applicable	Yes	No	Yes
Cutillo et al. - Safebook [6]	No	No	Not Applicable	Yes	Yes	Yes
Proposed Architecture	**No**	**No**	**Not Applicable**	**No**	**No**	**Yes**

Table 2. Vulnerability of the proposed architecture against known attacks

Attack	Proposed Architecture
Plain impersonation	Vulnerable.
Profile cloning	Vulnerable. But the attacker cannot rely on the OSN itself to create the cloned profile.
Profile hijacking	Secure, as the user profiles are not visible to external parties.
Profile porting	Vulnerable.
Profiling	Secure.
Fake requests	Vulnerable, since accepting (or rejecting) a fake request is a user's decision.
Crawling and harvesting	Secure.
Image retrieval and analysis	Secure, assuming that friends are trusted.
Communication tracking	Secure.
Fake profiles and Sybil attacks	Vulnerable
Censorship	Secure

Table 1 compares the proposed architecture to conventional OSNs and other architectures proposed by different scholars, in context of the faith placed on different entities involved in an OSN. It is noted as "Not Applicable" for architectures that do not have a centralized SNO and/or an external data storage provider. The proposed architecture provides the highest level of security in that sense. Some conventional OSNs (e.g. Facebook) provide certain privacy settings to be protected against external adversaries and non-friend OSN users. These settings, however, are effective only if a user opts to use them. Due to this, they are represented as 'Yes and No' in Table 1. It is evident that the proposed architecture is more secure as it has minimized the vulnerabilities and threat agents in the architecture itself.

The authors in [10] presented an attack spectrum in the context of OSN environments. Table 2 is an analysis of the proposed system's vulnerability against different types of attacks as presented in [10]. It shows that the proposed architecture is resilient against many attacks where the existing popular OSNs such as Facebook and LinkedIn are not. Also since user profile will only be shared at user discretion, the risk imposed by third party data users in the context of a conventional OSN is not present in the proposed architecture.

4.5   Functional and Performance Analyses

The supported functionality of an OSN varies from one SNO to another. Therefore, it is not possible to pinpoint a universally accepted comprehensive list of functions that should be supported by a particular OSN. However, by studying several OSNs we present some functions that are supported by different OSNs today, and whether these functions are supported in the proposed architecture (or are intended to be supported in future works) is presented in Table 3. While the proposed system currently fails to provide some functionality it still securely

Table 3. Functionality analysis of the proposed architecture

OSN Application/Functionality	Level of Support
Information sharing with trusted parties	Textual data and images are supported. Compressed videos are not supported. This is a future research direction.
User interactions on shared data (e.g. tagging, endorsing, commenting)	Supported.
User status updates	Supported
Communication (Messaging)	Supported.
Notifications	Notifications about friend requests are facilitated by emails and/or notifications at the Advertiser.
Connectivity expansion	Looking for friends and adding them securely is supported through the Advertiser and the public-key infrastructure. However, suggesting friends is not supported and is a future research direction.
Location based services	Not supported.
Advertising user profiles	Supported at the Advertiser.
Clustering friends and displaying relationships.	The proposed architecture so far supports only a single tier of connections at the same level of trust as friends.

supports most of the principal functions that are required for effective online interactions among different individuals in an OSN environment.

Assuming link bandwidths, the time taken for data pre-processing (e.g. image compression, image resizing, etc.) remains the same for the proposed architecture when compared to existing OSNs. The only time performance constraints in the proposed architecture are the time taken to create shares and time taken to reconstruct the original data. Experimental results in Section 5 provide an estimation about this overhead time complexity.

5 Experimental Results

A performance feasibility analysis of the proposed architecture was also carried out to ensure that it meets performance levels at an acceptable level with the extra burden on performance of higher security. We began our analysis by first recording the time it took to produce secret shares of a 175 kB bitmap image (Table 4). This image size was selected because most of the user-posted images (compressed) on popular OSNs like Facebook range between 50-200 kB. Next, we recorded the time it took to upload these shares to two different clouds (Table 4). The SN application was built in Java with a Java Server Faces (JSF) framework. It is backed by MySQL database for local DBs. The Advertiser is a restful web service built in java. For analysis purposes, and to achieve a general cloud average, we uploaded five shares to Google Drive, and five shares to Dropbox to determine an average time it would take to upload a share to any cloud service.

It should be noted that upload times may vary from one user to another based on internet connection speed and other third-party factors. The results recorded in this experiment were obtained with a 20Mbps download speed, and a 0.5 Mbps upload speed. The average share generation time for an image of size 175 kB was approximately 39.2 ms.

We found from our results that the average cloud upload time for an average image share size (50-200 kB) was approximately 5082.70 ms, or 5.0827 seconds per share. We gathered the same time measurements for an html file as well, and recorded the time it took to produce secret shares of a 175 byte html file (Table 4) and recorded how long it took to upload those shares to the two clouds. For the above observations it takes a total time of approximately 5.8018 seconds to produce shares for a user's profile home page (which included an image and an html file) and upload those shares to the corresponding cloud datacenters, which is at an acceptable level.

Table 4. Secret sharing and reconstruction times

Data	Time (ms)				
File	Share 1	Share 2	Share 3	Share 4	Reconstruction
Image (175 kB)	29.890	29.559	22.965	22.817	54.222
html file (175 B)	0.528	0.346	0.373	0.396	2.540

6 Limitations

This architecture exhibits the conventional tradeoff between user friendliness and information security. It is obvious that certain operations like adding friends to a user's OSN are not straightforward. Also, updating and viewing profiles may take slightly more time than for the existing OSNs. The user must be restricted to a single computing device when accessing the OSN. These limitations give rise to a lack of user friendliness in the proposed architecture when achieving higher user security and privacy. Video, especially in the compressed domain, is one of the major multimedia types being used extensively in mainstream OSNs. The proposed architecture, which is built upon secret sharing, lacks the support for compressed video for the time being. This is because studies on secret sharing schemes for compressed videos are still in early stages and are not yet sufficiently comprehensive to be considered in an application like this. As such, we did not try to address secret sharing of compressed videos, since it is outside of the scope of this study. Also, as mentioned above, the higher level of security and privacy, and availability of the proposed OSN may come at a certain cost to its users. The cost associated with data storage in cloud datacenters can be minimized by adopting computational secret sharing rather than perfect secret sharing.

7   Conclusions and Future Work

In this paper we have proposed and outlined a novel, decentralized architecture to build more private and secure social networks. This approach leverages the security and availability provided by secret sharing and the emerging global business model of cloud-based datacenters. The proposed architecture is an alternative to today's centralized OSNs. It addresses a requirement for highly secure and privacy-concerned online social networking.

This is the first step toward a more comprehensive architecture. There are multiple future enhancements that can be built upon the current work. Some suggestions for future research directions are provided in Table 3 in line with improving the functionality of the proposed architecture.

References

1. Lucas, M.M., Borisov, N.: Flybynight: Mitigating the Privacy Risks of Social Networking. In: 7th ACM Workshop on Privacy in the Electronic Society, pp. 1–8. ACM, New York (2008)
2. Atrey, P.K.: A Secret Sharing Based Privacy Enforcement Mechanism for Untrusted Social Networking Operators. In: 3rd International ACM Workshop on Multimedia in Forensics and Intelligence, pp. 13–18. ACM, New York (2011)
3. Anderson, J., Diaz, C., Bonneau, J., Stajano, F.: Privacy-enabling Social Networking Over Untrusted Networks. In: 2nd ACM workshop on Online Social Networks, pp. 1–6. ACM, New York (2009)
4. Shakimov, A., Lim, H., Caceres, R., Cox, L.P., Li, K., Liu, D., Varshavsky, A.: Vis-à-Vis: Privacy-preserving Online Social Networking via Virtual Individual Servers. In: 3rd International Conference on Communication Systems and Networks, pp. 1–10 (2011)
5. Baden, R., Bender, A., Spring, N., Bhattacharjee, B., Starin, D.: Persona: An Online Social Network with User-Defined Privacy. In: ACM SIGCOMM 2009 Conference on Data Communication, pp. 135–146. ACM, New York (2009)
6. Cutillo, L.A., Molva, R., Strufe, T.: Privacy Preserving Social Networking Through Decentralization. In: 6th International Conference on Wireless On-Demand Systems and Services, pp. 133–140. IEEE Press, New York (2009)
7. Shamir, A.: How to Share a Secret. Communications of the ACM 22(11), 612–613 (1979)
8. Alharthi, S., Atrey, P.K.: Further Improvements on Secret Image Sharing Scheme. In: 2nd ACM Workshop on Multimedia in Forensics, Security and Intelligence, pp. 53–58. ACM, New York (2010)
9. Atrey, P.K., Hildebrand, K., Ramanna, S.: An Efficient Method for Protection of Text Documents Using Secret Sharing. In: International Conference on Frontiers of Computer Science (2011)
10. Cutillo, L.A., Manulis, M., Strufe, T.: Security and Privacy in Online Social Networks. In: Furht, F. (ed.) Handbook of Social Network Technologies and Applications, pp. 497–522. Springer, Heidelberg (2010)

A Hybrid Algorithm for Image Watermarking against Signal Processing Attacks

Amit Kumar Singh[1], Mayank Dave[1], and Anand Mohan[2]

[1] Department of Computer Engineering,
National Institute of Technology, Kurukshetra, Haryana-India
[2] Department of Electronics Engineering,
Indian Institute of Technology BHU, Varanasi, Uttar Pradesh-India
amit_245singh@yahoo.com,
mdave67@gmail.com, amohan@bhu.ac.in

Abstract. In this paper, we have presented a hybrid image watermarking technique and developed an algorithm based on the three most popular trans form techniques which are discrete wavelet transforms (DWT), discrete cosine transforms (DCT), and singular value decomposition (SVD) against signal processing attacks. However, the experimental results demonstrate that this algorithm combines the advantages and remove the disadvantages of these three transform. This proposed hybrid algorithm provides better imperceptibility and robustness against various attacks such as Gaussian noise, salt and pepper noise, motion blur, speckle noise, and Poisson noise etc.

Keywords: mage watermarking, steganography, discrete wavelet transforms, discrete cosine transforms, singular value decomposition.

1 Introduction

Recently, with the explosive growth of information and communication technologies (ICT), various new opportunities emerged for the creation and delivery of content in the digital form which includes applications such as real time video and audio delivery, electronic advertising, digital libraries, telemedicine, e-commerce, e-governance, media forensics and web publishing [1]. However, these advantages have the consequent risks of data piracy, which motivate for the development of new protection mechanisms. One such effort that has been attracting interest is based on the digital watermarking techniques, which is a technique for inserting information into an image and later extracted or detected for variety of purposes including identification and authentication. With this technique, we can recognize the source, owner, distributor or creator of a document or an image. Simmons [2] has demonstrated a sample scenario where watermarking can be thought of in terms of the "Prisoner's problem". The data hiding technique in which the message signal is hidden in the cover signal without any perceptual distortion is a form of communication that depend on the channel used to transfer the host content. It is classified in to two categories: 1) steganography and 2) digital watermarking [3]. However, the former refers to hiding of a secret message inside another message in order to avoid the detection

S. Ramanna et al. (Eds.): MIWAI 2013, LNCS 8271, pp. 235–246, 2013.

and/or decoding it by others. It is used for spying in corporate and intelligence industries like for copyright purposes in entertainment industry [4]. On the other hand in later technique, a watermark signal is embedded into a host signal (image, audio, video or a text document) robustly and invisibly at the same time [5]. However, the watermarking techniques based on the type of document have been divided into the several categories such as text, image, audio and video watermarking [6]. Further, the image watermarking techniques are divided into two domain methods: spatial domain method and transform domain method [7]. In the spatial domain methods [8, 9], the data is embedded directly by manipulating the pixel values and bit stream or code values of the cover image. These methods are less complex, very simple and computationally straightforward, however, they are not robust against signal processing attacks, whereas on the transform domain watermarking techniques are more robust. In this method, the data has been embedded by modulating the wavelet coefficients of the image in transform domain such as DCT, DFT and DWT [10]. The wavelet transforms provides excellent spatial-frequency localization properties as discussed in detail in [11-13].

Jiansheng et al. [14] proposed an algorithm for the digital image watermarking based on two transform method (DCT and DWT). In embedding process, the host image is decomposed into 3rd level wavelet transform and the watermark information is embedding only in high frequency band information of DWT image. However, it has been DCT transformed before embed the watermark information. This algorithm has strong capability of embedding signal and anti-attack but the computing speed is lower. Lai and Tsai [15] proposed a watermarking scheme based on DWT and SVD. In embedding process, the cover image is decomposed into four subband and the SVD is applied to diagonally (H and V subband) only. After dividing the watermark image into two parts, singular values in H and V sub-band are modified with the half of the watermark image and then apply SVD to them, respectively.

Ahire and Kshirsagar [16] proposed a blind watermarking algorithm based on DCT and DWT that embeds a binary image into the gray image. In embedding process, DWT first applied to the cover image and it decomposed into 3rd level and DCT is applied on four selected subband of DWT. Now, the watermark information is embedded in all four selected DWT sub-band, where all four selected sub-band has been DCT. Umaamaheshvari and Thanushkodi [17] proposed a frequency domain watermarking method to check the integrity and authenticity of the medical images. In the embedding process, DCT first applied to original image to generate a resultant transformed matrix and a hybrid transformed image is obtained when Daubechies 4 wavelet transform applied on the resultant transformed matrix. The Daubechies 4 wavelet transform are useful for local analysis but it has higher computational overhead and is more complex.

2   Theoretical Background

The proposed work based on DWT, DCT, and SVD, which required certain theoretical considerations related to their application in image processing. Hence, a brief description of these concepts is discussed as follows.

2.1   Discrete Wavelet Transform (DWT)

Discrete wavelet transform is a system of filters that decomposes an image into a set of four sub bands that are non-overlapping multi-resolution [18], A (approximation/lower frequency sub-band), H (horizontal sub-band), V (vertical sub-band) and D (diagonal sub-band). The process can be repeated to obtain multiple scale wavelet decomposition.

2.2   Discrete Cosine Transform (DCT)

The discrete cosine transform (DCT) works by separating image into parts of different frequencies, low, high and middle frequency coefficients [19], makes it much easier to embed the watermark information into middle frequency band that provide an additional resistance to the lossy compression techniques, while avoiding significant modification of the cover image. The DCT has a very good energy compaction property. For the input image, I, of size $N \times N$ the DCT coefficients for the transformed output image, D, are computed using Equation (1). The intensity of image is denoted as I (x, y), where the pixel in row x and column y of the image. The DCT coefficient is denoted as D (i, j) where i and j represent the row and column of the DCT matrix.

$$D(i,j) = \frac{1}{\sqrt{2N}} C(i)C(j) \sum_{x=0}^{N-1} \sum_{y=0}^{N-1} I(x,y) \cos \frac{(2x+1)i\pi}{2N} \cos \frac{(2y+1)i\pi}{2N} \tag{1}$$

$$C(i), C(j) = \frac{1}{\sqrt{N}} \ for \ i,j = 0 \ and \quad C(i), C(j) = \sqrt{\frac{2}{N}} \ for \ i,j = 1,2, \dots \dots N-1$$

2.3   Singular Value Decomposition (SVD)

The singular value decomposition of a rectangular matrix A is as follows:

$$A = USV^T \tag{2}$$

where A is an $M \times N$ matrix, U and V are the orthonormal matrices. S is a diagonal matrix which consists of singular values of A. The singular values s1≥s2≥........≥sn≥0 appear in the descending order along with the main diagonal of S. However, these singular values have been obtained by taking the square root of the eigenvalues of AA^T and A^TA. These singular values are unique, however the matrices U and V are not unique. The SVD has two main properties from the viewpoint of image processing applications are: 1) the singular values of an image have very good stability, when a small perturbation is added to an image, its singular values do not change significantly, and 2) singular values represent the intrinsic algebraic image properties [20].

3   Performance Measures

However, the performance of the watermarking algorithm has been evaluated on the basis of its robustness and imperceptibility. A larger Peak Signal to Noise Ratio (PSNR) indicate that the watermarked image more closely resembles the original image meaning that the watermark more imperceptible. In general, the watermarked image with PSNR value greater than 28 is acceptable [21]. The PSNR is defined as:

$$PSNR = 10 \log \frac{(255)^2}{MSE} \qquad (3)$$

where the Mean Square Error (MSE) is defined as:

$$MSE = \frac{1}{X \times Y} \sum_{i=1}^{X} \sum_{j=1}^{Y} (I_{ij} - W_{ij})^2 \qquad (4)$$

where I_{ij} is a pixel of the original image of size X×Y and W_{ij} is a pixel of the watermarked image of size X×Y. The robustness of the algorithm is determined in term of correlation factor. However, the similarity and differences between the original watermark and extracted watermark is measured by the Normalized Correlation (NC). It value is generally 0 to 1. However, ideally its should be 1 but the value 0.7 is acceptable [21].

$$NC = \sum_{i=1}^{X} \sum_{j=1}^{Y} (W_{original\ ij} \times W_{recovered\ ij}) / \sum_{i=1}^{X} \sum_{j=1}^{Y} W_{original\ ij}^2 \qquad (5)$$

where $W_{original\ ij}$ is a pixel of the original watermark of size $X \times Y$ and $W_{recovered\ ij}$ is a pixel of the recovered watermark of size $X \times Y$.

4   Proposed Algorithm

In this paper, we have proposed a hybrid image watermarking algorithm, where second level DWT is performed on the cover image and first level DWT on watermark image. The two level DWT results in decomposition of the host image into four sub-bands $(A_{c1}, H_{c1}, V_{c1}, and\ D_{c1})$. However, among these four sub-bands, the H_{c1} sub-bands are chosen for watermark embedding. The decomposition of the watermark image into four sub-bands $(A_w, H_w, V_w, and\ D_w)$. Now, with the help of DCT and SVD, the singular value of the H_w is embedded into the singular value of H_{c1}. The watermark extraction process is same as the embedding process but in reverse order. The proposed algorithm has two parts, one is the watermark embedding and other is watermark extraction method as follows:

4.1   Watermark Embedding Algorithm

start:
STEP 1: Variable Declaration
Medical Image(Thorax): cover image
Leena: watermark image
C_w: read the cover image
W_w: read the watermark image
α : scale factor
DWT, DCT and SVD: Transform Domain Techniques
Wavelet filters: Haar

$A_c, H_c, V_c \text{ and } D_c$: First level DWT coefficients for cover image
$A_{c1}, H_{c1}, V_{c1} \text{ and } D_{c1}$: Second level DWT coefficients for cover image
$A_w, H_w, V_w \text{ and } D_w$: First level DWT coefficients for watermark image
D_c^1: DCT coefficients matrix for H_{c1}
H_w^1: DCT coefficients matrix for H_w
$U_c \text{ and } V_c^T$: orthonormal matrices for D_c^1
S_c: diagonal matrix for D_c^1
$U_w \text{ and } V_w^T$: orthonormal matrices for H_w^1
S_w: diagonal matrix for H_w^1
W_w^k: modified value of S_c
$U_{ww} \text{ and } V_{ww}^T$: orthonormal matrices for W_w^k
S_{ww}: diagonal matrix for W_w^k
W_{modi}: Modified DWT coefficient
W_{idct}: InverseDCT coefficients matrix
W_d: Watermarked Image

STEP 2: Read the Images
C_w← Thorax.bmp (Cover image of size 512*512)
W_w← Leena.bmp (Watermark image of size 256*256)

STEP 3: Perform DWT on Cover and Watermark image
Apply second level DWT on cover image and first level DWT on Watermark image
$[A_c, H_c, V_c, \text{ and } D_c]$ ← DWT (C_w, wavelet filter);
$[A_{c1}, H_{c1}, V_{c1}, \text{ and } D_{c1}]$ ←DWT (H_c, wavelet filter);
$[A_w, H_w, V_w, \text{ and } D_w]$ ←   DWT (W_w, wavelet filter);

STEP 4: Choice of subbands in Cover and Watermark image and obtain the DCT coefficients for the same
//Choose subband H_{c1} from cover image and H_w from watermark image
if (DCT on H_{c1})**then**
D_c^1 ← DCT (H_{c1});
endif;
if (DCT on H_w)**then**
H_w^1 ← DCT (H_w);
endif;

STEP 5: Compute the singular values of DCT coefficients for Cover and Watermark image
if (SVD on D_c^1)**then**
$U_c S_c V_c^T$ ← SVD (D_c^1)
endif;
if (SVD on (H_w^1) **then**
$U_w S_w V_w^T$ ← SVD (H_w^1)
endif;
STEP 6: Watermark Embedding
for $\propto$← 1: 5
$S_c + \propto S_w = W_w^k$;
  end;

STEP 7: Compute the singular values for W_w^k and obtain the modified DWT coefficients

if $(SVD \ on \ W_w^k)$then

$[U_{ww}S_{ww}V_{ww}^T \leftarrow SVD(W_w^k)$

endif;

//modified DWT coefficient

$W_{modi} \leftarrow U_c S_{ww} V_c^T$

Step 8: Obtain the Watermarked Image

$W_{idct} \leftarrow inverseDCT(W_{modi})$;

//Apply InverseDWT to $A_{c1}, H_{c1}, V_{c1}, and \ D_{c1}$ with modified coefficient

$H_c \leftarrow inverseDWT(A_{c1}, W_{idct}, V_{c1} and \ D_{c1}, wavelet \ filter)$;

//Apply InverseDWT to $A_c, H_c, V_c, and \ D_c$ with modified coefficient

$W_d \leftarrow inverse \ (A_c, H_c, V_c, D_c, wavelet \ filter)$;

end:

4.2   Watermark Extraction Algorithm

start:

STEP 1: Variable Declaration

α : scale factor

$A_c, H_c, V_c \ and \ D_c$: subbands for watermarked image

D_w^*: DCT coefficients matrix for H_c

U_w^* and V_w^{*T}: orthonormal matrices for D_w^*

S_w^*: diagonal matrix for D_w^*

S_c^k: diagonal matrix for DCT coefficients of cover image

S^{*k}: modified values

U_w^{*1} and V_w^{*1T}:   orthonormal matrices for S^{*k}

S_w^{*1}: diagonal matrix for S^{*k}

I_{cc}^*: modified DWT coefficients

I_{Wcc}^*: InverseDCT coefficients matrix

W_{EW}: Extracted watermark image

STEP 2: Perform DWT on Watermarked image (possibly distorted)

$[A_c, H_c, V_c, and \ D_c] \leftarrow \ DWT \ (W_d, wavelet \ filter)$;

STEP 3: obtain the DCT coefficients for H_c

if (DCT on H_c)then

$D_w^* \leftarrow DCT \ (H_c)$;

endif;

STEP 4: Compute the singular values for D_w^*

$U_w^* S_w^* V_w^{*T} \leftarrow SVD(D_w^*)$

end;

STEP 5: Perform the operation and then apply SVD

for $\alpha \leftarrow 1:5$

$S^{*k} = \dfrac{S_w^* - S_c}{\alpha}$

end;

$$U_w^{*1} S_w^{*1} V_w^{*1T} \leftarrow SVD(S^{*k})$$

STEP 6: Compute modified DWT coefficients

$$I_{cc}^* \leftarrow U_w S_w^{*1} V_w^T$$

STEP 7: Extract the watermark image

$$I_{Wcc}^* \leftarrow inverse(I_{cc}^*);$$
$$W_{EW} \leftarrow inverseDWT(A_w, I_{Wcc}^*, V_w, D_w, wavelet\ filter);$$

end:

5 Experimental Results

We have discussed the performance of the combined DWT-DCT-SVD watermarking algorithm. The gray–level images "Medical image (Thorax)" and "Lena" of size 256 ✕256 are used as cover and watermark image as shown in figure 1(a) and 1(b), respectively. We have evaluated the quality of watermarked image (as shown in figure 1(c)) and robustness of the proposed algorithm by the parameter PSNR and Normalized Correlation (NC) respectively. Also, we have compared the performance of the proposed algorithm with reported techniques by Jiansheng et al. [14], Lai et al. [15] and Nidhi et al. [22] against various kinds of attacks. The hybrid of DWT, DCT and SVD give the better results (see Table 1). However, our method needs less SVD computation than other methods. The effects of the attacks are shown in fig 2 to fig. 11. In the experiments, the value of the scale factor (α) are carried out from 1 to 5 and taking the average value of the PSNR and Normalized Correlation (NC), and the results are illustrated in Table 1. Without any noise attack the average PSNR obtained is 38.61dB and the NC value is 1. We found that the larger the scale factor, stronger the robustness and smaller the scale factor, better the image quality.

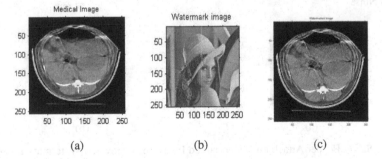

Fig. 1. (a) Cover Image (b) Watermark Image(c) Watermarked Image

Table 1. Comparison of robustness for Jiansheng [14], Lai [15], Nidhi [22] and our method

Various Attack	PSNR (Avg. Value) of proposed method	Jiansheng Method [14] Using DWT and DCT	Lai Method [15] Using DWT and SVD	Nidhi & Jani Method [22] Using DWT DCT and SVD	Proposed Method Using DWT, DCT and SVD
Gaussian Noise	18.51	Not Shown	0.97	0.9992	0.9997
Salt and Pepper Noise	17.94	0.8643	Not Shown	0.9995	0.9996
Motion Blur	17.02	Not Shown	Not Shown	Not Shown	0.9998
Low pass filter	18.88	0.9132	0.9597	0.9996	0.9996
Median filter	19.81	Not Shown	Not Shown	0.9995	0.9998
Speckle Noise	19.78	Not Shown	Not Shown	Not Shown	0.9996
Poisson Noise	25.61	Not Shown	Not Shown	Not Shown	0.9999
Contrast adjustment	18.84	Not Shown	0.9958	0.9995	0.9998
JPEG compression	18.12	0.8518	0.9761	0.9996	0.9996
Histogram equaliza-tion	19.21	Not Shown	0.9890	0.9991	0.9994

(a)   (b)

Fig. 2. (a) Gaussian Noise on Watermarked Image (b) Recovered Watermark Image from Gaussian Noise

(a)   (b)

Fig. 3. (a) Salt & Pepper Attack on Watermarked Image (b) Recovered Watermark Image from Salt & Pepper Attack

(a)   (b)

Fig. 4. (a) Motion Blurred on Watermarked Image (b) Recovered Watermark Image from Motion Blurred

(a)   (b)

Fig. 5. (a) Low pass filter on Watermarked Image (b) Recovered Watermark Image from Low pass filter

(a)   (b)

Fig. 6. (a) Median filter on Watermarked Image (b) Recovered Watermark Image from Median filter

(a)   (b)

Fig. 7. (a) Speckle Attack on Watermarked Image (b) Recovered Watermark Image from Speckle Attack

(a)   (b)

Fig. 8. (a) Poisson Attack on Watermarked Image (b) Recovered Watermark Image from Poisson Attack

(a)   (b)

Fig. 9. (a) Histogram equalization on Watermarked Image (b) Recovered Watermark Image from Histogram equalization

(a)   (b)

Fig. 10. (a) Contrast adjustment on Watermarked Image (b) Recovered Watermark Image from Contrast adjustment

(a)   (b)

Fig. 11. (a) JPEG Compression on Watermarked Image (b) Recovered Watermark Image from JPEG Compression

6   Conclusion

The three important parameters such as robustness, imperceptibility and capacity measure the quality of watermarking method, which need to be considered in the methods. There is some tradeoff between the imperceptibility, robustness and capacity, so there must be some balance among these requirements according to the applications.

The performance of watermarking methods depends on the overall watermarking method as well as embedding and detection techniques. In the transform domain, several efficient image watermarking techniques based on DFT, DCT, DWT and SVD have been developed. However, they do not have the directional information such as directional edges of the image. The discrete contourlet transform [23] is capable of capturing the directional information with multi-resolution representation. It makes use of Laplacian Pyramid for multi-resolution representation of the image followed by a directional decomposition on every band pass image using directional filters. However, this proposed hybrid scheme is robust and keeps the image quality very well. We would like improve the performance, which will be reported in future communication.

Acknowledgements. The authors are sincerely thankful to the potential reviewer's for their critical comments and suggestions to improve the quality of the paper.

References

1. Katzenbeisse, S., Petitcolas, F.A.P.: Information Hiding Techniques for Steganography and Digital Watermarking. Artech House, London (2000)
2. Simmons, G.J.: The Prisoners' Problem and the Subliminal Channel. In: Advances in Cryptology, Proceedings of CRYPTO 1983, pp. 51–67. Plenum Press (1984)
3. Bender, W., Gruhl, D., Morimoto, N.: Techniques for data hiding. IBM Systems Journal 35(3-4), 313–336 (1996)
4. Armstrong, T., Yetsko, K.: Steganography, CS-6293 Research Paper, Instructor: Dr. Andy Ju An Wang (2004)
5. Sequeira, A.: Enhanced Watermark Detection, M. Sc., thesis, University of Toronto, Canada (2003)
6. Mohanty, S.P.: Watermarking of Digital Images, M.S. Thesis, Indian Institute of Science, India (1999)
7. Wolak, M.C.: Digital Watermarking. Preliminary Proposal, Nova Southeastern University, United States (2000)
8. Nikolaidis, N., Pitas, I.: Digital Image Watermarking: An Overview. In: IEEE International Conference on Multimedia Computing and Systems, pp. 1–6 (1999)
9. Cox, I.J., Miller, M.L.: The First 50 Years of Electronic Watermarking (EURASIP). Journal on Applied Signal Processing, 126–132 (2002)
10. Singh, A.K., Dave, M., Mohan, A.: A novel technique for digital image watermarking in frequency domain. In: 2nd IEEE International Conference on Parallel Distributed and Grid Computing, JUIT, Solan-India (2012)
11. Meerwald, P., Uhl, A.: A Survey of Wavelet-Domain Watermarking Algorithms. In: Proceedings of the SPIE Security and Watermarking of Multimedia Contents, San Jose, pp. 505–516 (2001)

12. Hajjara, S., Abdallah, M., Hudaib, A.: Digital Image Watermarking Using Localized Bior-thogonal Wavelets. European Journal of Scientific Research 26(4), 594–608 (2009)
13. Paquet, A.H., Ward, R.K.: Wavelet-Based Digital Watermarking for Authentication. In: Proceedings of the IEEE Canadian Conference on Electrical and Computer Engineering, Winnipeg, pp. 879–884 (2002)
14. Jiansheng, M., Sukang, L., Xiaomei, T.: A Digital Watermarking Algorithm Based On DCT and DWT. In: International Symposium on Web Information Systems and Applications, Nanchang, P. R. China, pp. 104–107 (2009)
15. Lai, C.C., Tsai, C.C.: Digital Image Watermarking Using Discrete Wavelet Transform and Singular Value Decomposition. IEEE Transactions on Instrumentation and Measurement 59(11), 3060–3063 (2010)
16. Ahire, V.K., Kshirsagar, V.: Robust Watermarking Scheme Based on Discrete Wavelet Trans-form (DWT) and Discrete Cosine Transform (DCT) for Copyright Protection of Digital Images. International Journal of Computer Science and Network Security 11(8), 208–213 (2011)
17. Umaamaheshvari, A., Thanushkodi, K.: High Performance and Effective Watermarking Scheme for Medical Images. European Journal of Scientific Research 67(2), 283–293 (2012)
18. Mallat, S.: The Theory for Multiresolution Signal Decomposition: The Wavelet Represen-tation. IEEE Transactions on Pattern Analysis and Machine Intelligence 11(7), 654–693 (1989)
19. Al-Haj, A.: Combined DWT-DCT digital image watermarking. Journal of Computer Science 3(9), 740–746 (2007)
20. Liu, R., Tan, T.: An SVD-based Watermarking Scheme for Protecting Rightful Ownership. IEEE Transaction on Multimedia 4(1), 121–128 (2002)
21. Gunjal, B.L., Manthalkar, R.R.: An Overview of Transform Domain Robust Digital Image Watermarking Algorithms. Journal of Emerging Trends in Computing and Information Sciences 2(1) (2011)
22. Nidhi, H.D., Jani, N.N.: Image Watermarking Algorithm using DCT, DWT and SVD. International Journal of Computer Application (2012)
23. Do, M.N., Vetterli, M.: The Contourlet Transform: Directional Multiresolution Image Re-presentation. IEEE Transaction on Image Processing 14(12), 2091–2106 (2005)

Hindi Word Sense Disambiguation Using Semantic Relatedness Measure

Satyendr Singh[1], Vivek Kumar Singh[2], and Tanveer J. Siddiqui[3]

[1,3] Department of Electronics & Communication, University of Allahabad, Allahabad, India
[2] South Asian University, New Delhi, India
{satyendr,vivekks12}@gmail.com, jktanveer@yahoo.com

Abstract. In this paper we propose and evaluate a method of Hindi word sense disambiguation that computes similarity based on the semantics. We adapt an existing measure for semantic relatedness between two lexically expressed concepts of Hindi WordNet. This measure is based on the length of paths between noun concepts in an is-a hierarchy. Instead of relying on direct overlap the algorithm uses Hindi WordNet hierarchy to learn semantics of words and exploits it in the disambiguation process. Evaluation is performed on a sense tagged dataset consisting of 20 polysemous Hindi nouns. We obtained an overall average accuracy of 60.65% using this measure.

Keywords: Hindi Word Sense Disambiguation, Semantic Relatedness, Semantic Similarity, Semantic Distance.

1   Introduction

Polysemy is inherent to natural languages. Every natural language has a large number of words carrying multiple meanings. For instance, the English noun bark can mean sound made by a dog, a type of sailing vessel with three or more masts or the outermost layers of stems and roots of woody plants; similarly the Hindi word हल (hal) can mean जमीन जोतने का एक उपकरण (an instrument used to plough field) or समाधान/निबटारा (solution). Human beings are fairly apt in determining the correct sense of the word. But for machines this is a very difficult task. The task of computational identification of the correct sense of a word in a given context is called Word Sense Disambiguation (WSD). One of the widely used approaches in automatic WSD is to measure similarity between the textual context of an ambiguous word and dictionary definition of its senses and assigning appropriate sense based on the similarity value. However, this method requires an exact match of content words and fails to take into account semantic or conceptual similarity.

In this paper, we describe a method of word sense disambiguation based on the semantics. Our algorithm uses Leacock-Chodorow semantic relatedness measure [7] to compute semantic similarity between two lexically expressed concepts. It uses Hindi WordNet [2] hierarchy to compute distance between the two concepts. Instead of relying on direct overlap the algorithm uses Hindi WordNet hierarchy to learn

S. Ramanna et al. (Eds.): MIWAI 2013, LNCS 8271, pp. 247–256, 2013.

semantics of words and exploits it in the disambiguation process. The algorithm does not require any training corpus and can disambiguate any ambiguous word that is found in Hindi WordNet. The performance of semantic similarity based WSD algorithm is compared with Lesk-like direct overlap method [11].

A number of earlier work reports on the use of semantic similarity measures for WSD and other NLP task. Budanitsky and Hirst [3] evaluated five measures of semantic relatedness using WordNet. They compared the performance of these measures in detecting and correcting real world spelling errors. They found information content based measure proposed by Jiang and Conrath superior to those proposed by Hirst and St-Onge [4], Leacock and Chodorow [7], Lin [8], and Resnik [10]. Reddy et al. [9] evaluated and compared six semantic relatedness measures for Hindi semantic category labeling. They reported that adapted Lesk performed better than other measures. Sinha and Mihalcea [13] proposed an unsupervised graph-based WSD algorithm and reported results using six different measures on SENSEVAl-2 and SENSEVAL-3 dataset. Torres and Gelbukh [14] compared Lesk algorithm for original Lesk algorithm. They evaluated Jiang–Conrath, Lesk and combination of similarity measures using random sense and most frequent sense as a back-off procedure on senseval-2, Senseval-3, Semeval and Semcor corpus. The experimental results show that different measures performed better on different corpus and using different back-off procedure. In general, combination of similarity measures was more accurate than each measure separately. However, Lesk measure performs better for Senseval-3, Semval and Semcor corpus when back-off is random sense while Jiang–Conrath [5] measure shows good results with most frequent sense back-off is used. However, these results can not be generalized for Hindi. The work presented in this paper is an attempt to investigate the usefulness of existing semantic similarity measures for Hindi WSD task. To the best of our knowledge this kind of work is not reported earlier for Hindi WSD.

Other work on Indian languages includes [6, 9, 11, 12]. Sinha et al. [12] used an extended Lesk-like algorithm and performed contextual overlapping for Hindi WSD task. They explored extended sense definitions derived from Hindi WordNet and context of target polysemous word to perform contextual overlap between them. Winner sense is assigned as one which maximized the overlap. Khapra et al. [6] performed their study on domain specific WSD for nouns, adjectives and adverbs in a trilingual setting of English, Hindi and Marathi. Dominant senses of words in specific domains were used for performing disambiguation. Singh and Siddiqui [11] evaluated the effects of stemming, stop word removal and context window size on a Lesk like overlap based algorithm for Hindi WSD. They reported an improvement of 9.24% in precision after stop word removal and stemming over baseline.

The rest of the paper is organized as follows: In Section 2, an overview of semantic similarity is given and Leacock-Chodorow measure of semantic relatedness is discussed. Section 3 discusses the proposed algorithm. Details of the dataset and experiments conducted are provided in Section 4. Results are discussed in Section 5 and finally conclusions are drawn in Section 6.

2   Overview of Semantic Similarity

Semantic similarity is a kind of relatedness between two words. Semantic relatedness includes relationships between concepts including semantic similarity and other relations such as is-a-kind-of, is-a-part-of, is-a-specific-example-of, is-the-opposite-of, etc. A number of measures of semantic relatedness have been proposed [4, 5, 7, 8, 10]. The relatedness measure by Hirst and St-Onge [4] considers many other relations in WordNet and is not restricted to nouns. This measure assigns a relatedness score for word types rather than concepts. Jiang and Conrath [5] used information content proposed by Resnik [10]. They augmented it with a notion of path length between concepts. Resnik[10] introduced a measure of relatedness based on the information content. The information content is assigned to each concept in a hierarchy based on evidence found in the corpus. In Lin measure [8] of semantic relatedness, similarity between two concepts is measured by the ratio of the amount of information needed to state the commonalty of two concepts to the amount of information needed to describe them. In this work, we use the Leacock-Chodorow measure [7] of semantic relatedness to disambiguate a polysemous word. Relatedness is measured by computing the length of the shortest path between two concepts. We have used only nouns in this work. The underlying assumption is that the concepts of nearby nouns of target word are better indicator of the correct sense of target polysemous noun. The Leacock-Chodorow measure [7] of semantic relatedness is based on the length of paths between noun concepts in an is-a hierarchy. The shortest path between two concepts is the path having least number of intermediate nodes. The shortest path is scaled by the depth of the hierarchy. Depth is the length of the longest path from a leaf node to the root node of the hierarchy. The relatedness between two concepts $c1$ and $c2$ is computed as:

$$relatedness_{lch}(c1, c2) = [- \log (ShortestLength(c1, c2) / (2*D)]$$

ShortestLength $(c1, c2)$ is the shortest path length between two concepts and D is the maximum depth of the taxonomy. A hypothetical root node is created to join all the noun hierarchies. The maximum depth of Hindi WordNet is 12.

3   WSD Algorithm

We extend the idea of direct overlap method for performing disambiguation using the Leacock-Chodorow measure of semantic relatedness. Firstly, stop words are removed from test instances. A test instance is a context containing an ambiguous word. Then we extract all the nouns from the test instances irrespective of their usage in the context. The test instance is represented as a vector of words appearing in a window containing two nouns on either side of the target word. The sense definitions are also represented as a vector comprising of words appearing in the sense definitions. The similarity between these two vectors is computed. Instead of using a keyword-based similarity the Hindi WordNet hierarchy is utilized to get meaning-based similarity.

The algorithm takes as input synset id's of various senses of an ambiguous word and synset id's of nouns appearing in the test instance. For example, for Hindi noun सोना (sona), there are 3 senses listed in Hindi WordNet as shown in Fig. 1. The synsets pertaining to these 3 senses in Hindi WordNet along with their synset id are shown in Fig. 2.

1. सोना, स्वर्ण, कंचन, हेम, कनक, सुवरन, कांचन, सुवर्ण, अभ्र, हिरण्य, वरवर्ण, शातकुंभ, शातकुम्भ, शातकौंभ, शातकौम्भ, शुक्र, त्रिनेत्र, चामीकर, पुरुद, ज़र, वर्णि, अर्ह, अवष्टंभ, अवष्टम्भ, श्रीमत्कुंभ, श्रीमत्कुम्भ, रसविरोधक, रंजन, मनोहर, शतकुंभ, शतकुम्भ, शतकौंभ, शतकौंभक, शतखंड, शतखण्ड, भद्र, अश्मकर, अष्टापद, मरुत्, दत्र, आग्नेय, वसु, गारुड़, तामरस, तार्क्ष्य; एक बहुमूल्य पीली धातु जिसके गहने आदि बनते हैं ; "आज कल सोने का भाव आसमान छू रहा है/ चैतन्य महापुरुष के शरीर से स्वर्ण जैसी आभा निकलती थी"

2. शयन, सोना, सयन; सोने की क्रिया; "शयन के लिए हीरा त बनायी गयी है"

3. सोनापाठा, श्योनाक, टेंटू, सोना, सोनापाढ़ा, स्वर्णवल्कल, निसोथ, निसृता, निसौत, व्याघ्रादनी, पूतिपत्र, पूतिवृक्ष, त्रिमृत, त्रिमृता, शुक, शुकनास, पत्रोर्ण, अरलु, त्रिवेला, भूतपुष्प, धौतकोषज, पृथुशिंब, पृथुशिम्ब, भूमिपुत्र; एक प्रकार का ऊँचा पेड़; "सोना पाठा के बीज, छाल और फल दवा के रूप में काम आते हैं"

Fig. 1. Senses of सोना in Hindi WordNet

3045 - NOUN - [सोना, स्वर्ण, कंचन, हेम, कनक, सुवरन, कांचन, सुवर्ण, अभ्र, हिरण्य, वरवर्ण, शातकुंभ, शातकुम्भ, शातकौंभ, शातकौम्भ, शुक्र, त्रिनेत्र, चामीकर, पुरुद, ज़र, वर्णि, अर्ह, अवष्टंभ, अवष्टम्भ, श्रीमत्कुंभ, श्रीमत्कुम्भ, रसविरोधक, रंजन, मनोहर, शतकुंभ, शतकुम्भ, शतकौंभ, शतकौंभक, शतखंड, शतखण्ड, भद्र, अश्मकर, अष्टापद, मरुत्, दत्र, आग्नेय, वसु, गारुड़, तामरस, तार्क्ष्य]

8042 - NOUN - [शयन, सोना, सयन]

18571 - NOUN - [सोनापाठा, श्योनाक, टेंटू, सोना, सोनापाढ़ा, स्वर्णवल्कल, निसोथ, निसृता, निसौत, व्याघ्रादनी, पूतिपत्र, पूतिवृक्ष, त्रिमृत, त्रिमृता, शुक, शुकनास, पत्रोर्ण, अरलु, त्रिवेला, भूतपुष्प, धौतकोषज, पृथुशिंब, पृथुशिम्ब, भूमिपुत्र]

Fig. 2. Synset id's of सोना in Hindi WordNet

For our dataset, we have considered two senses of सोना, pertaining to gold and sleep sense. This consists of sense1 and 2 from Hindi WordNet. Thus, the synset id's of the target word सोना pertaining to sense 1 and 2 are {3045, 8042}. This constitutes the input sysnet id's of target word (सोना) used in this experiment.

The synset id's of the words in the test instance vector, excluding target word are computed. This forms output synset id's. For each synset id of the target word, semantic relatedness is computed for all output synset id's using the measure of Leacock Chodorow. Currently, we are not discriminating the senses of nouns appearing in the nearby context. The overall score for each sense is computed by summing the pair-wise semantic similarity between each of its senses. The sense pertaining to the synset id which maximizes the score is assigned as the winner sense. The steps in WSD algorithm are shown in fig. 3.

WSD Algorithm
1. Remove stop words from the test instances, extract all the nouns and then create context vector by extracting all the nouns to the left and right of the target word in the proximity of 2 words
2. Extract input synset id's of target word (Syn-input)
3. Extract output synset id's of the nouns in context vector, excluding target word (Syn-output)
4. For i = 1 to number of Syn-input do
      $Score_i \leftarrow$ similarity (Syn-input$_i$, Syn-output)
5. Return Syn-input$_i$ for which score is maximum.

Computing score:

similarity (Syn-input$_i$, Syn-output) // Syn-input$_i$ is the synset-id of the ith sense of target word and Syn-output is the synset id's of all nouns, excluding target words in context vector //
sense_score $\leftarrow 0$
      For each synset id a in Syn-output
relatedness$_{lch}$ (Syn-input$_i$, a) = [- log (ShortestLength (Syn-input$_i$, a) / (2*D)]
 // D is maximum depth of Hindi WordNet (12) //
sense_score $\leftarrow$ sense_score + relatedness$_{lch}$ (Syn-input$_i$, a)

      return sense_score

Fig. 3. WSD algorithm

4   Data Set and Experiment

4.1   Data Set

We evaluate our algorithm on a test dataset consisting of 20 Hindi polysemous nouns (Table 1). Sense inventory is derived from Hindi WordNet. Some of the senses having fine-grained sense distinctions are merged. Some of the senses for which instances were not easily available have been deleted. Instances were collected from Hindi Corpus [1] created by Centre for Indian Language Technology (CFILT), IIT Bombay. Instances were also collected from www.khoj.com and www.google.com by firing queries derived from sense definitions. Instances are from varying domains including news, literature, medical, sports, etc. Our dataset has a total of 710 instances. The average number of instances per word is 35.5. The average number of instances per sense is 14.48. The average number of senses per word is 2.45. The detail of the dataset is given in Table A1 in Appendix. The translation and transliteration of the dataset is given in Table A2 in Appendix. The performance evaluation is done in terms of precision and recall. Precision is defined as the ratio of the correctly answered instances and the total number of test instances answered for the target word. Recall is defined as the ratio of the correctly answered instances and the total number of test instances to be answered for the target word.

Table 1. Dataset

No of Senses	Words (Nouns)
2	कोटा, गुरु, चंदा, जेठ, डाक, तान, तीर, दाम, माँग, विधि, सोना, हल, हार
3	उत्तर, कुंभ, फल, वचन, संक्रमण, संबंध
4	मूल

4.2   Experiment

For evaluating the proposed algorithm test run for each target word is conducted on fixed window size of two. Precision and recall for 20 ambiguous Hindi words are shown in Table 2.We also conducted a test run using direct overlap method. The direct overlap method used for comparison is adapted from WSD algorithm after stop word removal from sense definitions and test instances (Case III) as given in [11].The precision and recall values averaged over a window size of 5, 10, 15, 20, 25 for direct overlap method are shown in Table 2. Table 3 summarizes the results of comparison.

Table 2. Precision and recall

Word	Semantic relatedness (Leacock-Chodorow Measure)		Direct Overlap Measure	
	Precision	Recall	Precision	Recall
उत्तर	0.9030	0.8898	0.7888	0.7772
कुंभ	0.5701	0.5701	0.6490	0.6490
कोटा	0.6416	0.6416	0.1940	0.1940
गुरु	0.6782	0.6782	0.5881	0.5881
चंदा	0.5687	0.4409	0.8265	0.7276
जेठ	0.6666	0.6666	0.2428	0.2428
डाक	0.5000	0.5000	0.1336	0.1336
तान	0.4166	0.4166	0.3500	0.3000
तीर	0.7500	0.7500	0.5615	0.5615
दाम	0.7368	0.7368	0.4798	0.4798
फल	0.6526	0.6288	0.4985	0.4737
माँग	0.6363	0.6363	0.3999	0.3999
मूल	0.9019	0.9019	0.4359	0.4272
वचन	0.3809	0.3809	0.2559	0.2559
विधि	0.5909	0.4090	0.3811	0.3681
संक्रमण	0.1809	0.1809	0.3919	0.3919
संबंध	0.3888	0.3055	0.6499	0.6499
सोना	0.7500	0.4732	0.4916	0.3196
हल	0.6114	0.6114	0.5070	0.5070
हार	0.6052	0.6052	0.3263	0.3263

Table 3. Comparision of Precision and recall

Method	Overall average Precision (over 20 words)	Overall average Recall (over 20 words)
Semantic relatedness (Leacock-Chodorow Measure)	0.6065	0.5711
Direct Overlap Measure	0.4576	0.4386

5   Results and Discussion

As shown in Table 3, the average precision and recall using WSD algorithm based on Leacock-Chodorow measure over 20 words is 60.65% and 57.11%.We obtained highest precision of 0.9030 for word उत्तर. The lowest precision of 0.1809 was observed for the word संक्रमण. A comparison of the proposed algorithm with Lesk like direct overlap method is shown in Table 3. For Direct Overlap method results were taken on window size of 5, 10, 15, 20 and 25 after stop word removal for every target word and then average is computed. The average precision over 20 words using direct overlap method is 45.76% as shown in Table 3.We obtained higher precision for many words using Leacock-Chodorow measure than direct overlap measure. The direct overlap method showed better precision than Leacock-Chodorow measure for words कुंभ , चंदा, संक्रमण and संबंध. This is due the availability of better content words in dictionary definitions for which contextual match is possible in the test instances. The use of semantic similarity measure results in an improvement of 32.53%. This is mainly due to the use of WordNet hierarchy in identifying similarity between the test instance and the sense definitions.

6   Conclusions and Future Work

In this paper, we proposed and evaluated Hindi Word Sense Disambiguation using Leacock-Chodorow measure of semantic relatedness. The experimental results demonstrate that the use of semantics increases the accuracy of disambiguation. A comparison with the direct overlap method yields an improvement of 32.53%. One of the drawbacks of our algorithm is that it does not consider part of speech other than noun. If words have multiple syntactic categories then the algorithm utilizes only the semantic similarity arising out of noun category of matching words.

References

1. Hindi Corpus, http://www.cfilt.iitb.ac.in/Downloads.html
2. Hindi WordNet, http://www.cfilt.iitb.ac.in/wordnet/webhwn/wn.php
3. Budanitsky, A., Hirst, G.: Evaluating WordNet-based Measures of Lexical Semantic Relatedness. Computational Linguistics 32(1), 13–47 (2006)

4. Hirst, G., St-Onge, D.: Lexical chains as representations of context for the detection and correction of malapropisms. In: Fellbaum, C. (ed.) WordNet: An Electronic Lexical Database, pp. 305–332 (1998)

5. Jiang, J., Conrath, D.: Semantic similarity based on corpus statistics and lexical taxonomy. In: Proceedings on International Conference on Research in Computational Linguistics, Taiwan, (1997)

6. Khapra, M., Bhattacharyya, P., Chauhan, S., Nair, S., Sharma, A.: Domain Specific Iterative Word Sense Disambiguation in a Multilingual Set-ting. In: Proceedings of International Conference on NLP (ICON 2008), Pune, India (2008)

7. Leacock, C., Chodorow, M.: Combining local context and WordNet similarity for word sense identification. In: Fellbaum, C. (ed.) WordNet: An Electronic Lexical Database, pp. 265–283 (1998)

8. Lin, D.: Using syntactic dependency as a local context to resolve word sense ambiguity. In: Proceedings of the 35th Annual Meeting of the Association for Computational Linguistics, pp. 64–71 (1997)

9. Reddy, S., Inumella, A., Singh, N., Sangal, R.: Hindi Semantic Category Labeling using Semantic Relatedness Measures. In: Proceedings of Global WordNet Conference (2010)

10. Resnik, P.: Using information content to evaluate semantic similarity in taxonomy. In: Proceedings of the 14th International Joint Conference on Artificial Intelligence, Montreal, pp. 448–453 (1995)

11. Singh, S., Siddiqui, T.J.: Evaluating Effect of Context Window Size, Stemming and Stop Word Removal on Hindi Word Sense Disambiguation. In: Proceedings of the International Conference on Information Retrieval & Knowledge Management, CAMP 2012, Malaysia, pp. 1–5 (2012)

12. Sinha, M., Kumar, M., Pande, P., Kashyap, L., Bhattacharyya, P.: Hindi Word Sense Disambiguation. In: International Symposium on Machine Translation, Natural Language Processing and Translation Support Systems, Delhi, India (2004)

13. Sinha, R., Mihalcea, R.: Unsupervised Graph-based Word Sense Disambiguation Using Measures of Word Semantic Similarity. In: Proceedings of the International Conference on Semantic Computing, ICSC 2007, pp. 363–369 (2007)

14. Torres, S., Gelbukh, A.: Comparing Similarity Measures for Original WSD Lesk Algorithm Advances in Computer Science and Applications. Research in Computing Science 43, 155–166 (2009)

Appendix

Table. A1. Statistics of Dataset

Word	No. of Senses	Instances			
		Sense1	Sense 2	Sense 3	Sense 4
उत्तर	3	10	23	12	
कुंभ	3	22	21	19	
कोटा	2	20	21		
गुरु	2	31	19		
चंदा	2	25	22		
जेठ	2	3	7		
डाक	2	19	20		
तान	2	6	7		
तीर	2	30	13		
दाम	2	19	6		
फल	3	30	25	5	
माँग	3	4	11		
मूल	4	3	17	12	12
वचन	3	7	9	8	
विधि	2	22	22		
संक्रमण	3	10	19	7	
संबंध	3	8	12	3	
सोना	2	21	8		
हल	2	8	23		
हार	2	10	19		

Table. A2. Dataset – Translation& Transliteration

Word	No of Senses	Sense1	Sense2	Sense3	Sense4
उत्तर (uttar)	3	Answer	North direction	A person's name	
कुंभ (kumbh)	3	Waterpot made of mud	A Sun Sign (Aquarius) in Hindi	A Holy event happing every 12 years in India	
कोटा (kota)	2	Reservation, quota	Name of a district in Rajasthan in India		
गुरु (guru)	2	teacher	Jupiter (name of a planet)		
चंदा (chanda)	2	moon	Financial contribution, subscription		
जेठ (jeth)	2	Name of a month in Hindi	Husband's elder brother, brother in law		
डाक (dak)	2	Bid, bidding	Post, postal service		
तान (taan)	2	Process of stretching	Music tone		
तीर (teer)	2	arrow	shore of river or sea		
दाम (dam)	2	Cost, price	Type of strategy or policy		
फल (phal)	3	fruit	result	Front sharp part of arrow or spear	
माँग (maang)	2	Requirement, need,	Parting of hairs on head where married Hindu woman put vermilion as a sign of marriage		
मूल (mul)	4	root of plant	Basic reason, fundamental	Time for a type of star	Capital/Principal money
वचन (vachan)	3	whatever one speaks or says, saying	Promise, commitment	Agent in Hindi grammar to denote singular or plural	
विधि (vidhi)	2	Way or process of doing something	law		
संक्रमण (sankraman)	3	Process of sun's transition from one star-sign to another	Process of disease infection	Process of transition from one place or state to another place or state	
संबंध (sambandh)	3	relation	Agent in Hindi grammar that shows relation between two words	marriage	
सोना (sona)	2	gold	sleep		
हल (hal)	2	solution	Instrument used to plough		
हार (har)	2	defeat	necklace, garland		

A Content-Based eResource Recommender System to Augment eBook-Based Learning*

Vivek Kumar Singh[1], Rajesh Piryani[1], Ashraf Uddin[1], and David Pinto[2]

[1] Department of Computer Science, South Asian University, New Delhi, India
vivek@cs.sau.ac.in, {rajesh.piryani,mdaakib18}@gmail.com
[2] Faculty of Computer Science,
Benemerita Universidad Autonoma de Puebla, Mexico
dpinto@cs.buap.mx

Abstract. This paper presents our experimental work to design a content-based recommendation system for eBook readers. The system automatically identifies a set of relevant eResources for a reader, reading a particular eBook, and presents them to the user through an integrated interface. The system involves two different phases. In the first phase, we parse the textual content of the eBook currently read by the user to identify learning concepts being pursued. This requires analysing the text of relevant part(s) of the eBook to extract concepts and subsequently filter them to identify learning concepts of interest to Computer Science domain. In the second phase, we identify a set of relevant eResources from the World Wide Web. This involves invoking publicly available APIs from Slideshare, LinkedIn, YouTube etc. to retrieve relevant eResources for the learning concepts identified in the first part. The system is evaluated through a multi-faceted process involving tasks like sentiment analysis of user reviews of the retrieved set of eResources for recommendations. We strive to obtain an additional wisdom-of-crowd kind of evaluation of our system by hosting it on a public Web platform.

Keywords: Concept Extraction, Learning Concepts, RDF, Recommender System, Semantic Annotation.

1 Introduction

With newer forms of digital storage devices, large screen readers and fast Internet access; eBooks are becoming a popular alternative to traditional printed books. The ease of creation and dissemination of content over electronic media and the World Wide Web (Web) is making eBooks more and more popular. The impact can also be understood from the fact that now publishers are transforming their popular books into eBooks and promoting their distribution. An important observation pertinent to our work, however, is that the eBooks are delivered and used in an electronic form over a eBook reader (like Kindle) or a

---

* This work is partly supported by an Indo-Mexican project funded jointly by DST, India and CONACYT, Mexico.

S. Ramanna et al. (Eds.): MIWAI 2013, LNCS 8271, pp. 257–268, 2013.

personal computer. This mode of usage of eBooks allow us to improve and enrich the learning environment of the readers. Imagine that while reading some section of an eBook (which describes some learning concept, say Topic Modeling), you are provided with some video tutorials and slides about the learning concept. Or perhaps a set of articles and names (and social profiles) of experts in that area. You can now pursue the additional learning resources for a better understanding and can also contact related experts through social media systems, in case of clarifications or doubts. This would no doubt improve both, the learning environment as well as learning outcome.

In this paper, we describe our experimental work on design of such a recommendation system, which can effectively improve the learning outcome by augmenting the learning environment with additional set of knowledge resources for the particular set of concepts described in that part of eBook. We call our system a content-based recommender system because it uses a text analytics approach to automatically identify learning concepts being pursued by a user, reading an eBook. The system designed has two identifiable components, which work in phases. In the first phase, a text analytics based formulation parses the text of the eBook section being read by the reader and extracts concepts. The extracted concepts are filtered and ranked, which then feeds as input to the second phase. In the second phase, different APIs are called and a set of eResources are identified and a ranked list is presented to the user. The recommended eResources include videos, slides, documents, web articles, twitter and linkedin ids of professionals working in the area. The figure 1 presents a block diagram of the proposed system, illustrating the system architecture, its components and an overview of its functioning.

The rest of the paper is organized as follows. Section 2 describes the process of parsing the eBook content, extraction of concepts from different sections, ranking the concepts by importance, and populating the RDF schema designed for eBooks. Section 3 explains the recommendation generation process, including identifying eResources from the Web and selection of relevant ones for presenting to the reader. Section 4 presents the dataset and experimental results. Section 5 presents evaluation of the the proposed system. The paper concludes with a short summary of usefulness of this work and future extension. A recent work which uses a similar concept extraction process used by us, though for the purpose of qualitative evaluation of eBooks, can be found in [1], [2]. We have used qualitative scores proposed in this work for ranking related eBook recommendation results.

2   Concept Extraction from eBook

The first phase of our recommendation system extracts learning concepts described a particular part of the eBook. This requires a number of tasks ranging from POS tagging to concept filtering. First of all we parse the textual contents of an eBook part and then use knowledge from linguistics to identify patterns that can represent concepts. The concepts so identified are subjected to a

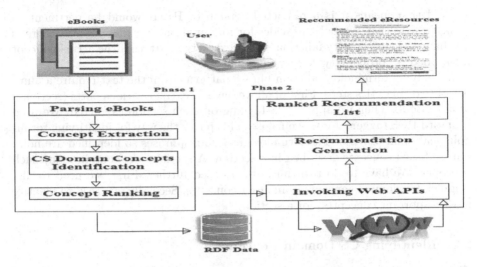

Fig. 1. Architectural Block Diagram of the System

filtering process for identifying Computer Science (CS) domain concepts. The CS domain concepts present in a section are then ranked in order of importance for use by the recommendation generation phase. For concept extraction, we had to first do multitude of text extractions from the eBook, that included extracting Table of Contents, Chapter and section texts. This was followed by POS tagging and terminological noun phrase identification.

2.1 Concept Extraction

We extracted concepts using the terminological noun phrase identification, a set of three kind of patterns known to represent important noun-phrase based concepts, based on the idea proposed in [3],[4].

$$P1 = C^*N \tag{1}$$

$$P2 = (C^*NP)?(C^*N) \tag{2}$$

$$P3 = A^*N^+ \tag{3}$$

where, N refers to a noun, P a preposition, A an adjective, and C = A or N. The pattern P1 represents a sequence of zero or more adjectives or nouns which ends with a noun. The pattern P2 is a relaxation of P1 that allows two such patterns separated by a preposition. Examples of the pattern P1 may include "probability density function", "fiscal policy", and "thermal energy". Examples of the pattern P2 may include "radiation of energy" and "Kingdom of Ashoka". The pattern P3 corresponds to a sequence of zero or more adjectives, followed by one or more nouns. In P3, an adjective occurring between two nouns is not allowed that means it is a restricted version of P1. It would be pertinent to mention here that symbol * provides for maximal pattern matches i.e., there is no chance to get "density function" as an extracted pattern if the actual concept mentioned is "probability density function".

Identifying terminological noun phrase patterns from the text require a number of text analytics steps. First of all we have to extract various parts (sections) of the eBook. Then we apply POS tagging on each section extracted. We used Stanford POS tagger [1] for this purpose. This paves the way for identifying terminological noun phrases. The terminological noun phrases so identified are noun phrase based concepts described in a section. A section may contain many such concepts. We have to do two things to proceed further. First, we need to distinguish CS domain concepts from other concepts. Secondly, we need to identify most important concept(s) for a section.

2.2 Identifying CS Domain Concepts

The terminological noun phrases extracted represent generic noun-phrase based concepts. Not all of them represent concepts belonging to CS domain. In order to identify relevant eResources to recommend, we need to know precisely what CS domain concepts are described in an eBook section. We have therefore tried to filter out the concepts not in the CS domain. For this, we have used a filtering list containing key concepts in CS domain. We understand that this list could not be an exhaustive list of CS domain concepts. This may result in loosing some CS domain concepts, however, the list is appropriate enough to identify key concepts in different subjects of study in CS domain. We have used ACM Computing Curricular Framework document (ACM CCF)[2] as our base CS domain concept list. We have augmented this list by incorporating in it terms from IEEE Computer Society Taxonomy[3] and ACM Computing Classification System[4]. The augmenting process involved merging the two later documents into the first one, while preserving the 14 categories it is divided into. The combined list is thus a set of 14 different sets of CS domain knowledge areas, each knowledge area containing key concepts (the important ones) worth learning in that area. We use this list of concepts as our filtering list.

Every concept identified through the terminological noun phrase identification process, is subject to this filtering. However, we can not do an exact term matching. For example, two terms "algorithm complexity" and "complexity of algorithm" will not be a match, if we go for exact matching scheme. Therefore, we have used Jackard similarity measure, which allows two concept phrases to result in a match even when the word order in the two are different, or there

---

[1] http://nlp.stanford.edu/software/tagger.shtml
[2] http://ai.stanford.edu/users/sahami/CS2013/
   ironman-draft/cs2013-ironman-v1.0.pdf
[3] http://www.computer.org/portal/web/publications/acmtaxonomy
[4] http://www.acm.org/about/class/2012

is an impartial match. The Jackard similarity equation is given in the equation below:

$$Similarity\,(A, B) = \frac{|A \cap B|}{|A \cup B|} \qquad (4)$$

where A and B stand for the two phrases. Here, $A \cap B$ is the set of common words in both phrases, $A \cup B$ is the set of union of words in both phrases and $|S|$ stands for the number of elements in the set S. We have to set a threshold value for deciding whether phrase A and B constitute a match. We empirically found a threshold between 0.5 and 0.6, works best for identifying CS domain concepts. A simple example could help in understanding the suitability of this threshold. Consider, a phrase A= "methods of numerical analysis" is an identified terminological noun phrase and a phrase B="numerical analysis methods" is a concept in the CS domain concept list. In this case we get the similarity score = 0.75, greater than threshold and confirming that A is a valid CS domain concept. Thus, we use the reference list and similarity scores for deciding about every terminological noun phrase extracted from an eBook for being a valid CS domain concept.

2.3 Ranking Concepts by Importance

Our implementation tells us that a typical section in an eBook may have occurrences of several valid CS domain concepts. Since, we have to recommend eResources for eBook reader pursuing a particular set of concepts, we need to select only the most important concepts as the input for generating recommendations. This means that if an eBook section results in 10 valid CS domain concepts, we can simply not generate recommendation of eResources for all the 10 concepts, since it would make the recommendations ineffective. We have to, therefore, restrict the list of concepts to be used as input for the process of recommendation generation. This is equivalent to try identifying most important concept(s) in a section. An ideal position will be if we have a scheme to figure out semantically the learning concept(s), a section is about. But, in the absence of such a scheme to identify semantic tags about concepts described in a section, the only option is to use statistical evidence about the concept importance in a section. We have used statistical measures of term occurrence in the concerned section and the entire eBook to rank the concept(s) in order of importance. The rank score (section-rank)of a concept C_i belonging to a particular section S_j is computed as follows:

$$RankScore\,(C_i, S_j) = Freq\,(C_i, S_j) + log\left(\frac{NOC}{GRank\,(C_i)}\right) \qquad (5)$$

where, Freq() gives the number of occurrences of a particular concept in a given section, NOC refers to the total number of CS domain concepts extracted from the eBook and GRank is the rank of a concept in the entire eBook (with highest occurring concept getting the rank 1). Thus, we have two ranks for each concept C, a section-rank and a global-rank. The equation makes it clear that we compute section-rank of a concept by combining its occurrence measures in the section and the entire eBook. If the concept C_i refers to the highest ranking concept (rank 1), the Freq(C_i,S_j) value is incremented substantially by addition of log normalized measure of its importance in the entire eBook. On the other hand, if the concept C_i refers to the a concept with lowest global rank (rank=no. of concepts), its log normalized measure value becomes zero (since rank is equal to the number of concepts in eBook) and the section-rank of this concept is only a measure of its occurrence in the concerned section. In this manner, we are able to compute importance of a concept in a given section (measured as section-rank). This is in a sense equivalent to attempting to find the key section (most important) for a concept [5].

2.4 Populating the RDF

The JAVA program designed to extract concepts, their ranks etc. produces a lot of other useful information from eBooks. We have designed an RDF (Resource Description Framework) schema to store the information produced for each eBook. All this information is generated and written automatically (through our program) in the RDF schema. The RDF schema contains rdfs:resources for the eBook metadata, concepts in a section and chapter, concept relations and eBook reviews obtained by crawling the Web. The eBook metadata comprises of eBook title, author, number of chapters, number of pages, eBook price, eBook rating, its main and two related categories as determined from augmented ACM CCF, coverage score, readability score and consolidated sentiment score profile. For each chapter node in the RDF, the entry consists of section and chapter titles, top concepts with ranks, and relations extracted for the chapter. The populated RDF structure contains a lot of other information for eBooks. We have used only some of this information for our recommendation generator component. The other information can be used for a number of purposes like querying about relevant information for the eBook, designing a concept locator in the eBook or designing a semantic annotation environment.

A sample example of RDF representation of eBook metadata is as follows:

```
<rdf:RDF
xmlns:rdf=http://www.w3.org/1999/02/22-rdf-syntax-ns#
xmlns:book="http://www.textanalytics.in/ebooks/
  Data_Mining_Concepts_and_Techniques_Third_Edition#" >
<rdf:Description rdf:about="http://www.textanalytics.in/ebooks/
  Data_Mining_Concepts_and_Techniques_Third_Edition#metadata">
<book:btitle>Data Mining Concepts and Techniques Third Edition
  </book:btitle>
<book:author>JiaweiHan,MichelineKamber,Jian Pei</book:author>
<book:no_of_chapters>13</book:no_of_chapters>
<book:no_of_pages>740</book:no_of_pages>
<book:bconcepts>rule based classication, resolution, support vector
  machines,machine learning,...</book:bconcepts>
<book:main_category>Intelligent Systems</book:main_category>
```

```
<book:main_cat_coverage_score>0.051107325</book:main_cat_coverage_score>
<book:related_category>Programming fundamen-tals</book:related_category>
<book:related_category>Information Management</book:related_category>
<book:googleRating>User Rating: **** (3 rating(s))</book:googleRating>
<book:readability_score>56 (Fairly Difficult)
</book:readability_score>
</rdf:Description>
```

In this representation, the category and related category refers to the two closest of the 14 classes defined in ACM CCF. Similarly, other important information include readability score, author(s), number of pages etc. The figure 2 shows the RDF Graph for a part of the eBook metadata.

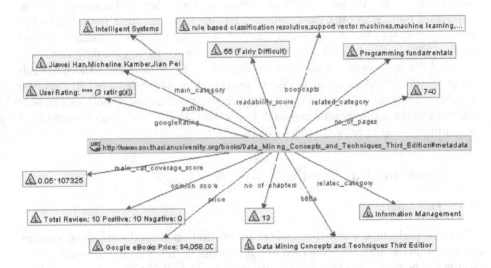

Fig. 2. RDF Graph For Book Metadata

The second key part of the information represented include information about concepts and their relations in the Chapter node of the RDF schema. A detailed discussion of the RDF schema and relation networks is available in [6].

3   Generating eResource Recommendations

After identifying important learning concepts presented in a section of eBook, we move to second phase of the system, which is to generate recommendations for relevant eResources for the learning concepts being pursued. While a section is being pursued by a reader, we have the key concepts in that section identified and ranked. The top three concepts then form input for the recommendation generation process. The design of the second part is fairly simple. First of all, we explored about what useful eResources may be readily available.

Thereafter, we wrote a JAVA code to invoke search APIs available for this purpose and integrate the results obtained. Our system returns a number of eResources, slides from Slideshare[5], web articles from Google Web Search[6], videos from YouTube[7], microblog posts in the area from Twitter[8], details of professionals working in the area from LinkedIn[9] and related documents from DocStoc[10].

The main objective of designing the recommender system for us was to identify and recommend additional set of eResources for eBook readers. While a reader is reading a particular section of an eBook, we want to provide him with additional learning resources as well as the set of professionals working in that area. While the first is aimed at improving the learning quality and pace; second is to provide an opportunity to the reader to connect to related professionals in the area. For learning concepts pursued by a reader, we generate a set of eResource recommendations. We have designed a web-based interface for this purpose. One important issue is to rank the recommendations based on their relevance to the learning concepts being pursued by the reader. The inherent ranking provided by the APIs invoked is one way to associate relevance to the learning concepts. These APIs use a sophisticated set of algorithms to retrieve only the most relevant results for a search query. We have, therefore, not attempted to rank the retrieved eResources afresh, except while recommending related eBooks (where we do rank the recommendations list). Our system design is thus a content-based recommendation system approach [7], [8].

4   Experimental Results

4.1   Dataset

We have performed our experimental evaluation on a moderate sized dataset collected on our own. We collected about 30 eBooks in CS domain from different sources. The text corresponding to various parts of a PDF eBook is extracted using the iText API[11] and programmatically reading the bookmarks. The different parts of an eBook are then parsed at a sentence level, starting with POS tagging and culminating in identification of learning concepts (denoted by terminological noun phrases).

4.2   Results

In the following paragraphs we present snapshot of some results produced at various stages of processing by our system. The snapshot of results shown correspond

---

[5] http://www.slideshare.net/about
[6] http://www.google.com
[7] http://www.youtube.com
[8] http://www.twitter.com
[9] http://www.linkedin.com
[10] http://www.docstoc.com/about/
[11] http://www.api.itextpdf.com

to a popular eBook on "Data Mining" that describes concepts and techniques of data mining and is a recommended eBook for graduate and research students. During phase 1 of system operation, we extract all probable learning concepts (measured as terminological noun phrases) from a section of the eBook. Then these concepts are filtered using the augmented ACM CCF reference document. For example, from the first chapter of the eBook having title "Introduction", we obtained 1443 concepts before filtering, out of which 96 concepts refer explicitly to the CS domain. Some example CS domain concepts from beginning portion of this chapter are:

```
business intelligence, knowledge management, entity relationship
models,information technology, database management system
```

After obtaining the filtered list of CS domain concepts in a section of the eBook, we rank them in order of importance. This required that both local (concept occurrence frequencies in the section) and global knowledge (concept ranking for the entire eBook) are available. Thus, we parse the entire dataset of eBooks, identify learning concepts in them and rank them in order of importance (assuming whole eBook as unit), beforehand. The concept occurrence frequencies in the currently accessed section are computed at the time of their actual use by the eBook reader. As stated earlier, all the information extracted is also written in an RDF schemea for future retrieval.

The second phase involves generation of recommendations for eresources relevant to the most significant learning concepts being pursued by the reader. Our recommendations list contain eResources of various kinds. The recommendation list generated by us include videos from YouTube, slides form Slideshare, documents from DocStoc, Web articles from Google Web search, profile ids of professionals working in the area from LinkedIn and some others. We present below a sample results for a concept "Data mining" from the first chapter of the eBook used as an example demonstration.

An example of recommended videos from YouTube for the concept are as follows:

```
Result for Concept: Data Mining
1. Thumbnail: http://i.ytimg.com/vi/UzxY1bK2c7E/hqdefault.jpg
URL: http://www.youtube.com/watch?v=UzxY1bK2c7E
2. Thumbnail: http://i.ytimg.com/vi/EUzsy3W4I0g/hqdefault.jpg
URL: http://www.youtube.com/watch?v=EUzsy3W4I0g
```

An example snapshot of recommended slides from SlideShare for the concept are as follows:

```
Result for Concept: Data Mining
1. Title:The Secrets of Building Realtime Big Data Systems
URL:http://www.slideshare.net/nathanmarz/
    the-secrets-of-building-realtime-big-data-systems
2. Title:Big Data with Not Only SQL
URL:http://www.slideshare.net/PhilippeJulio/big-data-architecture
```

A sample of recommended documents from DocStoc for the concept are as follows:

```
Result for Concept: Data Mining
1. Title: Data Mining
URL: http://www.docstoc.com/docs/10961467/Data-Mining
2. Title: Data Mining Introduction
URL: http://www.docstoc.com/docs/10719897/Data-Mining-Introduction
```

A snapshot of a part of recommended LinkedIn profiles for the concepts are as follows:

```
Result for Concept: Data Mining
1. Name: Peter Norvig
URL: http://www.linkedin.com/in/pnorvig?trk=skills
2. Name: Daphne Koller
URL: http://www.linkedin.com/pub/daphne-koller/20/3a8/405?trk=skills
```

It would be important to mention here that the results displayed are a very small part of the actual results obtained. More results can be seen at our text analytics portal[12]. Through a similar process of API invocation, we have also generated recommendations for top web links from Google Web Search and top profiles of persons writing on the topic on microblogging site Twitter. We have thus generated recommendations for a comprehensive set of eResources (in addition to identifying the most relevant eBook and its chapter) for a concept being pursued by a learner.

For a given important concept in a section, we also recommend related eBooks (ranked in order of their relevance). The recommended list of related eBooks are at present generated from our dataset collection itself. However, it is not a limitation and we can generate a list of related eBooks (related on the important learning concepts under consideration) from the Web. The list of related eBooks is ranked based on a computed sentiment score of their reviews obtained from Google book reviews and from Amazon. It was necessary to rank eBooks since the recommendation list of eBooks is not generated by an API having inherent ranking scheme, but by a concept-bases matching calculation. We want that the most popular eBooks (measured through wisdom-of-crowds) should be ranked at top and recommended. For this, we have collected user reviews of all the eBooks in the dataset by a selective crawling of Google Book review and Amazon sites. The textual reviews obtained for each eBook are then labeled as 'positive' or 'negative' through a sentiment analysis program designed by us [9], [10]. Thus for each candidate eBook, we compute sentiment labels and strengths of its reviews (between 10-50 reviews), normalize the strength score (by dividing with number of 'positive' or 'negative' reviews) and use it to rank the eBooks in order of their popularity. The figure 3 shows an example recommendation for the related eBooks recommended for concept "Data Mining".

---

[12] http://www.textanalytics.in

Fig. 3. Recommended eBooks for Concept: Data Mining

5   Conclusion and Future Work

We have presented our experimental work on design of concept-based eResource recommendation system. The system takes as input an eBook being currently read by a user and provides him with additional learning resources for the learning concepts being pursued by him. The system uses a text analytics approach and works in two phases. In first phase, it identifies the main learning concepts that a user is trying to understand. In the second phase, it generates a set of eResource recommendations that are relevant to the learning concept and provide the user with additional learning material on the concept in concern. The recommender system design proposed and demonstrated by us, appears to be useful for learners.

Evaluation of recommendations is a key parameter of study for recommendation system design. Here, we have used Web APIs for collecting and recommending eResources. These APIs are inherently known to retrieve most relevant results for an information need. There is no such previous system or benchmark against which we can evaluate our system. While the first phase of the system is tested to work appropriately, the results of second phase need some more evaluations for relevance. Our preliminary observation shows that the retrieved and recommended eResources for a learning concept are the most relevant and authoritative ones. We are, however, working towards a wisdom-of-crowd kind of evaluation of the relevance of the recommendation results. Since it is largely a manual effort, it will take some more time to collect user feedbacks from the system hosted on a in-house web portal and being used by volunteers.

There are some possible improvements and extensions of the current work. One of them is to work on a large dataset and explore our system's applicability on open source eBooks from the Web not only for CS but other domains as well.

Secondly, we are still working on an appropriate evaluation scheme for ascertaining the quality of recommendations generated. Though, wisdom-of-crowds seem the most natural way, other ways of evaluation may be explored. Thirdly, we wish to extend the system to a full-blown web-based learning resource recommendation system, which can automatically identify users information needs. Fourth, behavioural and user-based modeling studies may be carried out to evaluate usefulness of the system and to deduce lessons for information need modeling of IR systems. And lastly, the linguistics-based formulations for concept identification, refinement still have possibility of improvement.

References

[1] Relan, M., Khurana, S., Singh, V.K.: Qualitative Evaluation and Improvement Suggestions for eBooks using Text Analytics Algorithms. In: Klinov, P., et al. (eds.) ICECCS 2013. CCIS, vol. 389, pp. 324–335. Springer, Heidelberg (2013)
[2] Khurana, S., Relan, M., Singh, V.K.: A Text Analytics-based Approach to Compute Coverage, Readability and Comprehensibility of eBooks. In: Proceedings of the 6th International Conference on Contemporary Computing, Noida, India. IEEE Xplore (August 2013)
[3] Agrawal, R., Gollapudi, S., Kannan, A., Kenthapadi, K.: Data Mining for Improving Textbooks. ACM SIGKDD Explorations Newsletter 13(2), 7–19 (2011)
[4] Justeson, J.S., Katz, S.M.: Technical Terminology: Some Linguistic Properties of an Algorithm for Identification in Text. Natural Language Engineering 1(1) (1995)
[5] Agrawal, R., Gollapudi, S., Kenthapadi, K., Srivastava, N., Velu, R.: Enriching textbooks through data mining. In: ACM DEV (2010)
[6] Uddin, A., Piryani, R., Singh, V.K.: Information and Relation Extraction for Semantic Annotation of eBook Texts. In: Thampi, S.M., Abraham, A., Pal, S.K., Rodriguez, J.M.C., et al. (eds.) Recent Advances in Intelligent Informatics. AISC, vol. 235, pp. 215–226. Springer, Heidelberg (2014)
[7] Adomavicius, G., Tuzhilin, A.: Toward the next generation of Recommender Systems: A Survey of the state-of-the-art and possible extensions. IEEE Transactions on Knowledge and Data Engineering 17(6) (2005)
[8] Singh, V.K., Mukherjee, M., Mehta, G.K.: Combining a Content Filtering Heuristic and Sentiment Analysis for Movie Recommendations. In: Venugopal, K.R., Patnaik, L.M. (eds.) ICIP 2011. CCIS, vol. 157, pp. 659–664. Springer, Heidelberg (2011)
[9] Singh, V.K., Piryani, R., Uddin, A.: Sentiment Analysis of Movie Reviews and Blog Posts: Evaluating SentiWordNet with different Linguistic Features and Scoring Schemes. In: Proceedings of IEEE 3rd International Advanced Computing Conference, Ghaziabad, India, pp. 893–898. IEEE Xplore (2013)
[10] Singh, V.K., Piryani, R., Uddin, A., Waila, P.: Sentiment Analysis of Movie Reviews: A new Feature-based Heuristic for Aspect-level Sentiment Classification. In: Proceedings of International Multi Conference on Automation, Computing, Control, Communication and Compressed Sensing, Kerala, India, pp. 712–717. IEEE Xplore (2013)

Markov Decision Processes
with Functional Rewards

Olivier Spanjaard and Paul Weng*

LIP6, UPMC
firstname.lastname@lip6.fr

Abstract. Markov decision processes (MDP) have become one of the standard models for decision-theoretic planning problems under uncertainty. In its standard form, rewards are assumed to be numerical additive scalars. In this paper, we propose a generalization of this model allowing rewards to be functional. The value of a history is recursively computed by composing the reward functions. We show that several variants of MDPs presented in the literature can be instantiated in this setting. We then identify sufficient conditions on these reward functions for dynamic programming to be valid. In order to show the potential of our framework, we conclude the paper by presenting several illustrative examples.

1 Introduction

In sequential decision-making under uncertainty, Markov Decision Processes (MDPs) have received much attention as a natural framework both for modeling and solving complex structured decision problems [1–5]. In standard MDPs, scalar rewards – assumed to be additive – are granted along the process, and a policy is evaluated according to the probabilistic expectation of the cumulated rewards. Yet, it often happens that some of those assumptions have to be relaxed, as in the following cases already highlighted in the literature: rewards are not static, in the sense that they do not only depend on the current state and the action carried out: this is the case for instance in MDPs where rewards represent durations, and these durations vary over time (think about public transportation: the travel time depends on which bus you catch); rewards are not known with certainty: the determination of precise rewards remains a bottleneck in the specification of an MDP, and it is often advantageous to use imprecise or stochastic rewards; the evaluation of a policy is not performed via probabilistic expectation: this is the case for instance in high-stake planning situations, where one wishes to take the level of risk into consideration by using expected utility instead of probabilistic expectation. Each of these observations have encouraged researchers to provide different extensions of MDPs, together with dedicated solution procedures: Time-dependent MDPs [3], MDPs with imprecise rewards [5], MDPs with one-switch utility functions [4].

---

* Funded by the French National Research Agency under grant ANR-09-BLAN-0361.

S. Ramanna et al. (Eds.): MIWAI 2013, LNCS 8271, pp. 269–280, 2013.

In this article, we propose a general framework called *MDP with functional rewards* (abbreviated by FRMDP in the sequel) to handle most features of such variants, and we show how to recast them in our setting. Our approach is based on the use of functional rewards instead of scalar ones. This can be seen as a generalization of the work of Serafini [6], who searched for preferred paths in graphs with functional costs, and of the work of Liu and Koenig [4] who proposed functional value iteration in MDPs with *scalar* rewards. In FRMDPs, the value of a policy in a state can be defined in two ways: a *recursive definition* based on a functional version of the Bellman equations [7, 8] and an *explicit definition* based on the expectation of cumulated rewards. We first exhibit general conditions for which the backward induction procedure returns an optimal policy according to the recursive definition. We however emphasize that in some cases the optimal policy according to this recursive definition can differ from that according to the explicit definition. We then exhibit the condition under which both notions of optimality coincide. For the sake of simplicity, we only consider the finite horizon case in this paper, although the results could be extended in infinite horizon MDPs with goal states for instance.

2 Preliminaries and Motivating Example

A *Markov Decision Process* (MDP) can be defined as a collection of objects $\{S, A, p^a_{ss'}, r^a_s, g_s, h\}$ where S is a finite set of states, A is a finite set of actions, $p^a_{ss'}$ is the probability of reaching state s' after taking action a in state s, r^a_s is the immediate reward gained after taking action a in state s, g_s is the final reward received when the process stops at state s and $h \in \mathbb{N}$ is a finite horizon.

A *history* starting from state $s_h \in S$ corresponds to a sequence: $(s_h, a_h, s_{h-1}, a_{h-1}, \ldots, a_1, s_0)$ where $\forall i = 1, \ldots, h, (a_i, s_{i-1}) \in A \times S$. A *decision rule* $\delta : S \to A$ is a mapping from states to actions. By abuse of notation, transition probability $p^{\delta(s)}_{ss'}$ is denoted by $p^\delta_{ss'}$ and reward $r^{\delta(s)}_s$ by r^δ_s. The set of all decision rules is denoted by Δ. A *policy* π is a sequence of decision rules: $\pi = (\delta_h, \delta_{h-1}, \ldots, \delta_1)$ where $\forall t = 1, \ldots, h, \delta_t \in \Delta$ is the decision rule applied at the tth-to-last step.

The *value function* $v^\pi(s)$ of a policy $\pi = (\delta_h, \delta_h, \ldots, \delta_1)$ in a state s, defined as the expected cumulated reward obtained by executing policy π from state s, can be computed iteratively by using the *Bellman equations*:

$$v^\pi_0(s) = 0 \tag{1}$$
$$v^\pi_t(s) = r^{\delta_t}_s + \sum_{s' \in S} p^{\delta_t}_{ss'} v^\pi_{t-1}(s') \quad \forall t \leq h$$

A policy π is *preferred* to a policy π' if and only if $\forall s \in S, v^\pi(s) \geq v^{\pi'}(s)$. An *optimal* policy is a policy that is preferred to any other policy. The value function v^*_t of an optimal policy satisfies the *Bellman optimality equations*:

$$v^*_0(s) = 0 \tag{2}$$
$$v^*_t(s) = \max_{a \in A} r^a_s + \sum_{s' \in S} p^a_{ss'} v^*_{t-1}(s') \quad \forall t \leq h$$

```
1: ∀s ∈ S, v₀*(s) ← 0; t ← 0
2: repeat
3:     t ← t + 1
4:     for all s ∈ S do
5:         for all a ∈ A do
6:             qₜ(s, a) ← rₛᵃ + Σₛ'∈S pₛₛ'ᵃ vₜ₋₁*(s')
7:         end for
8:         vₜ*(s) ← maxₐ∈A qₜ(s, a)
9:     end for
10: until t = h
```

Fig. 1. Backward Induction

This problem can be solved using the *backward induction* algorithm (Fig. 1).

A Coffee Robot Example. As a motivating example for our new framework, we present an adaptation of a shortest path problem in a deterministic graph, proposed by Serafini [6], to the coffee robot environment [9]. Consider a mobile robot whose task is to bring a coffee to a sleepy researcher. For simplicity, assume that the laboratory map can be seen as a 3x3 grid. Each cell of this grid corresponds to an area of the laboratory. The coffee machine is located at cell $(1, 1)$ and the researcher waits for her coffee at cell $(3, 3)$, where coordinates (n_r, n_c) stand for the row and the column of a cell. As the robot can bump into walls, furniture or people, it could spill the cup of coffee. Spilling the coffee has not the same cost according to the place where it happens: for instance, it is more damaging in offices (where there are carpets) than in corridors (where the floor is covered by lino). In Figure 2, the left (resp. middle) grid corresponds to the cost (resp. probability) of spilling the coffee according to the location of the robot. Four actions are available for the robot, which simply consists in deciding in which direction to go: (N)orth, (S)outh, (E)ast or (W)est. Due to faulty connections in its circuits, the robot has a tendency to drift to the left of the expected direction (once in ten): for instance, instruction "East" at cell $(2, 2)$ sometimes leads to cell $(1, 3)$ instead of $(2, 3)$. The robot stops as soon as the coffee is delivered or spilled. The aim is to determine a policy minimizing the expected cost of spilling the coffee.

The formalism of MDPs seems well-suited for this problem of sequential decision making under uncertainty. It can be formalized as follows: $S = \{(n_r, n_c) : n_r \in \{1, 2, 3\}, n_c \in \{1, 2, 3\}\}$; $A = \{N, S, E, W\}$ (the possible directions); $p_{ss'}^a = 0.1$ if $s' = failure^a(s)$ or 0.9 if $s' = success^a(s)$ where state s equal (n_r, n_c) and state $success^a(s)$ (resp. $failure^a(s)$) is the state reached if action a succeeds (resp. fails, drift to the left of the expected direction) in state s; when action a brings the robot to the wall, we assume that $success^a(s) = failure^a(s) = s$, meaning that the robot does not move; $g_s = 0, \forall s \in S$;

It now remains to define the rewards from the matrices of Figure 2. We assume that cost r_s (resp. probability P_s) of spilling the coffee when performing an action in state s is equal to the value indicated at cell s in the left (resp. middle) matrix. Note that in the general case, those costs and probabilities may depend on the action and the destination state. The subtlety here is that a cost is incurred only

-1	-2	-1
-3	-3	-4
-1	-2	0

3%	5%	3%
2%	4%	2%
1%	5%	3%

S	E	S
S	E	S
E	E	

Fig. 2. The Coffee Example: costs (left) and probabilities (center) of spilling coffee, first decision rule (right) of π

if a spilling occurs, which does not seem to naturally fit the usual assumptions of MDPs. The most natural expression of v_t^π from v_{t-1}^π writes indeed, for any horizon h and $\pi = (\delta_h, \delta_{h-1}, \ldots, \delta_1)$:

$$v_0^\pi(s) = 0$$
$$v_t^\pi(s) = P_s r_s + (1 - P_s) \sum_{s' \in S} p_{ss'}^{\delta_t} v_{t-1}^\pi(s') \quad \forall t \leq h$$

by explicitly taking into account the fact that once the coffee is spilled the robot stops (and therefore no more cost is incurred in this case). For instance, $v_t^\pi(2,2) = 0.04(-3) + 0.96(0.9v_{t-1}^\pi(2,3) + 0.1v_{t-1}^\pi(1,3))$ for policy π whose first decision rule is indicated on the right of Figure 2. This is not the usual form of the Bellman equations due to P_s. In order to recover this usual form, one needs to redefine rewards as $c_s = P_s r_s$ and transition probabilities as $q_{ss'}^a = (1 - P_s)p_{ss'}^a$:

$$v_0^\pi(s) = 0$$
$$v_t^\pi(s) = c_s + \sum_{s' \in S} q_{ss'}^{\delta_t} v_{t-1}^\pi(s') \quad \forall t \leq h$$

We now show that MDPs with functional rewards provide a more intuitive setting for this problem, and encompass a wide class of MDP variants.

3   MDP with Functional Rewards

An *MDP with functional rewards* (FRMDP) is an MDP where each reward $r_{ss'}^a$ is replaced by a function $f_{ss'}^a : \mathcal{R} \to \mathcal{R}$ where $\mathcal{R}$ is a valuation space with a binary operator $\max^\mathcal{R}$ that defines a (not necessarily complete) order relation $\succeq_\mathcal{R}: \forall x, y \in \mathcal{R}, x \succeq_\mathcal{R} y \Leftrightarrow \exists z \in \mathcal{R}, x = \max^\mathcal{R}(y, z)$. Furthermore, each final reward $g_s \in \mathbb{R}$ is replaced by a value in $\mathcal{R}$. Function $f_{ss'}^a(x)$ measures the value of executing action a in state s, moving to state s', and assuming that $x \in \mathcal{R}$ has already been received. Such functions are called *reward update functions*. By abuse of notation, for any decision rule δ, function $f_{ss'}^{\delta(s)}$ is denoted $f_{ss'}^\delta$.

In order to value a policy from an initial state, we assume that set $\mathcal{R}$ is endowed with a mixture operation m that assigns an element $m(p, x, y) = px + (1 - p)y$ in $\mathcal{R}$ to each p in $[0, 1]$ and each ordered pair (x, y) in $\mathcal{R} \times \mathcal{R}$. Pair $(\mathcal{R}, m)$ is assumed to be a *mixture set* [10], i.e., the following properties hold:

M1. $1x + 0x = x$,
M2. $p_1 x + (1 - p_1)y = (1 - p_1)y + p_1 x$,
M3. $p_1[p_2 x + (1 - p_2)y] + (1 - p_1)y = (p_1 p_2)x + (1 - p_1 p_2)y$,

for all x, y in $\mathcal{R}$ and p_1, p_2 in $[0, 1]$. Note that the mixture of more than two elements, i.e., $\sum_i p_i x_i$, is defined by inductive application of property M3. Moreover, we impose the following *independence* condition:

I. $x \succeq_{\mathcal{R}} y \Leftrightarrow px + (1 - p)z \succeq_{\mathcal{R}} py + (1 - p)z$

for all x, y, z in $\mathcal{R}$ and p in $[0, 1]$. This condition well-known in decision theory [11] states that when comparing two elements, only the differing parts are important in choosing the preferred ones.

As mentioned in the introduction, preferences over policies can be defined in two ways, as the value function of a policy can be given two definitions: a recursive one and an explicit one. We present those two definitions in the following two subsections. We show that the recursive definition allows the use of dynamic programming under some conditions that we will specify. We then reveal which property the reward update functions have to satisfy in order for the explicit definition and the recursive definition to be equivalent.

Recursive Definition. The value function of a policy $\pi = (\delta_h, \delta_{h-1}, \ldots, \delta_1)$ in a state s can be defined recursively with the functional version of the Bellman equations:

$$v_0^{\pi}(s) = g_s \tag{3}$$

$$v_t^{\pi}(s) = \sum_{s' \in S} p_{ss'}^{\delta_t} f_{ss'}^{\delta_t} \left(v_{t-1}^{\pi}(s') \right) \quad \forall t \leq h$$

A policy π is *preferred* to a policy π' in state s at step t, denoted by $\pi \succsim_{s,t} \pi'$, if and only if $v_t^{\pi}(s) \succeq_{\mathcal{R}} v_t^{\pi'}(s)$. Note that preference relation $\succsim_{s,t}$ may be partial as $\succeq_{\mathcal{R}}$ is not necessarily complete.

Usually, valuation space $\mathcal{R}$ is simply taken as the real line $\mathbb{R}$. The mixture operation (resp. $\max^{\mathcal{R}}$) is then the usual convex combination between reals (resp. the usual max operator). For instance, this is the case of standard MDPs, which are FRMDPs where $f_{ss'}^a$ is defined from $r_{ss'}^a$ by $f_{ss'}^a(x) = r_{ss'}^a + x$. The coffee robot example can also be casted in this framework, by setting $f_{ss'}^a(x) = P_s r_{ss'}^a + (1 - P_s)x$ and $g_s = 0$.

Besides, FRMDPs include many previously proposed variants of MDPs as special instances when $\mathcal{R} \neq \mathbb{R}$. We detail here two examples:

Time-dependent MDPs [3]. FRMDPs can model problems where the rewards depend on time. For instance, consider a navigation problem where one wants to minimize the arrival time. The duration $d_{ss'}^a(t)$ of a transition following action a from state s to s' depends on the departure time t at state s. The value of a history or a policy from a state s, interpreted here as the arrival time, is thus a function of the departure time. This situation can be modeled with $\mathcal{R} = \mathbb{R}^{\mathbb{R}}$, which denotes the set of real functions. For any $f \in \mathcal{R}$, if the value $f(t)$ represents the expected final arrival time with t being the departure time from some state s', then $f_{ss'}^a(f)(t) = f\left(d_{ss'}^a(t) + t\right)$ represents the expected final arrival time when executing action a at time t from state s and arriving in state s' (at time $d_{ss'}^a(t) + t$). Function g_s is set to the identity function Id. The choice for g_s will become clear when we present the solving method. For two functions $f, g \in \mathcal{R}$, $\max^{\mathcal{R}}(f, g) = h$ where $\forall x \in \mathbb{R}, h(x) = \max(f(x), g(x))$. Then, clearly, $f \succeq_{\mathcal{R}} g$

if and only if $\forall x \in \mathbb{R}$, $f(x) \geq g(x)$. Here, note that $\mathcal{R}$ is only partially ordered by $\succeq_{\mathcal{R}}$. The mixture operation is simply the convex combination of functions. Note that, contrarily to the dedicated framework provided by [3], FRMDPs cannot accommodate problems where the transition function also depends on the departure time.

Risk-sensitive MDPs [12, 4]: In standard MDPs, the decision criterion for comparing policies in a state is simply the mathematical expectation. However, it is insufficient when one wants to take into account risk aversion for instance. In that aim, one can use *expected utility* (EU) instead. For a discrete random variable X (total reward) whose possible realizations are $x_1, \ldots, x_k$ in $\mathbb{R}$, with probabilities $p_1, \ldots, p_k$ respectively, it is formulated as follows: $\sum_{i=1}^{k} p_i u(x_i)$, where u is the *utility function*. An MDP using EU as a decision criterion is an FRMDP with $\mathcal{R} = \mathbb{R}^{\mathbb{R}}$, $\forall f \in \mathcal{R}$, $\forall x \in \mathbb{R}$, $f_{ss'}^a(f)(x) = f(r_{ss'}^a + x)$ and $g_s = u$, where $r_{ss'}^a$ is the immediate reward of the initial MDP. Here, again, the value function v^π of policy π in a state s is a real function. Its interpretation is the following: $v^\pi(s)(x)$ represents the expected utility of the total reward obtained by following policy π from state s, assuming that cumulated reward x has already been received. At the last time step, clearly the value function of any policy in a state s is $g_s = u$, which can then be applied on the cumulated rewards received before reaching the final time step. Operator $\max^{\mathcal{R}}$, relation $\succeq_{\mathcal{R}}$ and the mixture operation are defined as in the previous example.

In a close framework with $\mathcal{R} = \mathbb{R}$, Kreps and Porteus [7] axiomatically justified a preference system defined recursively for taking into account preferences on temporal resolution of uncertainty. They identified under which conditions preferences over policies can be represented by functions $f_{ss'}^a$ and showed that $f_{ss'}^a$'s have to be strictly increasing functions. In their framework, Kreps and Porteus [8] showed that backward induction can be used to determine an optimal policy in finite horizon. Those general results can naturally and easily be extended to FRMDPs[1].

Proposition 1. *If Condition I holds and the functions $f_{ss'}^a$'s are strictly increasing, then an optimal value function v_h^* can be computed recursively with a functional version of the Bellman optimality equations:*

$$v_0^*(s) = g_s \tag{4}$$

$$v_t^*(s) = \max_{a \in A}{}^{\mathcal{R}} \sum_{s' \in S} p_{ss'}^a f_{ss'}^a\left(v_{t-1}^*(s')\right) \quad \forall t \leq h$$

Note the importance of the existence of the operator $\max^{\mathcal{R}}$ as it entails the unicity of the optimal value function, even if $\mathcal{R}$ may be partially ordered (as when $\mathcal{R} = \mathbb{R}^{\mathbb{R}}$ for instance). The conditions of Proposition 1 hold for all previous examples. When $\mathcal{R} = \mathbb{R}^{\mathbb{R}}$, it is obvious from the fact that $\succeq_{\mathcal{R}}$ is nothing but the pointwise dominance relation between functions. From now on, we will assume that the validity conditions of Proposition 1 hold. Then, Prop. 1 implies that the *functional backward induction* (Fig. 3) can be used to find an optimal policy.

---

[1] For lack of space, we omit the proofs. They are available on a longer version of this paper at http://www-desir.lip6.fr/~weng/pub/miwai2013-1.pdf

```
1: ∀s ∈ S, v₀*(s) ← gₛ; t ← 0
2: repeat
3:     t ← t + 1
4:     for all s ∈ S do
5:         for all a ∈ A do
6:             qₜ(s, a) ← ∑_{s'∈S} p^a_{ss'} f^a_{ss'} (v*_{t-1}(s'))
7:         end for
8:         v*ₜ(s) ← max^R_{a∈A} qₜ(s, a)
9:     end for
10: until t = h
```

Fig. 3. Functional Backward Induction

Explicit Definition. Another natural way for constructing preferences over policies in FRMDPs is as follows. First, we define the value of a history $\gamma = (s_h, a_h, s_{h-1}, \ldots, a_1, s_0)$ by $r(\gamma) = f^{a_h}_{s_h s_{h-1}} \circ f^{a_{h-1}}_{s_{h-1} s_{h-2}} \circ \ldots \circ f^{a_1}_{s_1 s_0} \circ g_{s_0}$ where $\circ$ denotes the function composition operator. Note that $r(\gamma)$ is in $\mathcal{R}$. Then, as the application of a policy π in a state s induces a probability distribution, denoted P^π_s, over histories, the value of π in state s can be defined by:

$$\bar{v}^\pi_h(s) = \sum_\gamma P^\pi_s(\gamma) r(\gamma) \tag{5}$$

Using these value functions, we can compare policies. A policy π is preferred to a policy π' in state s at horizon h if and only if $\bar{v}^\pi_h(s) \succsim_\mathcal{R} \bar{v}^{\pi'}_h(s)$. Then, the value function of optimal policies in s at horizon h can be found as follows:

$$\bar{v}^*_h(s) = \max_\pi{}^\mathcal{R} \bar{v}^\pi_h(s) \tag{6}$$

Unfortunately, when one only assumes that $f^a_{ss'}$ is strictly increasing, (3) and (5) do not define the same functions in general, that is we do not have $v^\pi_h(s) = \bar{v}^\pi_h(s)$ for all policy π and all state s. Thus, optimal policies can be different in the two preference systems, which may be problematic. A linearity requirement on those reward update functions has to be enforced for the two preference systems to be equivalent. A function $f : \mathcal{R} \to \mathcal{R}$ is said to be *linear* (with respect to the mixture set) if and only if $\forall x, y \in \mathcal{R}, \forall p \in [0, 1]$:

$$f(px + (1 - p)y) = pf(x) + (1 - p)f(y)$$

The following proposition formally states this result:

Proposition 2. *If all reward update functions are linear functions, then:*

  (i) With (3), (5) can be computed recursively , i.e., $\bar{v}^\pi_t(s) = v^\pi_t(s)$ $\forall \pi, \forall s, \forall t$

  If moreover Condition I holds and all reward update functions are strictly increasing, then:

  (ii) With (4), (6) can be computed recursively, i.e., $\bar{v}^*_t(s) = v^*_t(s)$ $\forall s, \forall t$

For the previous examples, the linearity condition is satisfied. This is obvious for the coffee robot problem. When $\mathcal{R} = \mathbb{R}^\mathbb{R}$, for instance for risk-sensitive MDPs, we can check that $\forall f, g \in \mathcal{R}, \forall p \in [0, 1], \forall x \in \mathbb{R}$,

$$f^a_{ss'}(pf + (1 - p)g)(x) = (pf + (1 - p)g)(r^a_{ss'} + x)$$
$$= pf(r^a_{ss'} + x) + (1 - p)g(r^a_{ss'} + x)$$
$$= pf^a_{ss'}(f)(x) + (1 - p)f^a_{ss'}(g)(x)$$

```
1: ∀s ∈ S, V₀*(s) ← {gₛ}; t ← 0
2: repeat
3:    t ← t + 1
4:    for all s ∈ S do
5:       for all a ∈ A do
6:          Qₜ(s, a) ← max^{⪰ℛ}_{v∈V*_{t-1}} Σ_{s'∈S} p^a_{ss'} f^a_{ss'}(v(s'))
7:       end for
8:       Vₜ*(s) ← max^{⪰ℛ}_{a∈A} Qₜ(s, a)
9:    end for
10: until t = h
```

Fig. 4. Generalized Functional Backward Induction

Generalization When $\max^{\mathcal{R}}$ Does Not Exist. In this section, we relax the requirements imposed on $\mathcal{R}$. We do not assume anymore that there exists an operator $\max^{\mathcal{R}}$. The only assumption that we make on $\mathcal{R}$ is that it is partially ordered by an order relation $\succeq_{\mathcal{R}}$. For instance, in this setting, one can think of multicriteria problems with $\mathcal{R} = \mathbb{R}^k$ (where k is the number of criteria) and $\succeq_{\mathcal{R}}$ the Pareto dominance[2]. Clearly, in such an example, $\max^{\mathcal{R}}$ is not defined.

In this generalized framework, propositions similar to the previous ones can be proved. We first give a few notations. The set of maximal elements of a set with respect to an order relation $\succeq_{\mathcal{R}}$ is denoted $\forall Y \subseteq \mathcal{R}, \max^{\succeq_{\mathcal{R}}}(Y) = \{y \in Y : \forall z \in Y, \text{not}(z \succ_{\mathcal{R}} y)\}$. Furthermore, we denote by $\mathcal{P}^*(\mathcal{R}, \succeq_{\mathcal{R}})$ the set $\{Y \subseteq \mathcal{R} : Y = \max^{\succeq_{\mathcal{R}}}(Y)\}$. $\max^{\succeq_{\mathcal{R}}}$ can be seen as a binary operator, i.e. $\forall X, Y \in \mathcal{P}^*(\mathcal{R}, \succeq_{\mathcal{R}}), \max^{\succeq_{\mathcal{R}}}(X, Y) = \max^{\succeq_{\mathcal{R}}}(X \cup Y)$. Besides, for any function $f : \mathcal{R} \to \mathcal{R}$, for any $Y \subseteq \mathcal{R}$, $f(Y) = \{f(y) : y \in Y\} \subseteq \mathcal{R}$.

All the propositions that we have presented previously can now be written by replacing $\mathcal{R}$ by space $\mathcal{P}^*(\mathcal{R}, \succeq_{\mathcal{R}})$ which is endowed with $\max^{\succeq_{\mathcal{R}}}$ seen as a binary operator. We just write the equations for finding the maximal value functions, which are elements of the following sets:

$$V_0^*(s) = \{g_s\}$$
$$V_t^*(s) = \max_{a\in A}{}^{\succeq_{\mathcal{R}}}\left\{\sum_{s'\in S} p^a_{ss'} f^a_{ss'}(v(s')) : v \in V_{t-1}^*\right\}$$

where $V_{t-1}^* = \{v \in \mathcal{R}^S : \forall s \in S, v(s) \in V_{t-1}^*(s)\}$. In this more general setting, one can then formulate a *generalized functional backward induction* (Fig. 4)

Discussion. As noted before, FRMDPs may be reformulated as MDPs by state augmentation. However, this may not be a natural way for representing preferences in the sequential decision-making problem at hand. We argue that the general FRMDP framework is more suitable when preferences become a bit sophisticated. First, it allows more flexibility into the modeling of preferences by allowing functional rewards. Moreover, it clearly uncouples, when modeling a problem, the dynamic of the system and the preference structure.

In practice, for being able to apply efficiently the functional backward induction or the functional value iteration, one exploits specific properties of set $\mathcal{R}$, together with the nature of operator $\max^{\mathcal{R}}$ (or relation $\succeq_{\mathcal{R}}$). For instance, if $\mathcal{R}$

---

[2] $(x_1, \dots, x_k) \succeq_{\mathcal{R}} (y_1, \dots, y_k) \Leftrightarrow \forall i = 1, \dots, k, x_i \geq y_i$.

is the set of piecewise-linear real functions and $\max^{\mathcal{R}}$ is the pointwise maximum operation, Boyan and Littman [3] (in the setting of time-dependent MDPs) discuss the restrictions needed for exact computations, and suggest efficient data structures to represent and manipulate the subsequent (piecewise-linear) value functions. More generally, if $\mathcal{R}$ is the set of real functions, these real functions may be approximated by piecewise-linear functions, and algorithms dedicated to this latter type of functions can be applied to find approximate solutions by adapting the work of Liu and Koenig [13].

Interestingly, note that the class of piecewise linear functions is not the only one for which efficient algorithms can be designed. Consider a certain class $\mathcal{C}$ of real functions such that $\mathcal{C}$ is a real space vector. A real function f is called piecewise-$\mathcal{C}$ if the real line can be partitioned into intervals such that f restricted on each interval is in $\mathcal{C}$, i.e., $\exists n \in \mathbb{N}^*, \exists f_1, \ldots, f_n \in \mathcal{C}, \exists w_0, \ldots, w_n \in \mathbb{R} \cup \{-\infty, +\infty\}, w_0 < \ldots < w_n, \forall i = 1, \ldots, n, \forall x \in]w_{i-1}, w_i], f(x) = f_i(x)$. Obviously, any function in $\mathcal{C}$ is piecewise-$\mathcal{C}$. Let $PW\mathcal{C}$ denote the set of real functions that are piecewise-$\mathcal{C}$.

If we take $\mathcal{R} = PW\mathcal{C}$, value functions are elements of $PW\mathcal{C}^S$, the set of functions from S to $PW\mathcal{C}$. We define operator L from $PW\mathcal{C}^S$ to $PW\mathcal{C}^S$ by:

$$(Lv)(s) = \max_{a \in A}{}^{\mathcal{R}} \sum_{s' \in S} p_{ss'}^a f_{ss'}^a(v)$$

where $v \in PW\mathcal{C}^S$ and $s \in S$. With this operator, (4) can simply be rewritten:

$$v_0^*(s) = g_s$$
$$v_t^*(s) = L(v_{t-1}^*)(s) \quad \forall t \leq h$$

It is then easy to see that the real space vector $PW\mathcal{C}^S$ is closed under operator L, i.e., $Lv \in PW\mathcal{C}^S$ for any $v \in PW\mathcal{C}^S$.

This framework is a generalization of piecewise-linear functions. Other classes of functions may be convenient for efficient computations, e.g., the class of piecewise-linex function[3] [13]. Another interesting class is $\mathcal{C} = P_2$, the set of polynomials of degree 2, as the application of operator L on functions in PWP_2 can be computed efficiently. We detail how to proceed in the next first example.

4   Illustrative Examples

Our framework is very general as it encompasses many previously proposed models. We now present three new examples to illustrate the usefulness of our results. In standard MDPs, one wants to find a policy that maximizes the expected cumulated rewards. However, in high-stake planning situations, it is sensible to take also into account the variability of the total reward received from the initial state. A way to tackle this issue is to compute a policy optimizing the *expected utility* (EU) of the total reward. Expected utility is very popular in decision theory, as it enables to simply model risk aversion by concavity of the utility function $(\lambda u(x) + (1 - \lambda)u(x') \leq u(\lambda x + (1 - \lambda)x')$ for $\lambda \in [0, 1])$. If utility function u is quadratic, i.e., $u(x) = bx^2 + cx + d$, risk aversion is therefore modeled by setting coefficient b to a negative value.

---

[3] A linex function is the sum of a linear function and an exponential function.

Deterministic Rewards. In this first example, we show how one can optimize EU with a quadratic utility function and deterministic rewards. Let $P_2 = \{f(x) = bx^2 + cx + d : b, c, d \in \mathbb{R}\}$ be the set of polynomials of degree at most 2. Let PWP_2 be the set of functions that are piecewise-P_2.

In order to operationally implement operator L, three primitives are required: one primitive implementing function $f_{s,s'}^a$, one for the linear combination of functions in PWP_2, and one for the pointwise maximum of functions in PWP_2.

For any states s, s' and any action a, function $f_{ss'}^a$ is defined here by $\forall v \in PWP_2, \forall x \in \mathbb{R}, f_{ss'}^a(v)(x) = v(r_{ss'}^a + x)$. By definition, any piecewise-P_2 function v is defined by a sequence (w_i, b_i, c_i, d_i) for $i = 1, \ldots, n$ where $\forall x \in \mathbb{R}, \exists j = 1, \ldots, n, x \in]w_{j-1}, w_j], v(x) = b_j x^2 + c_j x + d_j$. Function $f_{ss'}^a(v)$ is given by the following sequence for $i = 1, \ldots, n, (w_i - r_{ss'}^a, b_i, 2b_i r_{ss'}^a + c_i, b_i r_{ss'}^a{}^2 + c_i r_{ss'}^a + d_i)$.

To obtain linear combinations of piecewise-quadratic functions, one first finds an interval partition of the real line such that on each interval, each PWP_2 function is in P_2, then computes linear combinations of coefficients b_i, c_i, d_i.

We now present how to compute the pointwise max of two functions in PWP_2. The case of more than two functions is obtained by induction. We present the computation in the general case, but this computation could be slightly simplified by using the fact that the functions are all increasing since they represent utility functions. Let $v, v' \in PWP_2$. We can assume that both functions are defined on the same interval partition, otherwise just take a finer interval partition. On each interval $]w, w']$, four cases can occur:

1. $v - v'$ has no root in $]w, w'[$, then the max on $]w, w']$ is v if $v - v'$ is positive in $]w, w'[$ and v' otherwise.
2. $v - v'$ has one root r in $]w, w'[$, then we obtain two P_2 functions (possibly identical), one defined on $]w, r]$, the other on $]r, w']$, equal to v or v' depending on the sign of $v - v'$ on these two intervals.
3. $v - v'$ has two roots $r_1 < r_2$ in $]w, w'[$, then we obtain three P_2 functions on the following three intervals $]w, r_1],]r_1, r_2]$ and $]r_2, w']$ equal to v or v' depending on the sign of $v - v'$ in those intervals.
4. $v - v' = 0$, then the max on $]w, w']$ is simply v.

Note that the computation of the roots is of course very simple since it amounts to solve quadratic equations by using discriminants.

For this setting, we implemented the functional value iteration with piecewise-quadratic functions. On the classic problem of navigation of an autonomous agent in a grid, our experiments show that the functional value iteration is between 10% to 50% slower than standard value iteration, depending on the size of the problem and the quadratic utility function. This experimental observation indicates that functional value iteration could be reasonably exploited when a suitable class of functions is chosen.

Random Rewards. The second problem we present is a generalization of an optimal path problem in graphs with stochastic weights investigated by Loui [14]. The difference here is that the consequences of actions are stochastic. More specifically, this problem corresponds to an MDP where rewards are not known precisely. For illustration, consider a stochastic shortest problem where the duration of an action is not known with certainty. In this setting, one could model the durations with independent random variables. If we were optimizing the expected duration time, one could replace the random variables by their mean and solve the problem as a standard MDP. However, if instead, we optimize the expected utility of the total duration time, in order to take into account risk attitude, the problem does not boil down to a standard MDP anymore.

Such a problem can be formalized in the setting of FRMDP as follows: $\mathcal{R}$ is the set of real random variables; $f^a_{ss'}(X) = R^a_{ss'} + X$ where $R^a_{ss'} \in \mathcal{R}$; $g_s = 0$ where 0 is the null random variable.

Without any assumption on the random variables, the determination of a policy optimizing EU in this setting can be hard in the general case. We assume from now on that all rewards are independent Gaussian random variables. This property has nice consequences from the computational viewpoint: the Gaussian distribution $\mathcal{N}(\mu, \sigma^2)$ is completely characterized by its mean μ and its variance σ^2, and furthermore $\lambda_1 X_1 + \lambda_2 X_2 \sim \mathcal{N}(\lambda_1\mu_1 + \lambda_2\mu_2, \lambda_1^2\sigma_1^2 + \lambda_2^2\sigma_2^2)$ for two independent random variables $X_1 \sim \mathcal{N}(\mu_1, \sigma_1^2)$ and $X_2 \sim \mathcal{N}(\mu_2, \sigma_2^2)$. Consequently, the sum of rewards obtained along a history and therefore the total reward obtained by applying a given policy are both Gaussian random variables. As underlined by [14], the expected utility of a Gaussian distribution $\mathcal{N}(\mu, \sigma^2)$ is equal to $d + c\mu + b\mu^2 + b\sigma^2$ if the utility function u is quadratic, i.e., $u(x) = bx^2 + cx + d$. As u is a utility function, we restrict its domain of definition to an interval on which it is increasing. Then, on that interval, the function $u' : \mathbb{R}^2 \to \mathbb{R}$ defined by $u'(x, y) = d + cx + bx^2 + by$ is increasing in its first argument. For risk aversion, b is negative, which means that u' is decreasing on its second argument. Consequently, for two Gaussian defined respectively by (μ, σ^2) and (μ', σ'^2), whenever $\mu \geq \mu'$ and $\sigma^2 \leq \sigma'^2$, one can conclude that the Gaussian defined by (μ, σ^2) is preferred. This observation suggests to use the following approach. The sequential decision-making problem can be formalized in the setting of FRMDPs as follows: $\mathcal{R} = \{(\mu, \sigma^2) : \mu \in \mathbb{R}, \sigma^2 \in \mathbb{R}^*_+\}$; $f^a_{ss'}(\mu, \sigma^2) = (\mu^a_{ss'} + \mu, \sigma^a_{ss'}{}^2 + \sigma^2)$ where $R^a_{ss'} \sim \mathcal{N}(\mu^a_{ss'}, \sigma^a_{ss'}{}^2)$; $g_s = (0, 0)$. We define $\succeq_\mathcal{R}$ as follows $(\mu, \sigma^2) \succeq_\mathcal{R} (\mu', \sigma'^2) \Leftrightarrow (\mu \geq \mu'$ and $\sigma^2 \leq \sigma'^2)$. Note that this order relation is partial. The mixture set on $\mathcal{R}$ is defined as follows: $p(\mu, \sigma^2) + (1-p)(\mu', \sigma'^2) = (p\mu + (1-p)\mu', p^2\sigma^2 + (1-p)^2\sigma'^2)$. Condition I holds in this context and one can easily check that functions $f^a_{ss'}$'s are linear and strictly increasing. Since order relation $\succeq_\mathcal{R}$ is partial, applying the Bellman optimality equations yields a set of maximal value functions. Finally, the value function that is optimal with respect to EU can be found by scanning this set.

Time-Dependent Rewards. As a final example, we simply note that various features of different previously proposed extensions of MDPs can be mixed and take into account simultaneously in FRMDPS. For instance, the optimization of expected utility in problems where rewards are time-dependent can naturally be expressed in our framework. However, to the best of our knowledge, no previous framework can model this kind of setting.

5   Conclusion

In this paper, we have proposed FRMDPs as a new general framework for modeling sequential decision-making problems under uncertainty when preferences are sophisticated. It generalizes many previously known propositions and encompasses new problems. We made explicit the conditions that allow the use of a functional backward induction algorithm. We showed in this paper that in some situations an FRMDP could be solved directly and efficiently. To illustrate our proposition, we showed its exploitation on three new problems that previous frameworks could not tackle.

References

1. Dean, T., Kaelbling, L., Kirman, J., Nicholson, A.: Planning with deadlines in stochastic domains. In: AAAI, vol. 11, pp. 574–579 (1993)
2. Littman, M.L., Dean, T.L., Kaelbling, L.P.: On the complexity of solving markov decision problems. In: UAI, pp. 394–402 (1995)
3. Boyan, J., Littman, M.: Exact solutions to time-dependent MDPs. In: NIPS, pp. 1026–1032 (2000)
4. Liu, Y., Koenig, S.: Risk-sensitive planning with one-switch utility functions: Value iteration. In: AAAI, pp. 993–999 (2005)
5. Regan, K., Boutilier, C.: Regret based reward elicitation for Markov decision processes. In: UAI, pp. 444–451 (2009)
6. Serafini, P.: Dynamic programming and minimum risk paths. European Journal of Operational Research 175, 224–237 (2006)
7. Kreps, D., Porteus, E.: Temporal resolution of uncertainty and dynamic choice theory. Econometrica 46, 185–200 (1978)
8. Kreps, D., Porteus, E.: Dynamic choice theory and dynamic programming. Econometrica 47, 91–100 (1979)
9. Boutilier, C., Dearden, R., Goldszmidt, M.: Stochastic dynamic programming with factored representations. Artificial Intelligence 121, 49–107 (2000)
10. Hernstein, I., Milnor, J.: An axiomatic approach to measurable utility. Econometrica 21, 291–297 (1953)
11. Fishburn, P.: Utility theory for decision making. Wiley (1970)
12. Denardo, E., Rothblum, U.: Optimal stopping, exponential utility and linear programming. Mathematical Programming 16, 228–244 (1979)
13. Liu, Y., Koenig, S.: Functional value iteration for decision-theoretic planning with general utility functions. In: AAAI, pp. 1186–1193 (2006)
14. Loui, R.: Optimal paths in graphs with stochastic or multidimensional weights. Communications of the ACM 26, 670–676 (1983)

Hand Gesture Segmentation from Complex Color-Texture Background Image

Vinay Kumar Verma[1], Rajeev Wankar[2], C.R. Rao[2], and Arun Agarwal[2]

[1] Institute of Development and Research in Banking Technology, Castle Hills, Road
No-1, Masab Tank, Hyderabad-57, India
[2] SCIS, University of Hyderabad, Gachibowli, Hyderabad-500046, India

Abstract. Gestures provide a rich, intuitive and natural form of interaction between human and other devices. In this paper an automatic
hand gesture segmentation technique from the complex color-texture Image is developed for segmentation of hand gesture with less false positive
rate(FPR). In this approach we propose a model for Skin Color Characterization and define a Potential of a Pixel (PoP) which are then used to
segment the hand gesture. This new skin segmentation technique takes
into account both the color-texture features for efficient segmentation. It
is observed that the classifier is robust with respect to usage of hand and
mode of hands like front or back side of hand. To evaluate the system
the hand gesture images have been acquired from set of students under
various complex background. The gesture segmentation technique has
false positive rate of nearly 5.7% and true positive rate near to 98.93%.

Keywords: Human-Device Interaction (HDI), Hand gesture, Segmentation, Feature Extraction, Classification.

1 Introduction

One of the attractive methods for providing natural human-computer interaction
is the use of the body movements as an input device rather than the cumbersome devices such as keyboards and mice, which need the user to be located
at a specific location to use these devices. Visual-based automatic body gesture recognition has acquired much attention. In this context strong efforts have
been carried out to develop intelligent and natural interfaces between users and
computer system based on body movements. The control of the robot is one of
the most important research fields. Gesture recognition is a topic in computer
science and language technology with the goal of interpreting human gestures
via mathematical algorithms.

Human-Device Interaction (HDI) can be considered as one of the most important Computer Vision domains. It has many applications in a variety of fields
such as: search and rescue, military battle, mine and bomb detection, scientific
exploration, law enforcement, entertainment and hospital care. HDI is the study
of interactions between people and Devices (example robots). Gestures are a
powerful means of communication among humans. In fact, gesturing is so deeply

S. Ramanna et al. (Eds.): MIWAI 2013, LNCS 8271, pp. 281–292, 2013.

rooted in our communication that people often continue gesturing when speaking on the telephone. Hand gestures provide a separate complementary modality to speech for expressing ones ideas. Information associated with hand gestures in a conversation is degree, discourse structure, spatial and temporal structure. So, a natural interaction between humans and devices can be achieved by using for example: hand gestures for communication between them. Hand gesture segmentation is very important and difficult task from the complex color-texture background image. For the accurate gesture recognition, gesture segment result should be good since segmentation result directly effect the recognition result. This paper proposed a hand gesture segmentation technique from the complex color-texture background image based on the skin detection[14].

There are various factors that affect the skin segmentation result like different type of skin color, luminance sensitivity of various color space, and object like skin color, different lighting condition etc. For reducing luminance sensitivity we have to select the color space that has less sensitivity for the luminance and lighting condition [1]. Jons and Rehg [2] shows that there is a significant overlap between skin and non skin pixel that reduce the classification accuracy. Traditional method used skin color detection for the Gesture segmentation in static image [5], color is the most robust and important cues for skin detection and also fast in the processing of the skin pattern, some other cues are shape, texture, geometry can be used for gesture segmentation. Zhao,Song and Li proposed a novel approach for the hand gesture contour segmentation based on the skin color elliptical model and GVF Snake model [3] it is showing excellent result in case of simple image it is showing high true positive rate. Another approach for skin Segmentation is Multi-pixel colour Clustering model, this method shows high true positive rate in the case of varying lighting condition and different type of skin, but drawback is that in the presence of similar object like skin, non-skin pixel also classified as skin pixel it has high false positive rate in the case of complex image [4]. Skin color detection model is also proposed based on the Neural Network, network may be symmetric or asymmetric classifiers. The asymmetric classifiers has two separate NN in output one for skin and other for non-skin pixel. In this approach first we have to train the NN by large no of skin pixel and non-skin pixel from the diverse race image, once NN is trained we can easily classify the skin and non-skin pixel [8].

In the video streaming for the hand gesture segmentation, Background Subtraction Technique(BST) is most widely used. In this model the major task is to build an explicit model for the background. Calculating the difference between current frame and previous frame we can extract the foreground and background of image and segmentation is easily performed[7][9]. BST are of two type Recursive and Non recursive. Recursive techniques include Running Gaussian average (RGA), Gaussian Mixture Model (GMM), Adaptive GMM (AGMM), Approximated median filtering (AMF)[10]. Another approach Histogram of Oriented Gradient and force field method are used for segmentation of gesture in video streaming. Locally normalise Histogram of Oriented Gradients shows excellent performance for the Gesture Segmentation. There are various type of color space they have there own advantage in different case but for the skin segmentation RGB, Normalise RGB, HSV,YCbCr,TSL, HSI and perceptually uniform color space like CIELAB and CIELUV color space are used. Most of the researcher believe that HSV and YCbCr color space are most appropriate color space for skin segmentation.

2   Skin Color Elliptical Model

Jain A K et al.[12] proposed a method that skin color distribution can be represented using elliptical model in the Cb'Cr' region by converting YCbCr color space to the non linear YCb'Cr' space and they proposed a method for face detection this shows good result for skin detection but for the complex background image this approach has high FPR. In the color space there is always a certain degree of non-linear dependency between the color and luminance value Y, so they proposed a method of non linear transform to eliminate the dependence. After a non linear fragmented color transformation, the color space is represented by YCb'Cr'. For making YCbCr color space to YCb'Cr' color space the conversion formula is given as follows:

$$
C_i' = \begin{cases} (C_i(Y) - \overline{C_i}(Y))\frac{Wc_i}{Wc_i(Y)} + \overline{C_i}(Y) & \text{if } Y \notin [K_l, K_h] \\ C_i & \text{if } Y \in [K_l, K_h] \end{cases} \tag{1}
$$

Where $K_l = 125$ and $K_h = 128$
And $Wc_i(Y)$ and $\overline{C_i}(Y)$ defined as follows...

$$
Wc_i(Y) = \begin{cases} WLc_i + \frac{(Y-Y_{min})(Wc_i-WLc_i)}{K_l-Y_{min}} & \text{if } Y < K_l \\ WHc_i + \frac{(Y_{max}-Y)(Wc_i-WHc_i)}{Y_{max}-K_h} & \text{if } Y > K_h \end{cases} \tag{2}
$$

$$
\overline{C_b}(Y) = \begin{cases} 108 + \frac{(K_l-Y)(118-108)}{K_l-Y_{min}} & \text{if } Y < K_l \\ 108 + \frac{(Y-K_h)(118-108)}{Y_{min}-K_h} & \text{if } Y > K_h \end{cases} \tag{3}
$$

$$
\overline{C_r}(Y) = \begin{cases} 154 + \frac{(K_l-Y)(154-144)}{K_l-Y_{min}} & \text{if } Y < K_l \\ 154 + \frac{(Y-K_h)(154-144)}{Y_{min}-K_h} & \text{if } Y > K_h \end{cases} \tag{4}
$$

In the formula C_i present C_b and C_r, $Y_{min} = 16$ $Y_{max} = 235$ (the minimum and maximum in color cluster region of the component Y), $Wc_b = 46.97$ and $Wc_r = 38.76$
$WLc_b = 23$ and $WHc_b = 14$
$WLc_r = 20$ and $WHc_r = 10$

(to gather statistics, Jain A.K. selected 137 image from the image library of Heinrlch-Hertz-Institute.)

In the transformed Cb'Cr' region, the color distribution was described using the elliptical model the elliptical formula can be expressed as follows:

$$\begin{bmatrix} x \\ y \end{bmatrix} = SM \times \begin{bmatrix} C_b^i - c_x \\ C_r^i - c_y \end{bmatrix} \tag{5}$$

$$\frac{(x - ec_x)^2}{a^2} + \frac{(y - ec_y)^2}{b^2} \leq 1 \tag{6}$$

where SM is a skin matrix.

Values for $c_x = 109.38$, $c_y = 152.02$, $ec_x = 1.60$ and $ec_y = 2.41$ are taken from [12].

After transforming YCb'Cr' color space, the Cb' and Cr' of each pixel is substituted in equation 5 to calculate $[x, y]^t$. We then substitute the value of $[x, y]^t$ in elliptical model 6, to check if pixel is within the ellipse. If it is then the pixel is the skin pixel otherwise it is non-skin pixel.

Retain the skin pixel (in the original RGB image as I') wherein non skin pixel are set to zero. In this way we can eliminate all the non skin pixels. It may happen that similar object with skin like color can also get classified as a skin pixel, which may lead to degradation in the recognition accuracy. For removing this type of non skin pixel we propose a new method called Potential of Pixel (PoP), explained in the section 4.

3   Estimation of a, b and SM

This section discusses methodology for estimating the values of a, b and SM to reflect the Indian skin properties. For this we have done several experiments on different samples of Indian skin. The experiment is conducted by acquiring samples from several students under different lighting conditions and background. Methodology followed is described below:

- Transform the RGB image to Cb'Cr' image. Take skin pixels in Cb'Cr' image for training
- Map the Cb'Cr' image to Cb'Cr' Cartesian space
- Initialize a, b and SM to value given in [12]
- Check if training skin pixels are getting mapped within the assumed ellipse. If no then modify the value of SM and re-estimate a and b. Repeat this till all the training skin pixels are correctly mapped into the assumed ellipse.

Graphical approach can also be adopted to estimate a, b and SM. In conclusion, it is seen that if we define the value of a and b tightly then some skin pixels may get classified as non-skin pixel. Therefore, we define these values within some tolerance so that such a situation can be avoided. The resultant values for SM, a and b after the above experimentation are:

$$SM = \begin{bmatrix} -0.6990 \ 0.8870 \\ -0.8990 \ 0.7905 \end{bmatrix} \quad and \quad a = 56.55 \quad and \quad b = 31.37 \tag{7}$$

4   Potential of Pixel: A New Approach

After several experiments we found that Lab color space is suitable for next stage of skin segmentation (i.e. Region based approach) as compared to other color spaces (for example HSV, TSL, YIQ etc). Therefore we transform the image I' into the Lab Color space image defined as I''. Lab color space has three plane **L**, **a** and **b**. L gives the gray value of the image this is much sensitive to lighting condition **a** and **b** gives the chromaticity value of the color space. So for making our approach less sensitive to lighting condition we taken only **a** and **b** value.

Let $I''(i, j)$ be defined as a central pixel of a region R. We now calculate (PoP) of $I''(i, j)$ which is defined as follows:

$$PoP = \frac{I''(i,j) * I''(k,l)}{r^2} \tag{8}$$

here we assume pixel region R to be of size 5×5 and $I''(k,l) \neq I''(i,j)$ and $\forall \; I''(k,l) \in R$. Value of r is computed using euclidean distance metric. Total potential of central pixel by its neighbour are defined as follows:

$$PoP_{total} = \sum_{\forall (k,l)} \frac{I''(k,l) * I''(i,j)}{r_{k,l}^2} \tag{9}$$

Apply Potential equation 9 on the transformed plane **a** and **b**. Let us call them as $aPoP$, $bPoP$. Define $cPoP$ as $(bPoP - aPoP)$. The plane passing through aPoP(i,j), bPoP(i,j) and cPoP(i,j) for every (i, j) is given as:

$$aPoP(i,j)x + bPoP(i,j)y + cPoP(i,j)z = d \quad \forall(i,j) \tag{10}$$

aPoP(i,j), bPoP(i,j) and cPoP(i,j) are the normal vector to the plan whereas x,y and z are the coordinate perpendicular to each other, and d is perpendicular drawn to from origin to the plane

The reflection of a point through plane 10 is given by Householder transformation [13]. The transformation matrix can be represented as:

$$\mathbf{HM} = \begin{bmatrix} 1 - 2aPoP^2(i,j) & -2aPoP(i,j)bPoP(i,j) & -2aPoP(i,j)cPoP(i,j) \\ -2aPoP(i,j)bPoP(i,j) & 1 - 2bPoP^2(i,j) & -2bPoP(i,j)cPoP(i,j) \\ -2aPoP(i,j)cPoP(i,j) & -2bPoP(i,j)cPoP(i,j) & 1 - 2cPoP^2(i,j) \end{bmatrix} \tag{11}$$

We convert the Equation 11 into a scalar value by using the equation 12. This will speed up the classification of pixel into skin or non-skin class by appropriate selection of the threshold value. Let $V(i,j)$ is the projection of the reflection matrix as a single point. $V(i,j)$ is defined as follows:

$$V(i,j) = \begin{bmatrix} aPoP(i,j) \\ bPoP(i,j) \\ cPoP(i,j) \end{bmatrix}^t HM \begin{bmatrix} \cos\theta \\ -\sin\theta \\ 1 \end{bmatrix} \tag{12}$$

Here $\theta = 0.34$ which was experimentally determined.

Hand gesture segmentation is divided in to four step process pre-processing, skin detection based on pixel based approach, skin detection based on texture based approach and post-processing. The block diagram for hand gesture segmentation based on two model given in figure 1.

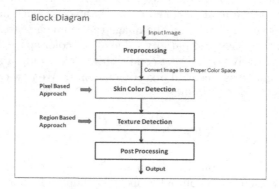

Fig. 1. Block Diagram for skin segmentation

5 Binarization

Binarization classifies each pixel into two classes: skin pixel or non-skin pixel. For a good binarization we have to choose a proper threshold value such that mis-classification is minimum. It is very difficult task for choosing proper threshold value. Many algorithms are proposed for defining good threshold like histogram based thresholding, optimal thresholding etc, but in the case of skin segmentation it is very difficult task to choose a proper thresholding algorithm, since there are various kind of skins like black, white, mixed etc and it is affected by varying lighting condition etc.

Our binarization is given by equation 13.

$$I'''(i,j) = \begin{cases} 1 \text{ if} & V(i,j) \in [th1, \quad th2] \\ 0 & otherwise \end{cases} \tag{13}$$

Where $th1$ and $th2$ are limits of the threshold band. One need to choose these threshold values aptly. The proposed approach was applied on several images. Under controlled lighting condition and on mixed skin samples the threshold range is found to be between 20 and 80.

6 Proposed Algorithm for Gesture Segmentation

Proposed algorithm is robust for skin segmentation in the case of complex background images. Following are the step of the proposed algorithm for skin segmentation:

Algorithm for Hand Gesture Segmentation

1. Convert the image into YCbCr color space, then using equation 1, 2, 3 and 4 convert into YCb'Cr' space.
2. Using the Cb' and Cr' value of the pixel, map it to the elliptical model using equation 5 and 6. If it is inside the ellipse retain the original RGB value of the pixel (also called skin pixel) otherwise make it zero.
3. The image obtained in step-2, convert the skin pixel in to Lab color space.
4. On the a-plane and b-plane apply PoP model (equation 9 and 10) and calculate aPoP and bPoP. Now adjust the mean value of aPoP and bPoP such that mean of aPoP is 4.60 and mean of bPoP is 6.20 and calculate value cPoP=(bPoP-aPoP).
5. Apply equation 11 for finding reflection of a point of plane passing through aPoP, bPoP and cPoP, convert reflection matrix into one dimension using equation 12.
6. Apply threshold equation 13 and put skin pixel as one(1) and non skin pixel as zero(0).
7. Apply post-processing step on image to get the segmented hand gesture image.

7   Post-processing

Post-processing is another important step for hand gesture segmentation to increase the classification accuracy. In this final step for the gesture segmentation we eliminate small region sizes. This is accomplished (after binarization) by calculating the size of each connected component and taking the biggest connected component (MBR) as a hand gesture.

Algorithm for Post-processing:

1. Apply median filter (5 x 5) on binary image. This will eliminate randomly distributed noisy pixels and thin lines that exist after binarization step.
2. Find connected components and retain the largest component, i.e. the "Hand Gesture".
3. If required small holes within the segmented hand gesture can be filled using a standard function available in MatLab.

8   Experimental Result and Discussion

This section discusses results of our new approach for gesture segmentation from color-texture background images. To demonstrate the efficacy of our method several images with varying complex backgrounds are considered. Figures shown below are the result of our proposed new hand gesture segmentation method. Fig 2, 3, 4, 5, 6 and 7 shows the results of different steps of the hand gesture

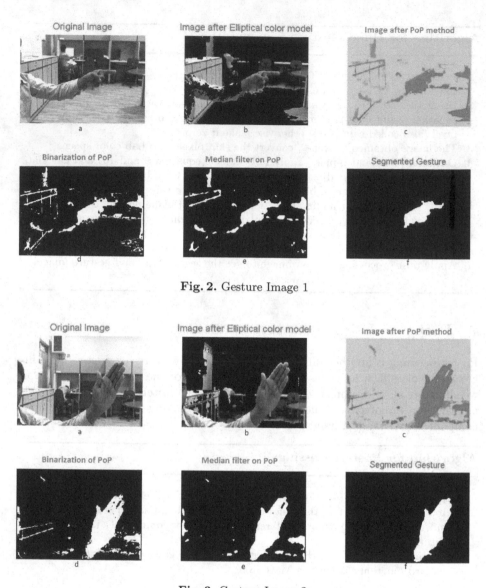

Fig. 2. Gesture Image 1

Fig. 3. Gesture Image 2

segmentation as was discussed in the previous section: **a**: Original Image (Input), **b**:Output of Elliptical Color Model, **c**: Image obtained after application of PoP method, **d**:Binarization of the PoP image, **e**:Post processing the binarized image using median filter and **f**:Final image of the segmented hand gesture.

For the quantitative analysis of skin classifier we use the following measures: False Positive Rate(FPR) and True Positive Rate(TPR). These are calculated

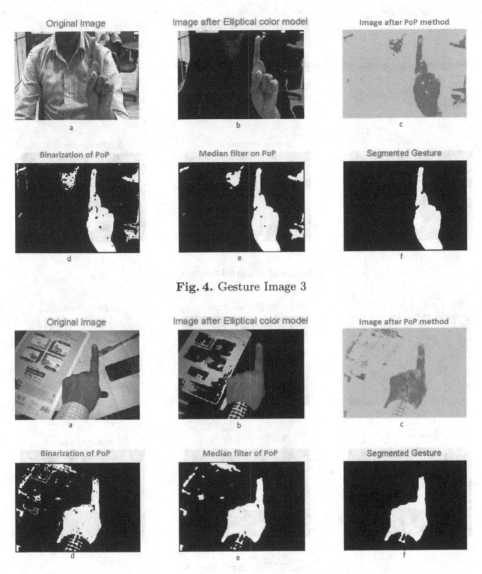

Fig. 4. Gesture Image 3

Fig. 5. Gesture Image 4

using pixel by pixel comparison of perfectly segmented image(ground truth image) with their respective segmented image using equation 14 and 15:

$$FPR = \frac{\text{No. of non-skin pixel classified as skin}}{\text{Total no. of non-skin pixel}} \qquad (14)$$

$$TPR = \frac{\text{No. of skin pixel correctly classified}}{\text{Total no. of skin pixel}} \qquad (15)$$

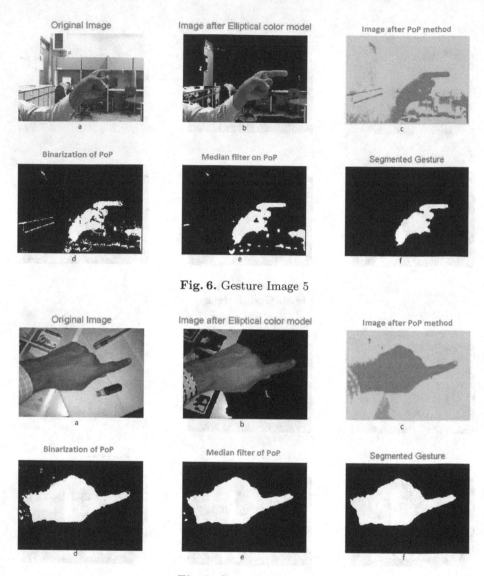

Fig. 6. Gesture Image 5

Fig. 7. Gesture Image 6

Quantitative Analysis of Image in terms of TPR and FPR is given in table 1.

To automate gesture segmentation process, a new approach is proposed and its quantitative measures such as False Positive Rate and True Positive Rate are computed. The proposed segmentation approach has FPR 5.7% and TPR 98.93% for the images having complex color-texture background. The proposed algorithm is showing excellent result on low variance of lighting condition. Experiment is done on Indian skins and it is found that the proposed approach is less sensitive for the different skin tones.

Table 1. Quantitative analysis of the proposed approach

Image	TPR(%)	FPR(%)
1	99.17	4.91
2	99.03	4.84
3	99.21	6.56
4	99.02	5.97
5	98.57	5.82
6	98.69	0.88

9   Conclusions

An automatic hand gesture segmentation technique from the complex image is developed for segmentation of hand gesture with less false positive rate. In this approach we have proposed a model for Skin Color Characterization and defined a Potential of a Pixel (PoP) which are then used to segment the hand gesture. This new skin segmentation technique takes into account both the color-texture features for efficient segmentation. It is observed that the classifier is robust with respect to usage of hand and mode of hands like front or back side of hand.

The drawback of the proposed approach is that in the case of high varying lighting condition and different type of skin color(black, white and mixed skin) the threshold value varied, so in future we can develop a automatic threshold selection method. To check the efficacy of the proposed segmentation method a hand gesture recognition system can be developed for the recognition of the hand gesture.

References

1. Kakumanu, P., Makrogiannis, S., Bporbakis, N.: A survey of skin-color modelling and detection methods. Patt. Recog. 40, 1106–1122 (2007)
2. Jones, M.J., Rehg, J.M.: Statistical color models with applications to skin detection. Int. J. Comput. Vis. 46, 81–96 (2002)
3. Shuying, Z., Song, X., Tan, W., Li, H.: A Novel Approach to hand Gesture Contour detection based on GVF Snake Model and Skin Color Elliptical Model. In: International Conference on Computer Application and System Modeling, vol. 5, pp. 381–84 (2010)
4. Naji, S.A., et al.: Skin Segmentation based on multi pixel color clustering models. Digital Signal Processing (2012),
   http://dx.doi.org/10.1016/j.dsp.2012.05.004
5. Hwang, C.L., Lu, K.-D., Pan, Y.-T.: Segmentation of Different Skin Colors with Different Lighting Condition by Combining Graph Cuts Algorithm with Probability Neural Network Classification, and its Application. Neural Process. Lett. 37, 89–109 (2013)
6. Sun, C., Talbot, H., Ourselin, S., Adriaansen, T.: Improved Automatic Skin Detection in Color Images. Digital Image Computing: Technique and Applications, vol. 7, pp. 10–12 (2003)

7. Hong, B., Xinggui, Z.: Study on Hand Gesture Segmentation. In: 2010 International Conference on Multimedia Technology (ICMT), pp. 1–4 (2010)
8. Bhoyar, K.K., Kakde, O.G.: Skin Color Detection Model Using Neural Network and its Performance Evaluation. Journal of Computer Science 6, 1549–3636 (2010)
9. Lee, D.-S., Hull, J.J., Erol, B.: A Bayesian Framework for Gaussian Mixture Background Modeling. In: International Conference on Image Processing, ICIP, pp. 973–976 (2003)
10. Calderara, S., Melli, R., Prati, A., Cucchiara, R.: Reliable background suppression for complex scenes. In: Proceedings of the 4th ACM International Workshop on Video Surveillance and Sensor Networks, vol. 4, pp. 211–214. ACM (2006)
11. Gupta, R.K.: A comparative Analysis of Segmentation Algorithms for Hand Gesture Recognition. In: Third International Conference on Computational Intelligence, Communication Systems and Networks (CICSyN), pp. 231–235 (2011)
12. Hsu, R.-L., Abdel-Mottaleb, M., Jain, A.K.: Face Detection In color Image. IEEE Trans. Pattern Analysis and Machine Intelligence 24(5), 696–706 (2002)
13. Householder, A.S.: Unitary Triangularization of a Non-symmetric Matrix. Journal of the ACM 5(4), 339–342 (1958), doi:10.1145/320941.320947. MR 0111128
14. V. K. Verma.: Hand Gesture Segmentation and Recognition From Complex Color-Texture Background Image. MTech dissertation, SCIS, University of Hyderabad (June 2013)
15. Ng, C.-M.P.: Skin Color Segmentation by Texture Feature Extraction and K-Mean Clustering. In: Third International Conference on Computational Intelligence, Communication Systems and Networks (CICSyN), pp. 213–218 (2011)

Distributed Query Plan Generation Using HBMO

T.V. Vijay Kumar, Biri Arun, and Lokendra Kumar

School of Computer and Systems Sciences,
Jawaharlal Nehru University, New Delhi-110067, India
tvvijaykumar@hotmail.com,
{biriarun,lokendrakashyap286}@gmail.com

Abstract. Processing a distributed query entails accessing data from multiple sites. The inter site communication cost, being the dominant cost, needs to be reduced in order to improve the query response time. This would require the query optimizer to devise a distributed query processing strategy that would, for a given distributed query, generate query plans involving fewer number of sites in order to reduce the inter site communication cost. In this paper, a distributed query plan generation algorithm, based on the honey bee mating optimization (HBMO) technique that generates query plans for a distributed query involving less number of sites and higher relation concentration in the participating sites, is presented. Further, experimental comparison of the proposed HBMO based DQPG algorithm with the GA based DQPG algorithm shows that the former is able to generate distributed query plans at a comparatively lesser total query processing cost, which in turn would lead to efficient processing of a distributed query.

Keywords: Distributed Query Processing, Swarm Intelligence, Honey Bee Mating Optimization.

1 Introduction

The objective of distributed query processing is to transform a high level distributed query, posed to a distributed database system, into an efficient distributed query execution plan in a low level language [16]. Query optimization is an important part of query processing. Since more than one execution plan could be the correct transformation of the same high level query, the one that minimizes resource consumption should be chosen [18][19]. Resource consumption is measured as the total cost of processing a distributed query where the total cost is the sum of the time spent in processing the operators of the query at various sites and the time used for inter site communication [16][19]. Another measure of resource consumption is the response time of a query [2][7]. Since operations at different nodes can be executed in parallel, the response time for a query could be substantially less than its total cost [16]. In a distributed database system, CPU costs, I/O costs and communication costs constitute the total cost that need to be minimized [19]. The CPU cost is the amount of time used for performing operations in the main memory. The I/O cost is the time

S. Ramanna et al. (Eds.): MIWAI 2013, LNCS 8271, pp. 293–304, 2013.

required for disk access. Reducing the number of disk accesses will decrease this cost. The communication cost is the time spent in data transfers between sites hosting the involved relations or fragments [19]. In centralized DBMSs, the I/O cost and the CPU cost are the only costs used for measuring the total cost. The communication cost component is the most important and dominant cost factor in distributed database systems. Most of the early proposals considering the total cost of a DDBS, ignored the I/O and the CPU cost [19]. Henceforth, minimization of inter site communication cost has become the aim of distributed query optimization. Local optimization can then be done independently using the known methods for centralized database systems. Due to faster communication networks now available, recent researches have come up with a weighted total cost [20]. Nevertheless, the overhead cost for inter site communication (e.g., software protocols) still renders the communication cost to be a major cost factor [19]. In order to reduce this communication cost, appropriate sites for performing joins need to be selected by a distributed query optimizer in the query optimization phase [2][5][21]. This optimizer would generate query plans and evaluate the cost of these plans. The query plans consuming the least amount of resources for executing a distributed query are selected for distributed query processing. This would require a trade-off between the query response time and the resource consumption.

This paper addresses the problem regarding the generation of efficient distributed query processing plans for a distributed relational query. The number of possible query processing plans increases exponentially with increase in the number of relations in a user query [11]. It becomes impossible for the query optimizer to exhaustively search all the query plans including those with moderate number of relations. The complexity of the problem increases further because of data fragmentation and replication at multiple sites in the distributed database system. In such an expansive search space, it is computationally impossible to search for the best query plan having the least cost. This, being a combinatorial optimization problem, has been addressed by randomized optimization techniques [11][13][17][23]. But the efficiency of these techniques depreciates in many cases [12]. In this paper, an attempt to generate near optimal distributed query processing plans, using the "close property" of a query plan as defined in [24], has been made. The close property of a query plan is defined by the number of sites involved in a query plan and the number (concentration) of relations, accessed by the query, at the involved sites [24]. Lesser the number of sites involved in a query plan, and greater the number/concentration of relations in these sites, the closer would be the query plan [24]. A closer query plan would involve lesser inter site communication. In addition, it would have more local join operations [22]. This would significantly improve the efficiency of distributed query processing [22].

For example, consider relations R1, R2, R3, R4, R5, R6, R7 and R8 accessed by a user query. Let there be eight sites in a DDBS, namely S1, S2, S3, S4, S5, S6, S7 and S8. The relation-site matrix is given in Figure 1. In this matrix, the indices of the cells with value 1 represent the relation and its corresponding host site.

Relations\Sites	S_1	S_2	S_3	S_4	S_5	S_6	S_7	S_8
R_1	0	0	1	1	1	0	1	1
R_2	1	1	1	1	1	1	1	0
R_3	0	0	0	0	1	0	1	1
R_4	0	1	1	1	1	0	1	1
R_5	0	0	0	1	1	0	0	1
R_6	1	0	1	0	1	0	1	0
R_7	1	1	0	1	1	0	1	0
R_8	0	1	1	1	1	1	0	1

Fig. 1. Relations with their sites

Consider the following distributed relational query:

Select B1, B2, B3, B4
From R1, R2, R3, R4, R5, R6, R7, R8
Where R1.B1=R2.B1 and R3.B2=R4.B2 and
         R5.B3=R6.B3 and R7.B4=R8.B4

Some valid query plans for above query are given in Figure 2.

QueryPlanNo	Relations								Query Plans
	R1	R2	R3	R4	R5	R6	R7	R8	
1	Site 3	Site 1	Site 5	Site 2	Site 4	Site 1	Site 1	Site 2	[3,1,5,2,4,1,1,2]
2	Site 4	Site 1	Site 7	Site 3	Site 5	Site 5	Site 2	Site 3	[4,1,7,3,5,5,2,3]
3	Site 7	Site 5	Site 8	Site 4	Site 5	Site 5	Site 4	Site 4	[7,5,8,4,5,5,4,4]
4	Site 8	Site 3	Site 7	Site 7	Site 8	Site 7	Site 7	Site 6	[8,3,7,7,8,7,7,6]

Fig. 2. Valid Query Plans

The number of possible valid query plans is computed as $(NS_1 \times NS_2 \times NS_3 \dots \times NS_n)$, where NS_i is the number of sites where relation 'i' is residing [24]. Since there are 8 relations distributed over 8 sites, the total number of possible query plans is 226800 ($5 \times 7 \times 3 \times 6 \times 3 \times 4 \times 5 \times 6$). The number of possible distributed query processing plans increases exponentially with increase in the number of relations in a user query. It becomes impossible for the query optimizer to exhaustively search all distributed query plans, even those with moderate number of relations. The complexity of the problem further increases, as the data is fragmented and replicated at multiple sites, in the distributed database system. In such an expansive search space, it is computationally impossible to search for the most 'close' distributed query plan having the least cost. This, being a combinatorial optimization problem, can be addressed by generating query plans that incur less query processing cost. Since the dominant cost in distributed query processing is the cost due to inter site communication, the aim would be to generate query plans involving lesser site-to-site communication cost. This cost, which is modeled as the query proximity cost (QPC) in [24], is used to define the 'close' query plan [24]. QPC is defined as given below:

$$QPC = \sum_{i=1}^{M} \frac{NS_i}{N} \left(1 - \frac{NS_i}{N} \right)$$

where M is the number of sites accessed by the query plan, NS_i is the number of times the ith site is used in the query plan and N is the number of relations accessed by the query.

Query plans having lower QPCs are more desirable than those with higher QPCs. The QPC [24] of the valid query plans, given in Figure 2, is computed whereafter the query plans are arranged in the increasing values of QPC as shown in Figure 3.

QueryPlanNo	Query Plan	QPC
4	[8,3,7,7,8,7,7,6]	0.66
3	[7,5,8,4,5,5,4,4]	0.68
1	[3,1,5,2,4,1,1,2]	0.75
2	[4,1,7,3,5,5,2,3]	0.81

Fig. 3. Query plans according to increasing value of QPC

The DQPG problem, discussed above, is a combinatorial optimization problem [24]. The size of the search space is large, as the number of possible query plans increases exponentially with increase in the number of relations accessed by the distributed query and also with the increase in the number of sites containing these relations [24]. The aim is to generate distributed query plans, from amongst all possible query plans, for a given distributed query at minimum QPC. This problem has been solved using the Genetic algorithm in [24]. This paper aims to solve this problem using the Honey Bee Mating Optimization (HBMO) [8][9], which is a meta-heuristic swarm intelligence [4][14] technique, used to solve combinatorial optimization problems. Two types of behaviour of honey bees have been imitated to solve combinatorial optimization problems. These are the foraging behaviour [3][6] and the marriage behaviour [1][3]. The proposed HBMO [8][9] based DQPG algorithm for generating 'close' query plans [24] for a given distributed query is based on the marriage behaviour of honey bees.

The paper is organized as follows: The HBMO based DQPG algorithm is discussed in section 2. Section 3 gives experimental results. The conclusion is given in section 4.

2   DQPG Using HBMO

As discussed above, the distributed query plan generation problem (DQPG) is a complex problem. In this paper, an attempt has been made to address this problem using HBMO [8][9]. An HBMO [8][9] based DQPG algorithm is presented that generates a query processing plan, for a given distributed query, which minimizes the QPC [24]. The HBMO [8][9] based DQPG algorithm is given in Figure 4. This algorithm generates distributed query plans based on the close property of a query plan, as defined in [24]. Inputs to the algorithm are the relation-site matrix, number of drone query plans, maximum initial speed of the queen query plan, minimum permissible speed of the queen query plan required to continue mating flights, the speed reduction scheme, size of the queen's query plan spermatheca (number of query plans to be taken for improvement), number of mating flights (iterations) to be undertaken and the number of worker bees (solution improvement heuristics). In this approach an initial population of bees, representing the initial set of drone query plans is randomly generated. This initial population, which acts as the input, is sought to be improved by the HBMO based DQPG algorithm to obtain a superior quality population of query plans.

Input: RS is Relation–Site Matrix, D is number of drone query plans, QS is Queen query plan speed, QS_{max} is maximum initial queen query plan speed, QS_{min} is Minimum permissible queen query plan speed, α is the speed reducing scheme, S is Size of queen's query plan spermatheca, W is the number of worker bees heuristic and F is Number of mating flights

Output: Top-K query plans

Method:

Step 1: Generate initial drone query plans D in accordance with RS

Step 2: Compute QPC of each drone query plan

$$QPC = \sum_{i=1}^{M} \frac{NS_i}{N} \left(1 - \frac{NS_i}{N} \right)$$

Where M is the number of sites accessed by the query plan, NS_i is the number of times the ith site is used in the query plan, N is the number of relations accessed by the query

Step 3: Select the drone query plan with least QPC as the queen query plan 'Q'

Step 4: Repeat

    Repeat // **Bee mating phase //**

    Select a drone query plan 'D$_i$' randomly from 'D'

    Compute $\Delta(QPC)=|QPC(Q)-QPC(D_i)|$

    Generate a random number 'r'

    $P(D_i)=1/exp(\Delta(QPC)/QS)$

    If $(P(D_i)) > $ 'r' then add D$_i$ to queen spermatheca S

        Decrease QS, QS=α*QS //α is speed reducing scheme

    Until S is full or QS≤QS_{min}

    Repeat // **Egg fertilization phase //**

        Select a drone query 'S$_i$' randomly from spermatheca S and use it to fertilize queen's query plan by applying crossover between S$_i$ and Q

    Until all drone query plans in spermatheca are used

    Mutate the newly generated fertilized queen query plans

    // **Brood generation phase //**

    Employ any one of the two worker bees heuristics Iterative Improvement or Simulated Annealing to feed the new brood query plans

    If (the fittest brood query plan is fitter than the queen query plan)

    Replace queen query plan with this fittest brood query plan

    Else

    Retain the old queen query plan

    Generate new population of drone query plans by discarding the old one

  Until all the mating flights F are completed

  Return Top-K brood query plans as Top-K query plans

Fig. 4. HBMO [8][9] based DQPG Algorithm

The query proximity cost (QPC), given in [24], is used to measure the fitness of these query plans. Lesser the value of QPC of a bee query plan, fitter the bee query plan is; and larger the value of QPC of a bee query plan, less fit the bee query plan is [24]. The bee query plan in the population, having maximum fitness, is the queen bee. So, the fitness of each drone query plan is computed and the one with least QPC is chosen to be the queen query plan of the initial population. The queen query plan undertakes a number of mating flights [8]. During each such mating flight, the queen query plan probabilistically mates with a number of drone query plans and collects their sperms (query plans) in her queen query plan spermatheca. This probability, as defined in [8] is given below:

$$P(D_i)=1/exp(\Delta(QPC)/QS)$$

Where $P(D_i)$ is the probability of adding the sperm of a drone query plan 'D_i' to the queen's spermatheca 'S', $\Delta(QPC)$ is the absolute difference between the QPC value of the drone query plan 'D_i' and the queen query plan 'Q', and 'QS' is the speed of the queen query plan in a mating flight (iteration). Initially, the speed of the queen query plan is assigned with a maximum value QS_{max}. In every mating flight, the queen query plan spends her energy in attempting to mate, as a result of which her speed gets reduced by a speed reducing scheme α as given below [8]:

$$QS=\alpha*QS$$

The mating process ends when the queen query plan spermatheca is full. The mating process also comes to a halt when the queen query plan reaches a speed less than the minimum permissible speed QS_{min}. In the presented approach the queen query plan has only one unique egg (i.e. the queen query plan). This queen's query plan is repeatedly used to mix with the collected sperms (i.e. drone query plans) in her spermatheca [8]. Fertilization is imitated by the crossover operator, which is used to exchange portions of the queen query plan and the drone query plan. The crossover operator used in the proposed approach causes only the portions of the drone query plans to be exchanged, while the queen query plan remain intact [9]. Fertilization is completed when portions of all the collected drone query plans are replaced by portions of the queen query plan. A very small percentage of the fertilized eggs (fertilized query plans) are mutated. Mutation is carried out by randomly selecting very few of fertilized query plans and replacing portions of these query plans with some valid portions in a random manner. The resultant new born brood query plans survive only when they are cared for and fed. These brood query plans are fed by worker bees (improvement heuristics). In this approach two search improvement heuristics namely, the iterative improvement [11][17] and simulated annealing [13][11][10][15], have been considered. After employing one of the two heuristics, the fittest brood query plan of the population is compared with the queen query plan. If this brood query plan is fitter than the queen query plan, then this brood query plan becomes the new queen query plan, replacing the old queen query plan. Thereafter a new swarm of drone query plans are generated by discarding the present population of query plans, with which the new queen query plan can mate to further improve the population of query plans. In this way fitter brood query plans are generated with the birth of new fitter queen query plan. In the event of there being no fitter brood query plan to replace the existing queen query plan, the queen query plan continues to be the queen and a new population of drone query plans is generated to mate with it. This cycle of mating, fertilization and brood generation continues until all the mating flights are undertaken. Thereafter, the Top-K query plans are produced as output.

An example that illustrates the use of the HBMO based DQPG algorithm to generate query plans for a distributed relational query is given next.

2.1   An Example

Consider a distributed database system (DDBS) with eight sites S1, S2, S3, S4, S5, S6, S7, and S8. Consider the following distributed query accessing eight relations R1, R2, R3, R4, R5, R6, R7, and R8

>Select   A1, A2, A3, A4
>From   R1, R2, R3, R4, R5, R6, R7, R8
>Where   R1.A1=R2.A1 and R3.A2=R4.A2 and
>   R5.A3=R6.A3 and R7.A4=R8.A4

Consider the relation-site matrix given in Figure 1. The objective is to generate the Top-5 distributed query plans, for the above distributed query, using the HBMO based DQPG algorithm given in Figure 4.

Using this relation-site matrix given in Figure 1, a pre-specified population of 20 drone query plans is randomly generated. These are shown in Figure 5(a). The fitness QPC[24], for each drone query plan is computed and is shown in Figure 5(a). The 8th drone query plan has the least QPC and thus is chosen as the queen query plan. The queen query plan then takes mating flights to probabilistically mate with the congregation of the drone query plans. After a successful mating the drone query plan is added to queen's spermatheca and then this particular drone query plan is removed to prevent it from mating twice. The mating phase ceases either when the queen query plan speed becomes slower than the required minimum speed or when queen query plan spermatheca is full. The drone query plans collected in the queen's spermatheca are shown in Figure 5(b). With the start of the fertilization phase, the queen query plan randomly selects a drone query plan from the spermatheca and uses it to fertilize the queen's query plan. Fertilization is accomplished through the application of a single point crossover operator between the queen's query plan and the drone's query plan. The resultant collection of query plans after the fertilization phase is shown in Figure 5(c). This collection of fertilized query plans is mutated. Mutation is carried by randomly selecting a site and replacing it with another valid site. In this iteration, the 20th query plan [4,1,7,7,5,5,5,6] is mutated to [4,5,7,7,5,5,5,6].

The newly hatched brood query plans are taken care of and fed by any of the two worker heuristics i.e. Iterative Improvement or Simulated Annealing. In this iteration, worker bee heuristic iterative improvement is selected and used to care for and feed the brood query plans. The population of brood query plans, after the application of iterative improvement, is shown Figure 6(a). Thereafter, a new fitter queen query plan is searched for, from amongst the generated brood query plans. If such a query plan is found, it replaces the old queen query plan. The fittest brood query plan (Bee 15) is fitter than the old queen query plan (Bee 8) and thus Bee 15, i.e. {8, 5, 5,5, 5, 5, 5, 8}, becomes the new queen query plan. The Top-5 query plans by the end of first iteration are shown in Figure 6(b). Thereafter, a new congregation of drone query plans is randomly generated. This continues for a pre-specified number of iterations. The Top-5 query plans, after 100 iterations, are shown in Figure 6(c).

(a) Initial Population of Query Plans

Bee	Query Plan	QPC
Bee 1	[3,1,5,2,4,1,1,2]	0.75
Bee 2	[4,1,7,3,5,5,2,3]	0.81
Bee 3	[7,5,8,4,5,5,4,4]	0.68
Bee 4	[8,3,7,7,8,7,7,6]	0.66
Bee 5	[8,1,7,7,8,1,7,8]	0.66
Bee 6	[4,6,5,2,8,5,1,2]	0.66
Bee 7	[7,6,5,5,4,7,7,4]	0.72
Bee 8	[4,5,5,5,5,5,5,6]	0.40
Bee 9	[4,3,5,4,4,6,4,4]	0.56
Bee 10	[8,6,8,8,8,7,7,8]	0.53
Bee 11	[7,1,5,2,4,1,1,2]	0.75
Bee 12	[7,6,8,8,8,7,7,8]	0.59
Bee 13	[7,3,5,4,4,6,4,4]	0.69
Bee 14	[7,5,5,5,5,5,5,6]	0.41
Bee 15	[8,5,5,5,5,5,5,6]	0.41
Bee 16	[8,6,5,2,8,5,1,2]	0.78
Bee 17	[3,6,5,2,8,5,1,2]	0.81
Bee 18	[3,1,7,3,5,5,2,3]	0.75
Bee 19	[4,1,7,3,5,5,2,3]	0.81
Bee 20	[4,1,7,7,8,1,7,8]	0.72

(b) Query Plans in Queen's Spermatheca

Bee	Query Plan	QPC
Bee1	[3,1,5,2,4,1,1,2]	0.75
Bee2	[4,1,7,3,5,5,2,3]	0.81
Bee3	[7,5,8,4,5,5,4,4]	0.68
Bee5	[8,1,7,7,8,1,7,8]	0.66
Bee7	[7,6,5,5,4,7,7,4]	0.72
Bee9	[4,3,5,4,4,6,4,4]	0.56
Bee11	[7,1,5,2,4,1,1,2]	0.75
Bee12	[7,6,8,8,8,7,7,8]	0.59
Bee15	[8,5,5,5,5,5,5,6]	0.41
Bee20	[4,1,7,7,8,1,7,8]	0.72

(c) Query Plans after Fertilization

Bee	Query Plan	QPC
Bee 1	[3,1,5,2,5,5,5,6]	0.69
Bee 2	[4,1,7,3,5,5,5,6]	0.78
Bee 3	[7,5,8,4,5,5,5,6]	0.68
Bee 5	[8,1,7,7,5,5,5,6]	0.75
Bee 7	[7,6,5,5,5,5,5,6]	0.53
Bee 9	[4,3,5,4,5,5,5,6]	0.66
Bee 11	[7,1,5,2,5,5,5,6]	0.69
Bee 12	[7,6,8,8,5,5,5,6]	0.72
Bee 15	[8,5,5,5,5,5,5,6]	0.41
Bee 20	[4,1,7,7,5,5,5,6]	0.75

Fig. 5.

(a) Brood Query Plans generated after applying Iterative Improvement

Bee	Query Plan	QPC
Bee1	[3,5,5,2,5,5,5,6]	0.56
Bee 2	[4,5,7,5,5,5,5,6]	0.56
Bee 3	[7,5,7,4,5,5,5,6]	0.65
Bee 5	[7,1,7,7,5,5,5,6]	0.68
Bee 7	[7,5,5,5,5,5,5,6]	0.40
Bee 9	[4,6,5,5,5,5,5,6]	0.53
Bee 11	[5,6,5,2,5,5,5,6]	0.53
Bee 12	[8,6,8,8,5,5,5,6]	0.66
Bee 15	[8,5,5,5,5,5,5,8]	0.37
Bee 20	[8,5,7,7,5,5,5,6]	0.66

(b) Query Plans after First Iteration

Bee	Query Plan	QPC
Bee 8	[4,5,5,5,5,5,5,6]	0.40
Bee 7	[7,5,5,5,5,5,5,6]	0.40
Bee 9	[4,6,5,5,5,5,5,6]	0.53
Bee1	[3,5,5,2,5,5,5,6]	0.56
Bee 2	[8,6,8,8,5,5,5,6]	0.56

(c) Query Plans after 100th Iteration

Bee	Query Plan	QPC
Bee11	[5,5,5,5,5,5,5,6]	0.21
Bee20	[5,5,7,5,5,5,5,6]	0.40
Bee13	[4,5,5,4,5,5,5,4]	0.46
Bee 4	[4,5,5,3,5,5,5,4]	0.53
Bee 5	[7,7,7,7,5,5,5,6]	0.59

Fig. 6.

3   Experimental Results

The GA and HBMO based DQPG algorithms were implemented in MATLAB 7.7 in a Windows XP environment. Experiments, for comparing the two algorithms, were carried out on an Intel based 2 GHz PC having 1 GB RAM. The comparisons were carried out on parameters like average QPC, the number of iterations and the Top-K query plans.

First, line graphs were plotted to compare the HBMO and the GA based DQPG algorithm on the average QPC against the number of iterations for selecting the Top-10 query plans. These graphs for 4, 8, 12 and 16 relations are shown in Figure 7. These graphs were plotted by varying the queen's spermatheca size (S = 10, 20) and the queen's speed (QS= 10, 20), represented as HBMO(QS, S). These graphs were compared with the observed crossover and mutation probabilities of the GA based DQPG algorithm (GA (Pc, Pm)), i.e. for observed values GA(0.6, 0.05) and GA(0.8, 0.1) given in [24].

It can be observed from the graphs that the HBMO based DQPG algorithm, in comparison to the GA based DQPG algorithm, is able to generate query plans having a significantly lower average QPC in each iteration for 4, 8, 12 and 16 relations. The HBMO algorithm performs best for the queen speed of QS=10 and spermatheca size of S=10 i.e. HBMO(10, 10).

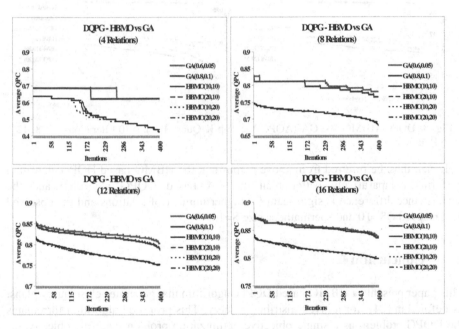

Fig. 7. DQPG–HBMO Vs. GA (AQPC Vs. Iterations)–Top-10 Query Plans–4, 8, 12, 16 Relations

Next, in order to ascertain the average QPC of the Top-K query plans over 400 iterations, line graphs were plotted for 4, 8, 12 and 16 relations and are shown in Figure 8. It can be observed from the graphs that the HBMO based DQPG algorithm, is able to generate Top-K query plans having significantly lower average QPC when compared with those generated by the GA based DQPG algorithm. This difference in average QPC is significant for higher number of relations accessed in the query. HBMO based DQPG algorithm again performed best for the queen speed QS=10 and the spermatheca size S=10 i.e. HBMO(10, 10).

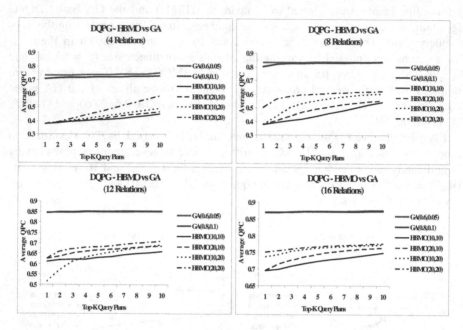

Fig. 8. DQPG–HBMO Vs. GA (AQPC Vs. Top-K Query Plans)–400 Iterations –4, 8, 12, 16 Relations

It can, thus, be inferred from above graphs that the HBMO based DQPG algorithm performs comparatively better than the GA based DQPG algorithm and the performance difference is significant for higher number of relations and the observed queen speed QS=10 and spermatheca size S=10.

4   Conclusion

This paper presents a HBMO based DQPG algorithm that generates close query plans, as defined in [24], for a given distributed query. This proposed algorithm formulates the DQPG problem as a single objective optimization problem with the objective of minimizing the QPC, as defined in [24]. The query plans so generated involve lesser number of sites, and higher concentration of relations, in the participating sites. Consequently, the query processing cost would be less, which in turn would lead to reduction in the response time of distributed queries.

Experiment based comparison of the HBMO based DQPG algorithm with the GA based DQPG algorithm shows that the former is able to generate Top-K query plans that have significantly lower average QPCs than those produced by the latter. This difference is significant when more relations are accessed in a distributed query. As a result, the query processing of a distributed query is faster, leading to reduced query response time. This in turn would facilitate the decision making process.

References

1. Abbass, H.A.: MBO: Marriage in Honey Bees Optimization a Haplometrosis Polygynous Swarming Approach. In: Proceedings of the 2001 Congress on Evolutionary Computation, pp. 207–214 (2001)
2. Apers, P.M.G., Hevner, A.R., Yao, S.B.: Optimization algorithms for distributed queries. IEEE Transactions on Software Engineering, SE-9, 57–68 (1983)
3. Batouche, M., Bitam, S.: A survey on bee colony algorithms. In: 2010 IEEE International Symposium on Parallel and Distributed Processing, pp. 1–8 (2010)
4. Blum, C., Merkle, D.: Swarm intelligence introduction and applications. Natural Computing Series. Springer (1998)
5. Cornell, D.W., Yu, P.S.: An optimal site assignment for relations in the distributed database environment. IEEE Transactions on Software Engineering 15, 1004–1009 (1989)
6. Davidović, T., Šelmić, M., Teodorović, D., Ramljak, D.: Bee colony optimization for scheduling independent tasks to identical processors. Journal of Heuristics 18(4), 549–569 (2012)
7. Epstein, R., Stonebraker, M., Wong, E.: Query processing in a distributed relational database system. In: Proc. ACM-SIGMOD Int. Conf. on Management of Data, pp. 169–180 (1978)
8. Fathian, M., Amiri, B., Maroosi, A., Application of Honey Bee Mating Optimization Algorithm on Clustering. Applied Mathematics and Computation (2007); Elsevier Inc. International Symposium on Parallel and Distributed Processing, pp. 1–8
9. Haddad, B.O., Afshar, A., Mariano, M.A., Adams, B.J.: Honey Bee Mating Optimization (HBMO) Algorithm: A New Heuristic Approach for Water Resources Optimization. Water Resources Management, 661–680 (2006)
10. Henderson, D., Jacobson, S.H., Johnson, A.W.: The Theory and Practice of simulated annealing, State-of-the-Art Handbook in Metaheuristic, pp. 287–319. Kluwer Academic Publishing, Norwell (2003)
11. Ioannidis, Y.E., Kang, Y.C.: Randomized Algorithms for Optimizing Large Join Queries. In: Proc. ACM-SIGMOD Intl. Conf. on Management of Data, Atlantic City, NJ, pp. 312–321 (1990)
12. Ioannidis, Y.E., Kang, Y.C.: Left-deep vs. bushy trees: An analysis of strategy spaces and its implementations on query optimization. In: SIGMOD International Conference on Management of Data, Denver, pp. 168–177 (1991)
13. Ioannidis, Y.E., Wong, E.: Query Optimization by Simulated Annealing. In: Proc. of the 1987 ACM- SIGMOD Conference on the Management of Data, San Francisco, CA, pp. 9–22 (May 1987)
14. Kennedy, J., Eberhart, R.C.: Swarm Intelligence. Morgan Kaufmann Publishers (2001)
15. Kirkpatrick, S., Gelatt Jr., C., Vecchi, M.: Optimization by Simulated Annealing. Science 220(4598), 671–680 (1983)
16. Kossmann, D.: The State of the Art in Distributed Query Processing. ACM Computing Surveys 32(4), 422–469 (2000)
17. Nahar, S., Sahani, S., Shragowitz, E.: Simulated Annealing and Combinatorial Optimization. In: Proceedings of the 23rd Design Automation Conference, pp. 293–299 (1986)
18. Ozsu, M.T., Valduriez, P.: Distributed Database Systems: Where are we now? IEEE Computer 24(8), 68–78 (1991)
19. Ozsu, M.T., Valduriez, P.: Principles of Distributed Database Systems, 3rd edn. Springer (2010)

304   T.V. Vijay Kumar, B. Arun, and L. Kumar

20. Page, T.W., Popek, G.J.: Distributed data management in local area networks. In: Proc. ACM SIGACT–SIGMOD Symp. on Principles of Database Systems, pp. 135–142 (1985)
21. Sacco, M.S., Yao, S.B.: Query optimization in distributed data base systems. In: Yovits, M. (ed.) Advances in Computers, vol. 21, pp. 225–273 (1982)
22. Segev, A.: Optimization of join operations in horizontally partitioned database systems. ACM Transactions on Database Systems 11(1), 48–80 (1986)
23. Swami, A., Gupta, A.: Optimization of large join queries. In: Proceedings of 1988 ACM-SIGMOD Conference, Chicago, pp. 8–17 (1998)
24. Vijay Kumar, T.V., Singh, V., Verma, A.K.: Distributed query processing plans generation using genetic algorithm. International Journal of Computer Theory and Engineering 3(1), 38–45 (2011)

Axiomatic Foundations
of Generalized Qualitative Utility

Paul Weng*

LIP6, UPMC
paul.weng@lip6.fr

Abstract. The aim of this paper is to provide a unifying axiomatic jus-
tification for a class of qualitative decision models comprising among others optimistic/pessimistic qualitative utilities, binary possibilistic utility,
likelihood-based utility, Spohn's disbelief function-based utility. All those
criteria that are instances of Algebraic Expected Utility have been shown
to be counterparts of Expected Utility thanks to a unifying axiomatization in a von Neumann-Morgenstern setting when non probabilistic decomposable uncertainty measures are used. Those criteria are based on
$(\oplus, \otimes)$ operators, counterpart of $(+, \times)$ used by Expected Utility, where
$\oplus$ is an idempotent operator and $\otimes$ is a triangular norm. The axiomatization is lead in the Savage setting which is a more general setting than
that of von Neumann-Morgenstern as here we do not assume that the
uncertainty representation of the decision-maker is known.

1 Introduction

The study of decision models is an important and necessary step in the task of
building autonomous agents and support systems. Indeed, in automatic decision-
making, an agent has to act "rationally" while in human assistance systems, to be
able to help, an agent has to model human preferences. For both cases, decision
theory can answer the critical question of which decision model to implement.

Decision models are generally axiomatically studied. An axiomatization reveals the exact properties of a decision model and shows what kind of preferences it can or can not describe. For instance, the axiomatization of Expected
Utility [1] allowed its limitations to be shown [2,3]. This axiomatic work can be
lead in two different settings. In a von Neumann-Morgenstern (vNM) setting,
the uncertainty representation is assumed to be known while in a Savage setting,
this assumption is relaxed. The second one is thus more general and interestingly, in this setting, the uncertainty representation can be revealed from the
decision-maker's preferences.

Although Expected Utility (EU) is used in many applications, this quantitative model is sometimes difficult to implement as one needs to elicit precise utilities and probabilities. When this information is not available or when
it is too costly to elicit, more qualitative decision models can be preferred.

---

* Funded by the French National Research Agency under grant ANR-10-BLAN-0215.

S. Ramanna et al. (Eds.): MIWAI 2013, LNCS 8271, pp. 305–316, 2013.

Besides as previously underlined by [4], probability is not always convenient to represent situations of partial ignorance. Possibility Theory can be considered in those situations and it is particularly justified in case of representing human beliefs as shown in psychological experiments lead by [5].

In this paper, we propose axiomatizations in a Savage setting for a class of decision models using possibilistic uncertainty representations: Generalized Qualitative Utilities, which have been proposed by [6]. Instances of this class are among others optimistic/pessimistic utilities [7], binary possibilistic utility [8], likelihood-based utility [9], Spohn disbelief function-based utility [10]. All those models have been axiomatized in a vNM setting and some of them have been axiomatized in a Savage framework [11,12].

This work allows us to justify axiomatically previously proposed models in a very general way and to unify them under a new decision model, generalized Sugeno integral. This unifying approach shows the similarities and the differences of all models belonging to the class of generalized qualitative utilities. This paper builds on the axiomatization proposed by [11].

In the next section, we give the used notations and recall the definitions of the many utilities that we work on. In section 3, we introduce generalized Sugeno integral and give its axiomatization. Building from this, we are then able to propose representation theorems of generalized qualitative utility and optimistic/pessimistic versions of likelihood based-utility. In section 4, we extend our results to binary utilities allowing us to provide an axiomatization of (binary) likelihood based-utility. Finally we conclude in the last section and show graphically how all those models are related to each other.

2   Decision Models

2.1   Definitions and Notations

In a problem of decision under uncertainty, a decision-maker has to choose an action, named act in the Savage setting, whose consequence is not known precisely. The decision problem can be formalized by defining: S a set of states; $X = \{x_1, \ldots, x_n\}$ a finite set of decreasingly ordered consequences (Elements $0_X = x_1$ and $1_X = x_n$ denote respectively the worst and the best consequences), $\mathcal{F}$ a finite set of acts, i.e., functions from the set of states to the set of consequences ($\mathcal{F} = X^S$). A state (or state of nature) describes a possible configuration of the environment. Uncertainty is defined as uncertainty about the actual state of nature. As the decision-maker does not know exactly in which state he is, he is uncertain about the consequences of his actions.

The decision maker's preference relation (which is simply a binary relation) over acts $\mathcal{F}$ is denoted by $\succsim_{\mathcal{F}}$, which reads "at least as good as". For any relation $\succsim$, $\succ$ and $\sim$ denote respectively the asymmetric and the symmetric part of $\succsim$.

For two acts f and g, an event A, fAg denotes the conditional act consisting in doing f if event A occurs and doing g otherwise, i.e., $\forall s \in A, fAg(s) = f(s)$ and $\forall s \in \overline{A}, fAg(s) = g(s)$. The constant act equal to $x \in X$ everywhere is denoted in bold $\mathbf{x}$. Relation $\succsim_{\mathcal{F}}$ restricted to constant acts induces a preference relation over consequences, denoted $\succsim_X$. Events, which are simply sets of states form a sigma-algebra, denoted by E. For an event A, binary act $1_X A 0_X$ is an act giving the best consequence 1_X when A occurs and the worst outcome 0_X otherwise. Preference relation $\succsim_{\mathcal{F}}$ restricted on those binary acts defines a plausibility relation over events, $\succsim_S$. Indeed for two events A and B, $1_X A 0_X \succsim_{\mathcal{F}} 1_X B 0_X$ means the decision-maker judges A at least as plausible as B. This is written $A \succsim_S B$. The complement of an event $A \in E$ is denoted $\overline{A}$.

The scale on which uncertainty (and sometimes preference) is measured is denoted by L which is totally ordered by $\geq_L$. Its lowest and greatest elements are denoted respectively by 0_L and 1_L. Operators max and min on scale L are respectively written $\vee$ and $\wedge$. The order reversing involution in L is denoted by n, i.e., $n(0_L) = 1_L, n(1_L) = 0_L$ and $\forall \lambda, \lambda' \in L, \lambda \geq_L \lambda' \Rightarrow n(\lambda) \leq_L n(\lambda')$. A *capacity* $\sigma : E \to L$ is a very general uncertainty representation. It is a measure that satisfies: $\sigma(\emptyset) = 0_L$, $\sigma(S) = 1_L$ and $\forall A, B \in E, A \subseteq B \Rightarrow \sigma(A) \leq_L \sigma(B)$. Probability, possibility, belief functions, kappa rankings and plausibility measures are all examples of capacities. We recall the definition of a possibility distribution. A capacity is a possibility distribution (denoted π) iff it additionally satisfies $\forall A \in E, \pi(A) = \bigvee_{s \in A} \pi(s)$. Distribution π can be viewed as an encoding of the ranking of the states in terms of plausibility. It can be interpreted as follows: $\pi(s) = 0_L$ means that state s is impossible, $\pi(s) = 1_L$ means that s is totally possible, when $0_L <_L \pi(s) <_L 1_L$, s is only possible to some extent, i.e., there are states more possible than s. We assume that there exists at least one state of possibility 1_L and of course, there could be several ones. A *basic utility assignment* which represents the decision-maker's preferences on consequences is a mapping that associates a value in a scale to each consequence in X. When scale L is used, a basic utility assignment is denoted $\mu : X \to L$. For clarity, we will change notations when the scale is different.

2.2 Qualitative Utilities

Optimistic/Pessimistic Utilities. We now present optimistic and pessimistic utilities, which have already been axiomatized in a vNM and a Savage settings [7,11]. For these criteria, uncertainty is assumed to be represented by a possibility distribution. Those qualitative utilities are functions from $\mathcal{F}$ to L. Optimistic utility U^+ and pessimistic utility U^- are defined by:

$$U^+(f) = \bigvee_{s \in S} \big(\pi(s) \wedge \mu(f(s)) \big) \qquad U^-(f) = \bigwedge_{s \in S} \big(n(\pi(s)) \vee \mu(f(s)) \big)$$

They are generalizations of respectively the maximax and the maximin criteria, which have been axiomatized by [13]. Using order reversing function $n : X \to X$, let $n \circ f$ be the act that is defined by $\forall s \in S, (n \circ f)(s) = n(f(s))$. Then the *dual* of a preference relation $\succsim_{\mathcal{F}}$ over $\mathcal{F}$, denoted $\succsim_{\mathcal{F}}^T$, is defined by $f \succsim_{\mathcal{F}}^T g \Leftrightarrow n \circ g \succsim_{\mathcal{F}} n \circ f$. Obviously the dual of the dual of a preference relation is the relation itself. By extension, we say that a decision model is the dual of another one if the preference relation defined by the former decision model is the dual of the relation defined by the latter. Pessimistic utility is the dual of optimistic utility.

Binary Possibilistic Utility. Binary[1] possibilistic utility (PU) introduced by [8] unifies the two previous qualitative decision models. It takes values in a particular scale L_2, which is a binary scale defined by the following set:

$$L_2 = \{\langle \lambda, \mu \rangle : \lambda, \mu \in L, \lambda \vee \mu = 1_L\}.$$

The scale is binary in the sense that a value of this scale is either in interval $[\langle 1_L, 0_L \rangle, \langle 1_L, 1_L \rangle]$ or in interval $[\langle 0_L, 1_L \rangle, \langle 1_L, 1_L \rangle]$. A pair in L_2 can be interpreted as a binary possibility distribution over two consequences : the best and the worst ones. The first element of the pair would be the possibility to get the best consequence and the second would be the possibility to get the worst consequence. Then a natural order relation can be defined on L_2:

$$\langle \lambda, \mu \rangle \geq_{L_2} \langle \lambda', \mu' \rangle \iff (\lambda \geq_L \lambda' \text{ and } \mu \leq_L \mu').$$

Indeed, the decision maker naturally prefers higher possibility of getting the best outcome and lower possibility of getting the worst one. As the order on scale L_2 is total, L_2 is in fact monodimensional.

Here, the basic utility assignment which takes values in binary scale L_2 is denoted by $u : X \to L_2$. Operator $\vee$ is extended as an operator on $L_2 \times L_2$:

$$\langle \lambda, \mu \rangle \vee \langle \lambda', \mu' \rangle = \langle \lambda \vee \lambda', \mu' \vee \mu' \rangle.$$

Note this operator is not operator max on L_2 (computed as $\max(\langle \lambda, \mu \rangle, \langle \lambda', \mu' \rangle) = \langle \lambda \vee \lambda', \mu \wedge \mu' \rangle$). Operator $\wedge$ is extended as an operator on $L \times L_2$:

$$\lambda' \wedge \langle \lambda, \mu \rangle = \langle \lambda' \wedge \lambda, \lambda' \wedge \mu \rangle.$$

In the same way as operator max, operator min on L_2 is given by $\min(\langle \lambda, \mu \rangle, \langle \lambda', \mu' \rangle) = \langle \lambda \wedge \lambda', \mu \vee \mu' \rangle$. PU is defined as a utility function from $\mathcal{F}$ to L_2:

$$PU(f) = \bigvee_{s \in S} \big(\pi(s) \wedge u(f(s)) \big). \tag{1}$$

As noted by [8], when the basic utility has the following restricted form: $\forall x \in X, \exists \lambda \in L, u(x) = \langle \lambda, 1_L \rangle$ then PU is an optimistic utility. Dually when we have $\forall x \in X, \exists \mu \in L, u(x) = \langle 1_L, \mu \rangle$, PU is a pessimistic utility. These remarks imply that optimistic and pessimistic utilities exploit only half of scale L_2.

PU is said to be *autodual* as it is its own dual. This property is shared by EU. Intuitively, it means that the decision model when stating a preference between two acts looks at how well and how bad these two acts do.

Sugeno Integral. We now present the Sugeno integral, which can be thought of as a very general decision criterion where uncertainty is assumed to be modeled by a capacity. It can be seen as the qualitative counterpart of Choquet

---

[1] This particular framework should not be confused with the bipolar one as stated by [14] for example. Here only one possibility distribution is used while in the bipolar setting two possibility distributions are exploited.

Expected Utility (CEU) [15,16], which is a generalization of EU. CEU has a higher descriptive power than that of EU. Being qualitative counterparts of EU, optimistic and pessimistic utilities are special cases of Sugeno integrals when the capacity is taken as respectively a possibility and a necessity distributions.

The utility of an act described by a Sugeno integral writes:

$$SU(f) = \bigvee_{x \in X} (\mu(x) \wedge \sigma(F_x)) \qquad (2)$$

where $F_x = \{s \in S : f(s) \succsim_X x\}$. While the quantitative models $(CEU$ or $EU)$ compute a weighted average, the Sugeno integral $SU(f)$ is the median of the utilities and uncertainty values in the following set [17]: $\{\mu(x) : x \in X\} \cup \{\sigma(F_x) : x \in X, x \succ_X 0_X\}$.

2.3 Axiomatizations of Qualitative Utilities

Sugeno Integral. [11] axiomatized this criterion using the following axioms:

Sav1 Preference relation $\succsim_{\mathcal{F}}$ over $\mathcal{F}$ is a total preorder.
WS3 $x \succsim_X y \Rightarrow \forall A \in E, \forall h \in \mathcal{F}, xAh \succsim_{\mathcal{F}} yAh$
Sav5 $\exists x, x' \in X, x \succ_X x'$
RCD $\forall f, g \in \mathcal{F}, \forall x \in X, g \succ_{\mathcal{F}} f$ and $\mathbf{x} \succ_{\mathcal{F}} f \Rightarrow g \wedge \mathbf{x} \succ_{\mathcal{F}} f$
RDD $\forall f, g \in \mathcal{F}, \forall x \in X, f \succ_{\mathcal{F}} g$ and $f \succ_{\mathcal{F}} \mathbf{x} \Rightarrow f \succ_{\mathcal{F}} g \vee \mathbf{x}$

Axiom $Sav1$ states that any two acts can be compared and the preference relation is transitive. Axiom $WS3$ declares that when there is a preference between two consequences, this preference can not reverse if considering two acts giving these consequences on a certain event and the same consequences on the complementary event. Axiom $Sav5$ is enforced in order to avoid triviality. Axiom RCD states that lowering the consequences of an act f, which is preferred to an act g, to a constant consequence that is also preferred to g still yields a better act than g. Finally axiom RDD is the dual of RCD. It says that an act f is preferred to an act g even if the worst consequences of g are improved to a constant consequence that is less preferred than f. As a side note, EU verifies neither RCD nor RDD [11]. Sugeno integrals have been characterized by [11]:

Theorem 1. *If $\succsim_{\mathcal{F}}$, a preorder over $\mathcal{F}$, satisfies Sav1, WS3, Sav5, RCD and RDD then there are a qualitative scale $(L, \geq_L)$, a basic utility assignment $\mu : X \to L$ and a capacity $\sigma : E \to L$ such that $f \succsim_{\mathcal{F}} f' \Leftrightarrow SU(f) \geq_L SU(f')$.*

Optimistic and Pessimistic Utilities. To enforce the optimistic utility, axiom RDD needs to be replaced by a stronger axiom:

OPT. $\forall f, g \in \mathcal{F}, \forall A \in E, f \succ_{\mathcal{F}} fAg \Rightarrow gAf \succsim_{\mathcal{F}} f$

This axiom says that if an act can be strictly downgraded in a certain event, this event is judged plausible enough to attract all the attention of the decision maker and what happens on the complementary event is not relevant.

The representation theorem for the optimistic utility as stated by [11]:

Theorem 2. *If $\succsim_{\mathcal{F}}$, a preorder over $\mathcal{F}$, satisfies Sav1, WS3, Sav5, RCD and OPT then there are a qualitative scale $(L, \geq_L)$, a basic utility assignment $v : X \to L$ and a possibility distribution $\pi : E \to L$ s.t. $f \succsim_{\mathcal{F}} f' \Leftrightarrow U^+(f) \geq_L U^+(f')$.*

For the pessimistic utility, RCD is to be replaced by:

PES. $\forall f, g \in \mathcal{F}, \forall A \in E, fAg \succ_{\mathcal{F}} f \Rightarrow f \succsim_{\mathcal{F}} gAf$

This axiom, which is the dual of OPT states that if an act is strictly improved in a certain event then it can not be strictly improved in the complementary event. And the pessimistic utility [11] is characterized by:

Theorem 3. *If $\succsim_{\mathcal{F}}$, a preorder over $\mathcal{F}$, satisfies Sav1, WS3, Sav5, RDD and PES then there are a qualitative scale $(L, \geq_L)$, a basic utility assignment $v : X \to L$ and a possibility distribution $\pi : E \to L$ s. t. $f \succsim_{\mathcal{F}} f' \Leftrightarrow U^-(f) \geq_L U^-(f')$.*

Binary Possibilistic Utility. PU is a Sugeno integral [12] with uncertainty modeled by *autodual possibilistic* measures, i.e., couples $\langle \pi(A), \pi(\overline{A}) \rangle$.

For any act $h \in \mathcal{F}$, $h^{\prec_{\mathcal{F}}}$ denotes the set of acts $\{f \in \mathcal{F} : f \prec_{\mathcal{F}} h\}$ and $h^{\succsim_{\mathcal{F}}} = \{f \in \mathcal{F} : f \succsim_{\mathcal{F}} h\}$. They are respectively the set of acts that are less preferred than h and its complement. The following axioms are required to state our theorem:

AD. Preference relation $\succsim_{\mathcal{F}}$ over $\mathcal{F}$ is autodual
OPT'. $\exists h \in \mathcal{F}, 1_{\mathbf{X}} \succ_{\mathcal{F}} h \succ_S 0_{\mathbf{X}}$ s.t. $\forall f, g \in h^{\succsim_{\mathcal{F}}}, \forall A \in E, f \succ_{\mathcal{F}} fAg \Rightarrow gAf \succsim_{\mathcal{F}} f$

Axiom AD states that the preferences of the decision-maker is autodual. Axiom OPT' states that the set of acts can be divided into two parts, where the set of less preferred acts satisfies axiom OPT.

Theorem 4. *If $\succsim_{\mathcal{F}}$, a preorder over $\mathcal{F}$, satisfies Sav1, WS3, Sav5, RCD, OPT' and AD then there are a qualitative scale $(L, \geq_L)$, a basic utility assignment $u : X \to L_2$ and a possibilistic distribution $\pi : E \to L$ such that $f \succsim_{\mathcal{F}} f' \Leftrightarrow PU(f) \geq_{L_2} PU(f')$.*

2.4   Semi-qualitative Utilities

Binary Likelihood-Based Utility. Consider the statistical problem described by $(Y, \Theta, P_{\theta \in \Theta})$ where Y is a random variable whose probability distribution is unknown, Θ is a parameter set and $P_{\theta \in \Theta}$ is a family of probability distributions. The real probability distribution of Y is assumed to be one of $P_{\theta \in \Theta}$. When observing a new data $Y = y$, one can compute the likelihood of a parameter θ by $P_\theta(Y = y)$. Thus one can define a likelihood function valued on $L = [0, 1]$ by $l(\theta) = P_\theta(Y = y)$. As noted by [9], likelihood can be viewed as a possibility measure. So from now on, we will denote likelihood l simply by π.

When using a likelihood function as a uncertainty representation, [9] justified axiomatically binary likelihood-based utility (LU) in a vNM setting.

LU is then defined as a utility function from $\mathcal{F}$ to L_2:

$LU(f) = \bigvee_{s \in S} \big(\pi(s)u(f(s))\big)$ where $u : X \to L_2$ is a basic utility assignment.

This criterion can be considered as an intermediate model between qualitative model such as PU and quantitative model such as EU. To the best of our knowledge, it has not yet received a Savagean axiomatization. One of the aims of this paper is to fill this gap.

Spohn's Disbelief Function-Based Utility. A Spohn's disbelief function that is a uncertainty representation is defined by a set function $\kappa : E \to \mathbb{N} \cup \{\infty\}$ satisfying $\kappa(S) = 0$, for event $A \neq \emptyset$, $\kappa(A) = \min_{s \in A} \kappa(\{s\})$ and $\kappa(\emptyset) = \infty$. The value of κ measures the surprise of the decision-maker.

Here scale $(L, \geq_L)$ can be defined by $(\mathbb{N} \cup \{\infty\}, \leq)$. Note that as the order is reversed, $\vee$ and $\wedge$ are exchanged. Spohn's disbelief function-based utility (κU) has been axiomatically justified in a vNM setting by [10]. κU is then defined as a utility function from $\mathcal{F}$ to L_2:

$$\kappa U(f) = \bigvee_{s \in S} \big(\kappa(s) + u(f(s))\big). \tag{3}$$

where $u : X \to L_2$ is a basic utility assignment.

This criterion is related to other decision models. It can be obtained from EU when using orders of magnitude of probabilities and utilities [10]. Besides it is isomorphic to a binary likelihood-based utility when equivalence classes of consequences and equivalence classes of events are countable.

2.5   Generalized Qualitative Utility

[7] have axiomatized, in a vNM setting, generalized qualitative utility (GU), which is a class of decision models that use a possibilistic uncertainty representation. GU generalizes optimistic and pessimistic qualitative utilities.

Before presenting GU, let us recall the definition of a it triangular norm or *t-norm*, denoted $\top$, which are operators defined on $[0,1] \times [0,1]$ into $[0,1]$. Operator $\top$ is a t-norm iff for all $x, y, z \in [0,1]$:

T1 Commutativity: $x \top y = y \top x$
T2 Associativity: $x \top (y \top z)) = (x \top y) \top z$
T3 Monotony: $x \top y \geq x \top z$ if $y \geq z$
T4 Limit condition: $x \top 1 = x$

min, $\times$ are two classic examples of t-norms. For a given t-norm $\top$, one can define a *triangular conorm* or *t-conorm*, denoted $\perp$, which are operators defined on $[0,1] \times [0,1]$ into $[0,1]$ by $x \perp y = 1 - \big((1-x) \top (1-y)\big)$. Thus, operator $\perp$ is a t-conorm iff it satisfies $T1$ to $T3$ and for all $x \in [0,1]$:

S4 Limit condition: $x \perp 0 = x$

Two examples of t-conorms are max and probabilistic sum defined by $x \perp y = x + y - xy$. GU takes its value on the same scale as the possibility scale and thus has two versions: generalized optimistic and generalized pessimistic utilities, which are functions from $\mathcal{F}$ to $[0,1]$. Generalized optimistic utility GU^+ and generalized pessimistic utility GU^- write for any t-norm $\top$ and its associated t-conorm $\perp$:

$$GU^+(f) = \bigvee_{s \in S} \left(\pi(s) \top v(f(s)) \right) \qquad GU^-(f) = \bigwedge_{s \in S} \left(n(\pi(s)) \perp v(f(s)) \right)$$

It is easy to see that those criteria are generalizations of U^+ and U^- when $\top = \wedge$ and generalizations of LU^+ and LU^- when $\top = \times$.

While optimistic/pessimistic utilities (and binary possibilistic utility) are special cases of Sugeno integrals, the last two criteria GU^+ and GU^- can be viewed as instances of a generalized Sugeno integral that we introduce now.

3   Proposed Axiomatizations

Generalized Sugeno Integrals. A utility function defined as a generalized Sugeno integral (GSU) is a function from $\mathcal{F}$ to L:

$$GSU(f) = \bigvee_{x \in X} \left(\mu(x) \top \sigma(F_x) \right) \tag{4}$$

where $\top$ is a t-norm[2] on L, $F_x = \{s \in S : f(s) \succsim_X x\}$ and $\sigma : E \to L$ is a monotone set function representing the belief of the decision-maker. Note that we do not define GSU as a function taking values in $[0,1]$ as we want to be general and because it is needed afterwards. Thus GSU could be defined on various scales, L_2 for example.

We now present the axiom that we will use in the axiomatization of GSU.

EX $(xA0_X \sim_{\mathcal{F}} 1_X B0_X$ and $yA0_X \sim_{\mathcal{F}} 1_X C0_X) \Rightarrow xC0_X \sim_{\mathcal{F}} yB0_X$

Assume that $x \succsim_X y$. We say that binary act $1_X B0_X$ is preferred to $1_X C0_X$ with a premium defined by the couple (x, y). This axiom states that if a binary act is preferred to another one with a certain premium (x, y), this preference can be cancelled by switching the two consequences (x, y).

We also use axiom $WS4$ introduced by [18].

WS4 $xAx' \succsim xBx' \Rightarrow yAy' \succsim yBy'$

where $x \geq_P y >_P y' >_P x', A, B \subseteq S$. This axiom entails some coherence in the preferences between binary acts. Changing to less extreme consequences in equivalent binary acts keep those acts equivalent. This axiom is not needed in the previous axiomatizations as it is implied by the set of axioms used in the

---

[2] To simplify the exposition, we call an operator on L a t-norm iff it satisfies $T1$ to $T4$ expressed on L. The same convention will be taken for t-conorm.

representation theorems. However, here as we work in a relaxed framework, this axiom has to be specified.

Thanks to these axioms, we can characterized[3] generalized Sugeno integrals:

Theorem 5. *If $\succsim_{\mathcal{F}}$, a preorder over $\mathcal{F}$, satisfies $Sav1$, $WS3$, $WS4$, $Sav5$, EX and RDD then there are a totally ordered scale L equipped with a t-norm $\top$, a basic utility assignment $\mu : X \to L$ and a capacity $\sigma : E \to L$ such that $f \succsim_{\mathcal{F}} f' \Leftrightarrow GSU(f) \geq_L GSU(f')$.*

Generalized Qualitative Utility.

We can now axiomatize GU^+ as it is a generalized Sugeno integral when using a possibility distribution by enforcing OPT instead of RDD. To better understand OPT, we prove a result that shows how strong it is as it imposes that only one pair of consequence and event is important in determining the value of an act.

Proposition 1. *If $\succsim_{\mathcal{F}}$ satisfies $Sav1$, $WS3$ and OPT then for any act $f \in \mathcal{F}$, which can be written $f = x_1 A_1 x_2 A_2 \ldots x_n A_n$, $f \sim_{\mathcal{F}} \max_{i=1,\ldots,n} x_i A_i 0_X$.*

The representation theorems for the optimistic and the pessimistic cases can be written as follows:

Theorem 6. *If $\succsim_{\mathcal{F}}$, a preorder over $\mathcal{F}$, satisfies $Sav1$, $WS3$, $WS4$, $Sav5$, EX and OPT then there are a basic utility assignment $\mu : X \to [0,1]$ and a possibility distribution $\pi : E \to [0,1]$ such that $f \succsim_{\mathcal{F}} f' \Leftrightarrow GU^+(f) \geq GU^+(f')$.*

If $\succsim_{\mathcal{F}}$, a preorder over $\mathcal{F}$, satisfies $Sav1$, $WS3$, $WS4$, $Sav5$, EX and PES then there are a basic utility assignment $\mu : X \to [0,1]$ and a possibility distribution $\pi : E \to [0,1]$ such that $f \succsim_{\mathcal{F}} f' \Leftrightarrow GU^-(f) \geq GU^-(f')$.

To obtain instances of the class of generalized qualitative utilities, one needs to enforce more restrictive axioms. Doing so, the differences of two instances can be highlighted by comparing their two sets of axioms. In the following section, we propose extra axioms to be added to get likelihood-based utilities.

Optimistic/Pessimistic Likelihood-Based Utilities.

Likelihood-based utility as presented in section 2.4 is a binary utility. One can restrict it to its optimistic and pessimistic versions. If basic utility assignment u only takes values in $\{\langle \lambda, 1_L \rangle : \lambda \in L\}$ then LU can be said to be an optimistic utility. Dually when u only takes values in $\{\langle 1_L, \mu \rangle : \mu \in L\}$ then LU can be qualified pessimistic. The two thus defined criteria would be counterparts of optimistic and pessimistic qualitative utilities (Section 2.2).

For a generalized optimistic qualitative utility to be an optimistic likelihood-based utility, one has to enforce this extra axiom that implies that the t-norm used in GU^+ is a strict t-norm:

ST $A \succ_S A' \Rightarrow \forall x \neq 0_X, xA0_X \succ_{\mathcal{F}} xA'0_X$

C1 $\forall n \in \mathbb{N}, xA_n0_X \succsim_{\mathcal{F}} f \Rightarrow x(\lim_{n\to\infty} A)0_X \succsim_{\mathcal{F}} f$

C2 $\forall n \in \mathbb{N}, x_nA0_X \succsim_{\mathcal{F}} f \Rightarrow (\lim_{n\to\infty} x_n)A0_X \succsim_{\mathcal{F}} f$

---

[3] For lack of space, we omit the proofs. They are available on a longer version of this paper at http://www-desir.lip6.fr/~weng/pub/miwai2013-2.pdf.

Axiom ST has a natural interpretation: Binary acts that gives with stronger belief the better outcome are strictly preferred. Axioms $C1$ and $C2$ are two continuity conditions. The representation theorems for the optimistic and pessimistic cases can then be stated as follows:

Theorem 7. *If $\succsim_F$, a preorder over F, satisfies $Sav1$, $WS3$, $WS4$, $Sav5$, EX, OPT, ST, $C1$ and $C2$ then there are a basic utility assignment $\mu : X \to [0,1]$ and a possibility distribution $\pi : E \to [0,1]$ such that $f \succsim_F f' \Leftrightarrow LU^+(f) \geq LU^+(f')$.*

If $\succsim_F$, a preorder over F, satisfies $Sav1$, $WS3$, $WS4$, $Sav5$, EX, PES, ST, $C1$ and $C2$ then there are a basic utility assignment $\mu : X \to [0,1]$ and a possibility distribution $\pi : E \to [0,1]$ such that $f \succsim_F f' \Leftrightarrow LU^-(f) \geq LU^-(f')$.

4   Extension to Binary Utilities

Generalized Binary Qualitative Utility. Generalized optimistic/pessimistic qualitative utilities can be easily unified in the same fashion as optimistic and pessimistic utilities with binary possibilistic utility. We now present this quick development and call the unifying criterion the generalized binary qualitative utility (GPU).

In this section, scale L is taken as $[0,1]$ on which a t-norm $\top$ is defined. T-norm $\top$ is extended as an operator on $L \times L_2$: $\lambda' \top \langle \lambda, \mu \rangle = \langle \lambda' \top \lambda, \lambda' \top \mu \rangle$.

GPU is then defined as a function from F to L_2:

$$GPU(f) = \bigvee_{s \in S} \left(\pi(s) \top u(f(s)) \right). \tag{5}$$

When the basic utility assignment has the following restricted form: $\forall x \in X, \exists \lambda \in L, u(x) = \langle \lambda, 1_L \rangle$ then GPU becomes GU^+. Dually when we have $\forall x \in X, \exists \mu \in L, u(x) = \langle 1_L, \mu \rangle$, GPU becomes GU^-.

[19] has proposed an axiomatization for GPU in a vNM setting as GPU is in fact an instance of algebraic expected utility. But no Savagean axiomatization for this criterion has been proposed yet. We now propose such an axiomatization:

Theorem 8. *If $\succsim_F$, a preorder over F, satisfies $Sav1$, $WS3$, $WS4$, $Sav5$, EX, OPT' and AD then there are a basic utility assignment $\mu : X \to L_2$ and a possibility distribution $\pi : E \to L$ such that $f \succsim_F f' \Leftrightarrow GPU(f) \geq_L GPU(f')$.*

Binary Likelihood-based Utility. Finally with the help of the previous results, a representation theorem is stated for binary likelihood-based utility.

Theorem 9. *If $\succsim_F$, a preorder over F, satisfies $Sav1$, $WS3$, $WS4$, $Sav5$, EX, OPT', ST, $C1$, $C2$ and AD then there are a basic utility assignment $\mu : X \to L_2$ and a possibility distribution $\pi : E \to L$ such that $f \succsim_F f' \Leftrightarrow LU(f) \geq_L LU(f')$.*

As a side note, one can check easily that Spohn's disbelief function-based utility satisfies all the axioms and in the construction of LU^+ and LU^-, one can use the additive generator of the t-norm [20] to build that criterion. However one extra technical axiom needs to be added to ensure that the scale is countable.

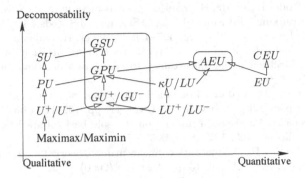

Fig. 1. Generalized Qualitative Utilities

5   Conclusion

In this paper, we have proposed axiomatizations in a Savage framework for previous proposed decision models that are instances of generalized qualitative models. Doing so, a better picture of how those criteria relate to each other can be drawn (Fig. 1). Models represented in a frame are general classes of decision models. An arrow can be read as "is an instance of". Except for those general models, decision criteria on the left hand-side are more qualitative in the sense that they require less information to be implemented whereas as we move on the right, models become more quantitative. Models on the bottom of the figure use a uncertainty representation that is decomposable while on the top, we have more general models that relax the hypothesis of decomposability.

Our work reveals the common properties that all those decision models share and in the same time underlines their differences from a decision-theoretic point of view. Besides the proposed axioms can be used to test if a decision-maker uses such qualitative utilities as decision criteria and represents his beliefs with possibility measures, likelihood functions or Spohn's disbelief functions. Finally as a by-product, during the process of axiomatizing those utilities, we introduced a generalization of Sugeno Integrals.

References

1. Fishburn, P.: Utility theory for decision making. Wiley (1970)
2. Allais, M.: Le comportement de l'homme rationnel devant le risque: critique des postulats de l'école américaine. Econometrica 21(53), 503–546
3. Ellsberg, D.: Risk, ambiguity, and the Savage axioms. Quarterly Journal of Economics 75, 643–669 (1961)
4. Dubois, D., Prade, H.: An introduction to possibilistic and fuzzy logics. In: Readings in Uncertain Reasoning, pp. 742–761. Morgan Kaufmann (1990)
5. Raufaste, E., da Silva Neves, R., Mariné, C.: Testing the descriptive validity of possibility theory in human judgements of uncertainty. Artificial Intelligence 148, 197–218 (2003)

6. Dubois, D., Godo, L., Prade, H., Zapico, A.: Advances in qualitative decision theory: Refined rankings. In: Monard, M.C., Sichman, J.S. (eds.) IBERAMIA-SBIA 2000. LNCS (LNAI), vol. 1952, pp. 427–436. Springer, Heidelberg (2000)
7. Dubois, D., Godo, L., Prade, H., Zapico, A.: Making decision in a qualitative setting: from decision under uncertainty to case-based decision. In: KR, vol. 6, pp. 594–607 (1998)
8. Giang, P., Shenoy, P.: A comparison of axiomatic approaches to qualitative decision making using possibility theory. In: UAI, vol. 17, pp. 162–170 (2001)
9. Giang, P., Shenoy, P.: Decision making on the sole basis of statistical likelihood. Artificial Intelligence 165, 137–163 (2005)
10. Giang, P., Shenoy, P.: A qualitative linear utility theory for spohn's theory of epistemic beliefs. In: UAI, vol. 16, pp. 220–229 (2000)
11. Dubois, D., Prade, H., Sabbadin, R.: Qualitative decision theory with Sugeno integrals. In: UAI, vol. 14, pp. 121–128 (1998)
12. Weng, P.: An axiomatic approach to qualitative decision theory with binary possibilistic utility. In: ECAI, vol. 17, pp. 467–471 (2006)
13. Brafman, R., Tennenholtz, M.: On the axiomatization of qualitative decision criteria. In: AAAI, vol. 14, pp. 76–81 (1997)
14. Benferhat, S., Dubois, D., Kaci, S., Prade, H.: Bipolar possibilistic representations. In: UAI, vol. 18, pp. 45–52 (2002)
15. Gilboa, I.: Expected utility with purely subjective non-additive probabilities. Journal of Mathematical Economics 16, 65–88 (1987)
16. Schmeidler, D.: Subjective probability and expected utility without additivity. Econometrica 57, 571–587 (1989)
17. Dubois, D., Prade, H.: Fuzzy sets and systems: theory and applications. Academy Press (1980)
18. Dubois, D., Prade, H., Sabbadin, R.: Decision-theoretic foundations of qualitative possibility theory. European Journal of Operational Research 128, 459–478 (2001)
19. Weng, P.: Axiomatic foundations for a class of generalized expected utility: Algebraic expected utility. In: UAI, vol. 22, pp. 520–527 (2006)
20. Klement, E., Mesiar, R., Pap, E.: Triangular norms. Kluwer Academic Publishers (2000)

Computing Semantic Association: Comparing Spreading Activation and Spectral Association for Ontology Learning

Gerhard Wohlgenannt, Stefan Belk, and Matthias Schett

Vienna University of Economics and Business, Augasse 2-6, 1090 Wien, Austria
{gerhard.wohlgenannt,stefan.belk,matthias.schett}@wu.ac.at
http://www.wu.ac.at

Abstract. Spreading activation is a common method for searching semantic or neural networks, it iteratively propagates activation for one or more sources through a network – a process that is computationally intensive. Spectral association is a recent technique to approximate spreading activation in one go, and therefore provides very fast computation of activation levels. In this paper we evaluate the characteristics of spectral association as replacement for classic spreading activation in the domain of ontology learning. The evaluation focuses on run-time performance measures of our implementation of both methods for various network sizes. Furthermore, we investigate differences in output, i.e. the resulting ontologies, between spreading activation and spectral association. The experiments confirm an excessive speedup in the computation of activation levels, and also a fast calculation of the spectral association operator if using a variant we called *brute force*. The paper concludes with pros and cons and usage recommendations for the methods.

Keywords: spreading activation, spectral association, ontology learning.

1 Introduction

Ontologies are the vocabulary and thereby the backbone of the Semantic Web. Constructing ontologies manually is an expensive and cumbersome process, and relies on highly specialized human labor [7]. (Semi-)automatic ontology learning from sources such as text supports the ontology engineer in his or her work.

Spreading activation is a method to search semantic networks, which can be used in ontology learning for example for finding semantically related concept candidates for existing concepts [14]. In the ontology learning system used as the foundation and test bed of this paper spreading activation is an essential tool used for the selection of domain-relevant concept candidates from a big semantic network as well as for positioning the new concepts in the ontology [16]. Spreading activation helps us to pick the most relevant concepts and associations from a vast number of evidence (candidate concepts and relations) generated from heterogeneous input sources.

S. Ramanna et al. (Eds.): MIWAI 2013, LNCS 8271, pp. 317–328, 2013.

In highly dynamic domains, and especially if the evolution of ontologies is of interest, new versions or updates of ontologies need to be generated in regular intervals. This makes the run-time performance for the ontology learning algorithms an important factor. Spreading activation is an iterative process and can be time-consuming to calculate, therefore Havasi et al. suggest a method called *spectral association* to compute the resulting activation levels after a number of steps of spreading activation in one step [11]. Spectral association approximates spreading activation using spectral decomposition of the concept matrix C, for details see Section 3.

This paper compares the two methods, i.e. the iterative spreading activation method and spectral association, in a specific environment for learning lightweight domain ontologies. We evaluate various aspects, including the run-time performance of both techniques and the impact on relevance of the resulting ontologies, discuss the implementation, and provide a general reflection of pros and cons of the methods observed as well as hints on when to apply which technique. The datasets used or domain is not important for the runtime of the evaluated algorithms – only the size of the semantic network.

Section 2 provides an overview of related work. The methods of spreading activation and spectral association are described in detail in Section 3. Section 4 introduces the ontology learning framework which applies the methods. Section 5 evaluates the run-time performance and other characteristics of the methods described earlier, and finally Section 6 summarizes the findings and gives an outlook on future work.

2   Related Work

This section starts with a presentation of a few selected ontology learning systems with respect to the question how evidence from algorithms or sources is integrated into the ontology. Text2Onto [3] uses a Probabilistic Ontology Model (POM) to aggregate results and represent the ontology. The results from various algorithms generate requests for changes of the POM; compared to the system for learning lightweight ontologies which this work is based on, Text2Onto aims at learning ontologies which are more expressive. Abraxas [18] takes quite a different approach, which is iterative and open-ended. The system identifies terms from a seed corpus and extracts ontological knowledge in the form of triples using lexico-syntactic patterns. Then it detects gaps in the collected ontological knowledge and tries to cover those with the help of external repositories such as the Web. KnowItAll [8] is not exactly an ontology learning system, but rather a system for the large scale extraction of facts. KnowItAll tests the plausibility of the candidate facts using pointwise mutual information (PMI) statistics computed by treating the Web as a massive corpus of text and thereby associates a probability with every fact. In their presentation of a generic architecture for ontology learning, Cimiano et al. stress the importance of methods of evidence integration especially when multiple and heterogeneous sources are combined, and state that a lot of further research is needed in that direction [2].

Spreading activation was first introduced by Collins et al. as a theory of human semantic processing [4]. Spreading activation is frequently used in information retrieval, Crestani provides a survey of spreading activation techniques on semantic networks in associative information retrieval [5]. The conclusions of the survey are positive, spreading activation is capable of providing good results.

Hasan proposes a system to apply spreading activation for information access within organizations [9]. The system integrates (i) documents, (ii) statistically derived information such as terms and entities extracted from those documents and (iii) a precise knowledge in form of an organizational ontology into a spreading activation network. User feedback on relevance of query results adapts the weights in the spreading activation network (learning).

Katifori et al. outline a framework which applies spreading activation over personal ontologies in the context of a personal interaction management system [13]. The goal of spreading activation is context inference depending on the users' recent actions and a populated personal information ontology to support user actions, for example generating suggestions when filling a Web form. The method is extended by augmenting the personal ontology with cached data from external repositories [6]. The selection of external data relates to the (spreading) activation level of entities already in the personal ontology or cached data.

Spectral association is an approximation technique for spreading activation networks, first presented in [11]. The paper describes the Colorizer application, which hypothesizes color values that represent given words and sentences. The common sense reasoning application determines the colors depending on physical descriptions of objects and emotional connotations. Spectral association interpolates colors for unknown concepts based on semantic relatedness to (in terms of color) known concepts. To ease computational complexity, Colorizer uses the spectral association approximation as a measure of semantic relatedness.

Spectral association is a variant of the AnalogySpace representation [15]. AnalogySpace addresses the problem of reasoning over large common sense knowledge bases with the characteristics of noisy and subjective data. The goal is to find rough conclusions based on similarities and tendencies, as traditional proof procedures are not feasible in such an environment. AnalogySpace forms analogical closures of a semantic network through dimensionality reduction.

A number of tools already utilize spectral association. Sentic Corner [1] is an application that dynamically collects audio, video, images and text related to and suitable for a user's current mood and displays this content on a multi-faceted classification Website. The system detects user mood via the analysis of semantics and sentics on the user's microblog (e.g. Twitter) postings. Spectral association is one of the fundamental methods used in the system, it generates semantically related concepts from the descriptions of music, films, images and text itself.

The Glass Infrastructure [10] allows the discovery of latent connections between people, projects, and ideas in an organisation. The system uses spectral association to generate a "semantic space" from project information, where closeness in the space signifies similarity of people, projects and ideas.

3   Methods

As this paper focuses on comparing spreading activation and spectral association, this section will describe them in some detail.

Spreading activation is a technique for searching associative, neural and semantic networks. The method requires a network structure with numeric or discrete relations between the network nodes. In the beginning the activation level of all nodes in the network is set to zero. The process starts by labeling source nodes, that is setting the activation level of one or more nodes to a value higher than the firing threshold (F). Every unfired node with an activation level $> F$ fires to all its connected nodes, the energy propagated results from a multiplication of the energy of the source, the weight of the connection, and the decay factor D. In further iterations unfired nodes that received activation in the previous step will fire to their neighbors. The lower the decay factor D, the less activation is spread to nodes further away from the original source nodes. The process terminates when no more unfired nodes (with activation above the firing threshold F) exist in the network. As a result of the process, the activation levels of all nodes in the network give measures of association to the source nodes, i.e. the higher the activation level of a node in the network at the end, the stronger the association to the source node(s).

Equation 1 outlines the activation level adaption of a single node A_j when receiving energy from node A_i via a connection with weight $W_{i,j}$, reduced by the static decay factor D. Obviously, this energy accumulation is applied for every incoming (and firing) node of A_j.

$$A_j = A_j + (A_i \cdot W_{i,j} \cdot D) \tag{1}$$

As already mentioned, spreading activation can be time-consuming to calculate [11]. Spectral association approximates many steps of spreading activation. The spectral association method starts with the transformation of the spreading activation network into a square symmetric matrix C of concepts. The rows and columns of C are labeled with the concepts from the network, and the values in the matrix represent the relation strength between the concepts. The relation strength of a concept to itself is 1. If there was no connection in the spreading activation network between two concepts the value 0 results. Step two is to scale rows and columns to unit vectors. Applying C to a vector of activations (of one or more concepts) yields the result of one round of spreading activation. The operator e^C to simulate any number of spreading activation rounds (with diminishing returns) is calculated by Equation 2, see [11]:

$$e^C = 1 + C + \frac{C^2}{2!} + \frac{C^3}{3!} + \dots \tag{2}$$

As C is square symmetric, it can be decomposed to $C = V \Lambda V^t$. In this eigendecomposition, V is the orthogonal real matrix of eigenvectors, whereas Λ is the diagonal matrix of eigenvalues. We can raise this expression to any power, and cancel anything but the power of Λ. What follows is that

$$e^C \approx V e^{\Lambda} V^t \tag{3}$$

To further ease computation, one can apply *dimensionality reduction* [11], i.e. keeping only the largest eigenvalues and their corresponding eigenvectors.

4   The Ontology Learning System

This section describes the system for learning domain-specific ontologies (T-box) underlying the work presented in this paper. The framework evolved since 2005 and has been presented in various publications. The original system [14] learns from domain text only and already includes spreading activation as the major building block to integrate evidence. Weichselbraun et al. evaluate information from social media as additional evidence source [16] and present novel methods for learning non-taxonomic relations [17]. Finally, Wohlgenannt et al. discusses data structures and algorithms for the fine-grained optimization of the system from feedback collected with games with a purpose [19].

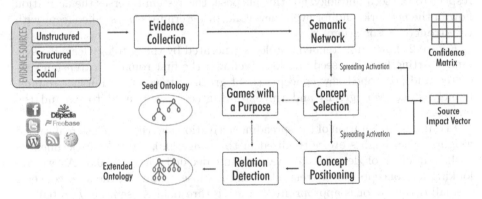

Fig. 1. Ontology Learning System Architecture Diagram [19]

The description given here is limited to a basic overview of the system needed to better understand the remainder of the paper. Figure 1 gives a graphical illustration of the major building blocks. The initial input to the process is a typically very small seed ontology (a few concepts and relations) in a particular domain. The system starts by collecting evidence for new concepts and relations in heterogeneous sources, i.e. domain text, social sources and structured sources (e.g. WordNet or online ontologies). A semantic network stores the gathered data. This semantic network is then transformed into a spreading activation network, from which new concept candidates are computed. Another round of spreading activation positions each of the new concepts in the seed ontology. Games with a purpose verify the relevance of new concepts and their position, as well as help to optimize the system performance. Finally the extended ontology acts as new seed ontology for the next step of ontology extension. After a pre-configured number of extension steps the system halts.

So what is spreading activation actually used for? The evidence acquisition phase collects a large number of concept candidates and relations to the seed concepts, the number of candidate terms can easily exceed a few thousand. The system includes a big number of so-called *evidence sources*, for example co-occurring keywords in domain text of US media, UK media, related tags from twitter, disambiguated hyper-/hyponyms from WordNet, and many others. Each of the evidence sources generates a number of new candidate terms (in this context we use *term* and *concept* synonymously) for any seed concept – all this information gets collected in the *semantic network*. A transformation algorithm creates the spreading activation network from the semantic network, the evidence source which suggested the relation influences the weight of the link. Spreading activation detects the most relevant n concepts (where $n = 25$ for example) from all the candidates. Roughly, candidates which are supported by different sources and by sources with a higher link weight ("source impact") are more likely to be selected.

Spreading activation is further used to position the selected concepts with respect to the seed ontology. For this purpose, the system reverses the activation flow in the network, and for every new concept determines the seed concept with the strongest association.

Figure 2 shows an extended ontology generated by the ontology learning system. Starting from the seed ontology (yellow), the first round of extension generated and positioned new concepts (light-green, "stage one"). Further rounds, namely stage two (green) and three (dark-green), were used to expand the ontology.

In our implementation of the spreading activation algorithm we use real valued weights. The weights are normalized in the range [0.0, 1.0]. We experimented with a number of decay factor values, and decided on $D = 0.3$. As we are looking for concepts closely related to and in close vicinity of the seed concept, a small decay factor is appropriate. No firing threshold is used, i.e. $F = 0.0$.

The implementation of the spectral association approximation of spreading activation starts with building the square symmetric concept matrix. Next a normalization step takes place. Havasi et al. propose to use unit vectors for rows and columns [11], but this led to increasing returns in Equation 2 when increasing n in $\frac{C^n}{n!}$ – which is not the expected behavior. Using the *determinant* of the matrix instead for normalization gave results as expected.

We calculate the operator e^C in two ways. The first variant uses spectral decomposition (eigendecomposition) to ease the approximation of e^C, in line with the description in Section 3 of the seminal paper [11]. This variant is referenced as *spectral association* or *SPECTRAL* in the evaluation section (Section 5). For the computation of eigenvectors and eigenvalues we use NumPy's `linalg.eigh` module [12], which is geared towards the generation of eigenvectors for symmetric matrices. NumPy[1] is a package for scientific computing in Python. The second

---

[1] http://www.numpy.org

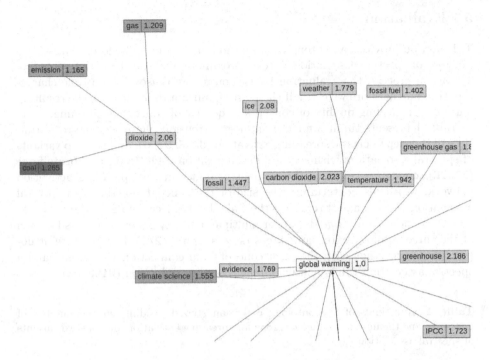

Fig. 2. An extended ontology (clipped) – concept labels and relations

variant, called *brute force* or *BRUTE* just calculates e^C according to Equation 2 – without the use of eigendecomposition. This makes the matrix multiplications more complicated, but saves the computation of eigenvectors. We use *brute force* (i) to validate the correct implementation of the *spectral association* method, and (ii) as a baseline when evaluating the run-time characteristics of spectral association.

Finally, as soon as e^C is computed, it can be used to simulate the activation of nodes simply by multiplying (using the dot-product) e^C with a concept vector where the values for the concepts to activate are set to 1, else 0.

The current version of the spectral association method can be found as an open source package on the Web[2] – free for use and modification. It is written in the Python programming language and uses numpy for all matrix operations. The most important parameter for run-time performance tuning is APPROX_DEPTH. In Equation 2, the implementation calculates C as far as the power of APPROX_DEPTH. We currently use the value of 10 – which proved to be more than sufficient in our experiments, as the factor gets very small (close to zero) with higher powers. The same parameter is used in the approximation of e^Λ, too.

---

[2] http://wwwai.wu.ac.at/~wohlg/spectral_association

5   Evaluation

This section provides an extensive evaluation of the methods described. As the purpose of spectral association is to approximate and speed up the spreading activation process, the evaluation focuses on a comparison of run-time characteristics. Furthermore we check if there are significant differences in the resulting ontologies regarding quality of concepts or quality of concept positioning.

Table 1 presents timings for the ontology extension phase, i.e. the total run-time of the step of classic spreading activation (SPREADING), and the two variants of spectral association: with spectral decomposition (SPECTRAL) and brute force (BRUTE). concepts refers to the number of nodes in the spreading activation network, which also determines the size of the concept matrix C in spectral association. The connections are the links between concepts in the network. The table includes *average* data over multiple ontology generation runs for each of the three ontology extension stages (Avg stage-1/2/3). Furthermore, it depicts the single stage with the most concepts and connections (biggest) and a spectral association run which uses dimensionality reduction (With DR).

Table 1. Run-times of the ontology extension step depending on the number of concepts and the number of connections for spreading activation and the two variants of spectral association

Run	concepts	connections	SPREADING	SPECTRAL	BRUTE
Avg stage-1	2495	3303	00:00:10 (1)	00:05:56	00:00:11
Avg stage-2	3843	9054	00:40:56 (91)	00:30:05	00:00:57
Avg stage-3	6101	18655	00:38:47 (86)	01:22:40	00:01:54
Biggest	6842	22342	00:44:35 (109)	01:43:46	00:02:35
With DR	6842	22342	–	00:35:33	–

It was very surprising to see that the *brute force* method performed best by far overall, for networks with > 5000 connections 20-40 times faster than the other two. Classic spreading activation sometimes outperformed spectral association, especially for small networks, and if no dimensionality reduction was applied.

We performed a thorough analysis to investigate *where time is spent*:

- *SPREADING activation*: As expected, almost all of the computing time is consumed by applying the activation algorithm to the network in concept detection and concept positioning.
- *BRUTE force* spends most of its time (about 70%) generating the operator e^C according to Equation 2.
- *SPECTRAL association* uses almost all of its time in the generation of e^C. In doing so, NumPy consumes about 33% for generating the eigenvectors and -values. The matrix multiplication in Equation 3 consumes the remaining computation time. This single matrix multiplication is surprisingly costly, keeping in mind that we have around 10 matrix multiplications (with matrices of same size) in Equation 2 when computing e^C *brute force* – depending

on the APPROX_DEPTH. Additional investigation showed that the matrices in Equation 3 are very dense, and that the concept matrix C in Equation 2 is typically sparse – an important point further covered below.

Another surprising observation is the huge difference between *Avg stage-1* (2672 concepts, 11 seconds) and *Avg stage-2* (3154 concepts, 26 minutes) for the method *spreading activation*. This is caused by two factors: (i) In stage-1 the network depth is very low (most nodes are connected to the seed concepts directly), whereas in stage-2 the network depth increases – which results in more steps of activation propagation. (ii) The number of activation processes differs. In stage-1 the system just needs to activate the seed concepts, while in stage-2 it positions the new concepts from stage-1 and activates the network. This results in about 60-110 activations, the exact number is given in parentheses in Table 1. A single activation in stage-2 and stage-3 has a runtime of about 20-30 seconds.

The eigenvector variant of spectral association can be approximated further (and thereby sped up) by *dimensionality reduction* (DR, see Section 3). Our implementation has an (optional) parameter to set the requested number of dimensions, we used a value of 50 in the experiments. As can be seen in Table 1, line With DR, applying DR reduces runtime to about 35%. The 35% largely result from calling NumPy for the calculation of eigenvectors (see above), so this variant is very effective to speed up the process. The use of DR had no impact on the resulting ontologies, the approximation works as expected.

Table 2 depicts memory usage for the three methods depending on the network size. The presented data reflects the maximum amount of memory allocated during runtime of the process.

Table 2. Maximum memory usage of the ontology extension step depending on the stage and algorithm used

Run	concepts	connections	SPREADING	SPECTRAL	BRUTE
stage-1	2672	3531	0.5 GB	0.4 GB	0.4 GB
stage-2	3154	12243	1.0 GB	2.0 GB	1.7 GB
stage-3	6842	22342	1.2 GB	3.3 GB	2.7 GB

We did not focus on optimizing memory usage at all, the goal of Table 2 is to give rough comparison of the methods, and to observe scaling regarding memory consumption. As expected, memory usage tends to scale linearly with the number of concepts and connections. In variant *SPECTRAL*, memory usage is highest during NumPy's computation of eigenvectors and -values. If required, there will be a number of options to decrease memory usage, for example experimenting with Numpy's sparse matrix types or specific optimizations regarding memory in the Python code. The machine we ran the experiments on had enough memory to prevent any swapping to disk.

Complementing the evaluation of run-time performance, we also compared the quality of results. Spectral association is intended as approximation of spreading activation, so results should roughly be the same, although there are various parameter settings available. We did three ontology extension runs for March, April and May 2013 which extended the seed ontology with 25 candidate concepts in each of the three extension iterations, summing up to 225 candidate concepts. Manual evaluation by domain experts showed that there was no significant difference in the relevance of candidate concepts. The relevance was about 50% for all three methods (50% BRUTE, 50% SPECTRAL, 49% SPREADING). We also investigated the effect of the application of the three methods on the quality of concept positioning; no significant differences were found.

The experiments conducted suggest the following *pros and cons*, as well as *areas of application* for the tree methods:

- *Spreading activation*: Among the pros of spreading activation is the interpretability of the activation process, which can be traced easily, whereas spectral association is rather a black box model. Spreading activation provides parameters for tuning the activation process to fit ones specific needs (mainly with the firing threshold F and decay D), while we do not see a simple way how to apply these tuning parameters in the spectral association process. As stated above, the *depth* of the network, i.e. the number of propagation steps necessary, strongly affects runtime performance. The more propagation steps, the slower the process. Naturally, the number of propagation steps also depends on decay factor D and firing threshold F.

  The application of the method is advisable for small networks and situations in which runtime performance is not an issue, in the (unlikely) case of only one or a few activations needed, or if parameter tuning is required.
- *Variants of spectral association*: In contrast to spreading activation, in both spectral association variants the main factor is the generation of e^C, once this is done activation is very fast - it's just a simple *matrix ∘ vector* multiplication. This is a crucial point, activation processes can be done almost instantly with this method. If the activation process has to be further quickened, one can apply dimensionality reduction by using only the largest eigenvalues and their corresponding eigenvectors [11]. Dimensionality reduction also speeds up the the generation of e^C, in our experiments by a factor of approx. 3. However, it is not applicable in the *brute force* variant (as we do not compute eigenvectors).

  The sparser the concept matrix, i.e. the less connections between concepts, the faster the computation of e^C with the *brute force* method. We also tested with a dense matrix (almost all values $\neq 0$), in that case *brute force* is consistently slower (around factor 2) than *SPECTRAL* overall. But in a typical scenario with a sparse concept matrix the computations in Equation 2 are very efficient.
  - *Spectral association (with eigendecomposition)*: The variant allows the application of dimensionality reduction to further reduce and approximate e^C, and thereby enables even faster activation processes. It is well suited for very dense networks and if the runtime of activation is critical, and not so much the generation of e^C.

- *Brute force*: In our run-time analyses which have to be further confirmed in other environments and programming languages, the described *brute force* approach is surprisingly efficient. *Brute force* is preferable when the number of activations is limited and the time spent in both the generation of e^C as well as activations is critical – as in our ontology learning framework, where the method was performing best by far.

6   Conclusions

This paper compares a classic method for searching networks, i.e. spreading activation, with spectral association. Spectral association approximates the computationally intensive activation process using a matrix operator called e^C. The generation of this matrix via spectral decomposition is complex for large spreading activation nets, therefore spectral association is particularly well suited where e^C is used for many activation processes. In situations where instant results for an activation (e.g. in interactive applications) are needed, spectral association is the only option. We also tested a simpler way to generate the operator e^C, named *brute force*, which – under the circumstances we investigated – provides far superior runtime performance in generating e^C.

The main contributions of this paper are (i) extensively evaluating spreading activation vs. spectral association in the domain for ontology learning regarding various sizes of concept networks, (ii) showing that the *brute force* method is very efficient (20–40 times faster than spreading activation) in our experiments, (iii) the provision of an (open-source) implementation of spectral association in Python, (iv) giving hints under what circumstances to apply which method.

Future work will include experimenting with a wider range of network sizes and parameter settings, using other libraries for example to compute eigenvectors, and the application of other programming languages to verify the results.

Acknowledgements. The work presented in this paper was developed within DIVINE (www.weblyzard.com/divine), a project funded by the Austrian Ministry of Transport, Innovation & Technology (BMVIT) and the Austrian Research Promotion Agency (FFG) within FIT-IT (www.ffg.at/fit-it). The work has also been supported by uComp (www.ucomp.eu), a project in EU's ERA-NET CHIST-ERA programme.

References

1. Cambria, E., Hussain, A., Eckl, C.: Taking refuge in your personal sentic corner. In: Bandyopadhyay, S., Okumurra, M. (eds.) Proceedings of IJCNLP, Workshop on Sentiment Analysis where AI meets Psychology, pp. 35–43. Chiang Mai, Thailand (2011)
2. Cimiano, P., Mädche, A., Staab, S., Völker, J.: Ontology learning. In: Staab, S., Studer, R. (eds.) Handbook on Ontologies. International Handbooks Information System, pp. 245–267. Springer, Heidelberg (2009)

3. Cimiano, P., Völker, J.: Text2onto. In: Montoyo, A., Muñoz, R., Métais, E. (eds.) NLDB 2005. LNCS, vol. 3513, pp. 227–238. Springer, Heidelberg (2005)
4. Collins, A.M., Loftus, E.F.: A spreading-activation theory of semantic processing. Psychological Review 82(6), 407–428 (1975)
5. Crestani, F.: Application of spreading activation techniques in information retrieval. Artificial Intelligence Review 11(6), 453–482 (1997)
6. Dix, A.J., Katifori, A., Lepouras, G., Vassilakis, C., Shabir, N.: Spreading activation over ontology-based resources: from personal context to web scale reasoning. International Journal of Semantic Computing 4(1), 59–102 (2010)
7. Drumond, L., Girardi, R.: A survey of ontology learning procedures. In: de Freitas, F.L.G., Stuckenschmidt, H., Pinto, H.S., Malucelli, A., Corchoo, Ó. (eds.) Proceedings of the 3rd Workshop on Ontologies and their Applications (WONTO), Brazil. CEUR Workshop Proceedings, vol. 427, CEUR-WS.org, Salvador (2008)
8. Etzioni, O., Cafarella, M., Downey, D., Kok, S., Popescu, A.M., Shaked, T., Soderland, S., Weld, D.S., Yates, A.: Web-scale information extraction in knowitall (preliminary results). In: WWW 2004: Proceedings of the 13th International Conference on World Wide Web, pp. 100–110. ACM, New York (2004)
9. Hasan, M.M.: A spreading activation framework for ontology-enhanced adaptive information access within organisations. In: van Elst, L., Dignum, V., Abecker, A. (eds.) AMKM 2003. LNCS (LNAI), vol. 2926, pp. 288–296. Springer, Heidelberg (2004)
10. Havasi, C., Borovoy, R., Kizelshteyn, B., Ypodimatopoulos, P., Ferguson, J., Holtzman, H., Lippman, A., Schultz, D., Blackshaw, M., Elliott, G.T.: The glass infrastructure: Using common sense to create a dynamic, place-based social information system. AI Magazine 33(2), 91–102 (2012)
11. Havasi, C., Speer, R., Holmgren, J.: Automated color selection using semantic knowledge. In: AAAI Fall Symposium Series. Arlington, Texas (2010)
12. Jones, E., Oliphant, T., Peterson, P., et al.: SciPy: Open source scientific tools for Python (2001), http://www.scipy.org/
13. Katifori, A., Vassilakis, C., Dix, A.J.: Ontologies and the brain: Using spreading activation through ontologies to support personal interaction. Cognitive Systems Research 11(1), 25–41 (2010)
14. Liu, W., Weichselbraun, A., Scharl, A., Chang, E.: Semi-automatic ontology extension using spreading activation. Journal of Universal Knowledge Management (1), 50–58 (2005)
15. Speer, R., Havasi, C., Lieberman, H.: Analogyspace: Reducing the dimensionality of common sense knowledge. In: Fox, D., Gomes, C.P. (eds.) AAAI, pp. 548–553. AAAI Press (2008)
16. Weichselbraun, A., Wohlgenannt, G., Scharl, A.: Augmenting lightweight domain ontologies with social evidence sources. In: Tjoa, A.M., Wagner, R.R. (eds.) 9th International Workshop on Web Semantics, 21st International Conference on Database and Expert Systems Applications (DEXA 2010), pp. 193–197. IEEE Computer Society Press, Bilbao (2010)
17. Weichselbraun, A., Wohlgenannt, G., Scharl, A.: Refining non-taxonomic relation labels with external structured data to support ontology learning. Data & Knowledge Engineering 69(8), 763–778 (2010)
18. Wilks, Y., Brewster, C.: Natural language processing as a foundation of the semantic web. Foundations and Trends in Web Science 1(3-4), 199–327 (2009)
19. Wohlgenannt, G., Weichselbraun, A., Scharl, A., Sabou, M.: Dynamic integration of multiple evidence sources for ontology learning. Journal of Information and Data Management (JIDM) 3(3), 243–254 (2012)

Evolution of Self-interested Agents:
An Experimental Study

Naoki Yamada* and Chiaki Sakama

Department of Computer and Communication Sciences
Wakayama University, Sakaedani, Wakayama 640-8510, Japan
sakama@sys.wakayama-u.ac.jp

Abstract. In this paper, we perform an experimental study to examine the evolution of self-interested agents in cooperative agent societies. To this end, we realize a multiagent system in which agents initially behave altruistically by sharing information of food. After generations of a genetic algorithm, we observe the emergence of selfish agents who do not share food information. The experimental results show the process of evolving self-interested agents in resource-restrictive environments, which is observed in nature and in human society.

Keywords: evolution, genetic algorithm, multiagent system, self-interested agents.

1 Introduction

In nature, animals communicate in many ways to share information. For example, ants inform each other about the location of food using scent, and sheep alert others about predator attacks by making a bleating sound. On the other hand, animals compete with one another for limited resources, such as food, space and mates. In his book "*The Selfish Gene*" [1], Richard Dawkins says: "Any altruistic system is inherently unstable, because it is open to abuse by selfish individuals, ready to exploit it." Animals are inherently self-interested and often behave dishonestly to have their own benefit. In [6], Searcy and Nowicki argue that "The predominant view nowadays, however, is that selection acts largely at the level of the individual, so that behavior evolves toward what is best for the individual performing the behavior, and not toward what is best for the group. If behavior is commonly selfish, in this sense, then it is not always obvious why animals should exchange information cooperatively. Instead, one might expect many instances in which signalers would attempt to profit individually by conveying dishonest information." Some studies are devoted to modelling evolution of selfish or dishonest behaviors of animals. Wade and Breden [8] provide a population generic model and examine necessary conditions for the spread of genes that determine selfish and cheating behaviors. Sober [7] provides a simple model which explains that lying and credulity are behaviors that evolved by natural selection. Rowell *et al.* [5] develop a game-theoretic model of animal communication in which animals effectively use deceptive signals as strategies.

---

* Current Address: Shima Seiki MFG., LTD., 85 Sakata Wakayama 641-8511, Japan.

S. Ramanna et al. (Eds.): MIWAI 2013, LNCS 8271, pp. 329–340, 2013.

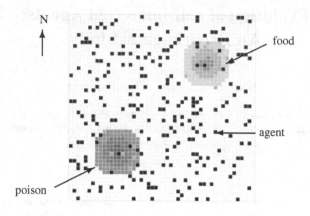

Fig. 1. An environment

Recent studies show the evolution of selfishness in robot communication. Floreano *et al.* [2,4] show that robots which compete for food learn to conceal food information. In their studies, a group of robots has the task of finding a food source in a field. Once a robot found the food, it stays nearby and emits blue light. This informs other robots of the location and results in overcrowding around the food. After a few generations, robots become more secretive and learn to conceal food information for their own survival. The study shows the possibility of designing artificial agents which would acquire selfish or dishonest attitudes in their environment.

In this study, we consider an environment similar to [2,4] and observe how self-interested behaviors evolve in an artificial society. In contrast to [2,4], we do not use robots but realize software agents who can communicate and move in an environment. We implement a multiagent system in which agents initially behave cooperatively by sharing information of food. After generations of a *genetic algorithm*, we observe the evolution of self-interested agents who do not share food information. We analyze experimental results and see the adaptation of agents in a resource-restrictive environment. The rest of this paper is organized as follows. Section 2 presents an agent society which is considered in this paper. Section 3 provides experimental results and considerations. Section 4 discusses related issues and Section 5 concludes the paper.

2 Agent Society

2.1 Agents

We set the *environment* as a two-dimensional grid of 50×50 cells where 250 *agents* are living. Each agent stays at one cell and two different agents cannot stay at a cell at the same time. The environment contains 132 cells of food and 132 cells of poison. The spacial constraints on the food allow a maximum of 132 agents to be fed simultaneously. The maximum amounts of food or poison in each cell are initially given. The range of maximal values is from 3 to 1 in order of the depth of a color (Figure 1). If the amounts

Input Neurons   Output Neurons

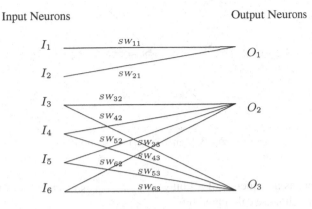

Fig. 2. Neural Network

of food or poison in each cell become less than its maximum value by consumption, they are automatically supplemented in every fixed period. Agents act synchronously in discrete time steps. A *generation* consists of 200 time steps. Each agent can move to neighbor cells (horizontally or vertically adjacent cells), send a signal, and obtain food or poison at every step. If an agent stays at a cell where food or poison is located, then it is counted as one at a step. The numbers of food or poison which an agent obtains in one generation are counted.

Each agent can send a signal when they find food or poison. An agent can recognize signals on the 360° field and decides a direction to move at the next step based on the amount and direction of signals. Each agent has a simple *neural network* which consists of 6 input neurons I_i ($1 \leq i \leq 6$) and 3 output neurons O_j ($1 \leq j \leq 3$) through 10 synaptic weights SW_{ij} representing the strength of connections between the input neuron I_i and the output neuron O_j (Figure 2).[1]

The value $v(I_i)$ of the input neuron I_i ($i = 1, 2$) is decided by the location of an agent as follows.

$$v(I_1) = \begin{cases} 1 & \text{if there is food on the cell where an agent is staying;} \\ 0 & \text{otherwise} \end{cases}$$

$$v(I_2) = \begin{cases} 1 & \text{if there is poison on the cell where an agent is staying;} \\ 0 & \text{otherwise} \end{cases}$$

The values of I_3, I_4, I_5 and I_6 are decided by the amount of signals which an agent perceives from each direction as follows.

$$v(I_3) = \frac{S_w}{S}, \quad v(I_4) = \frac{S_e}{S}, \quad v(I_5) = \frac{S_s}{S}, \quad v(I_6) = \frac{S_n}{S}. \tag{1}$$

Here, S_w, S_e, S_s and S_n respectively represent the amount of signals from the four sections (west, east, south and north) of 90° each and $S = S_w + S_e + S_n + S_s$ (Figure 3).

---

[1] In [2,4], a similar but more complicated neural network is used which consists of 11 input neurons connected to 3 output neurons through 33 synaptic weights.

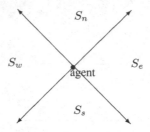

Fig. 3. Signals from four sections

Each agent has a sequence of 80-bit genes $\langle g_i \rangle_{1 \le i \le 80}$ as a binary digit. The values of synaptic weights are calculated by

$$SW_{ij} = \frac{D_{ij} \times 2}{255} - 1 \tag{2}$$

where D_{ij} represents the decimal number corresponding to 8-bit genes as follows: $D_{11} = (g_1 \ldots g_8)_{10}$, $D_{21} = (g_9 \ldots g_{16})_{10}$, $D_{32} = (g_{17} \ldots g_{24})_{10}$, $D_{33} = (g_{25} \ldots g_{32})_{10}$, $D_{42} = (g_{33} \ldots g_{40})_{10}$, $D_{43} = (g_{41} \ldots g_{48})_{10}$, $D_{52} = (g_{49} \ldots g_{56})_{10}$, $D_{53} = (g_{57} \ldots g_{64})_{10}$, $D_{62} = (g_{65} \ldots g_{72})_{10}$, $D_{63} = (g_{73} \ldots g_{80})_{10}$. For instance, when the 8-bit gene is 00000000 (resp. 11111111), it becomes $D_{ij} = 0$ and $SW_{ij} = -1$ (resp. $D_{ij} = 255$ and $SW_{ij} = 1$) by (2). Thus, a synaptic weight takes a value of $-1 \le SW_{ij} \le 1$.

The values of output neurons are computed using the values of input neurons and synaptic weights as follows:

$$O_1 = \tanh \left(\sum_{k=1}^{2} (v(I_k) \times SW_{k\,1}) \right),$$

$$O_n = \tanh \left(\sum_{k=3}^{6} (v(I_k) \times SW_{k\,n}) \right) \quad (n = 2, 3).$$

O_j is expressed using the hyperbolic function and takes a value between -1 and 1. The values of output neurons are used for deciding action of an agent.

2.2   Action Rules

An agent can take two different actions at each step: sending a signal or moving to neighbor cells. First, an agent sends a signal if the value of the output neuron is $O_1 > 0$. This may happen when an agent finds food ($I_1 > 0$) or poison ($I_2 > 0$), but whether $O_1 > 0$ or not depends on the values of the synaptic weights. If $O_1 \le 0$, then an agent does not send any signal even if it obtains food or poison. Next, a move of an agent is decided by the values of output neurons O_2 and O_3. A move of an agent is expressed by (dx, dy) where dx (resp. dy) represents a movement in the x-axis (resp. y-axis) direction. Each movement is defined by O_2 and O_3 as follows:

$$dx = \begin{cases} -1 & (O_2 < -\frac{1}{3}) \\ 0 & (-\frac{1}{3} \leq O_2 \leq \frac{1}{3}) \\ 1 & (\frac{1}{3} < O_2) \end{cases} \qquad dy = \begin{cases} 1 & (O_3 < -\frac{1}{3}) \\ 0 & (-\frac{1}{3} \leq O_3 \leq \frac{1}{3}) \\ -1 & (\frac{1}{3} < O_3) \end{cases} \qquad (3)$$

where the values of O_2 and O_3 are divided into three.

We consider a few agents who may not act properly in an environment. This is realized by making an agent move randomly at the probability of 0.2, regardless of the values of output neurons.

The movement of an agent depends on the values of its synaptic weights. As stated before, the synaptic weights of an agent are calculated by a sequence of 80-bit genes of the agent. The initial genes are randomly generated, and then evolve under a fitness condition using the *genetic algorithm*. The fitness of a gene is computed by

$$f = F - P \qquad (4)$$

where F and P respectively represent the numbers of food and poison which an agent obtained in one generation. By definition, the fitness increases if an agent obtains more food, while the fitness decreases if an agent obtains more poison. In each generation, agents having higher fitnesses are selected, and their genes are modified (by *crossover* and *mutation*) to form a new generation. The selection is made using the fitness proportionate selection in which the i-th individual is selected based on the probability:

$$p_i = \frac{f_i}{\sum_{k=1}^{n} f_k}$$

where $n = 250$ is the total number of agents and f_i is the fitness of the i-th individual ($1 \leq i \leq 250$). At the end of each generation, the 250 agents are ranked based on their fitness and the best 20% are selected. From these selected agents, two individuals are randomly chosen and paired to perform crossovers and mutations to create a new generation of 250 individuals. Here we use the *uniform crossover* which evaluates each bit in the parent strings for exchange with a probability of 0.5. We also set mutation with a rate of 0.01. The individual which has the highest fitness is also retained in the next generation (*elitism*).

3 Experiments

We perform experiments in three different situations: (i) the field contains food only; (ii) the field contains poison only; and (iii) the field contains both food and poison. In each case, the location of food and poison are fixed; food are located in the north-east corner of the field and poison are located in the south-west corner of the field (Figure 1).

Initially, each agent is assigned a sequence of 80-bit genes which is randomly generated. The initial location of 250 agents is also decided randomly and the evolution of agents is observed in 300 generations. The fitness value of an agent depends on its initial location. To reduce such biases, a group of agents in each generation is produced based on the average of fitness values in 20 trials in which each trial consists of 200 steps of transactions by the same agents. With these settings, we observe the following

Fig. 4. Snapshot of an experiment

changes over generations: (i) the number of signals in the field, and (ii) the average values of the synaptic weights among all agents. A snapshot of an experiment is shown in Figure 4 where some (but not all) agents send signals around food.

3.1   Food and Signal

We first observe how the number of signals by obtaining food changes in generations. When the field contains food only, the number of signals increases at first and arrives at the peak around the 20th generation (Figure 5). Then, the number of signals suddenly decreases and keeps low values in subsequent generations, mostly less than 5000. This phenomenon is explained as follows. At the initial stage some agents at the location of food send signals which attract other agents. Then the number of agents monotonically increases around the food and the number of signals increases accordingly. In several generations, however, it results in the crowd around food. Those agents who actively send signals around the food cannot obtain food as before (because they cannot move neighbors where other agents stay), which results in the decrease of fitness values of those agents. In contrast, those agents which do not actively send signals would have relatively high fitness values. (Note that some agents would not send signals at food due to the negative synaptic weights given initially.) Then, the probability of selecting agents who actively send signals around the food reduces, which results in the decrease of signaling agents in the next generation. This is observed by Figure 6 in which the average synaptic weight SW_{11} increases at first, while it decreases after the peak around the 20th generation. After the 50th generation, SW_{11} mostly takes negative values which indicates that most agents do not send signals around the food. However, agents who send signals around the food do not die out. This is because reduction of signals has the effect of solving overpopulation around the food, which results in weakening the selection pressure on secretive agents.

When the field contains both food and poison, the number of signals also decreases in Figure 5 but the values sharply oscillate compared with the case of food only. This is because agents around the poison also send signals, which eliminates the effect of

number of signals

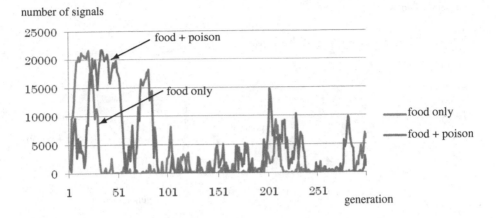

Fig. 5. Number of signals by obtaining food

value

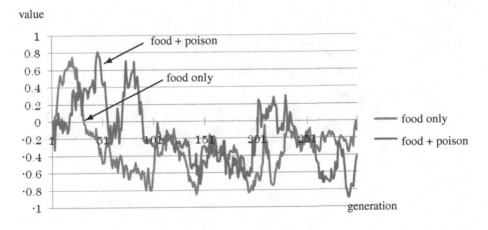

Fig. 6. Evolution of SW_{11}

decrease of signals in the field. The average synaptic weight SW_{11} also oscillates between positive and negative values in Figure 6.

3.2   Poison and Signal

We next observe how the number of signals by obtaining poison changes in generations. When the field contains poison only, the number of signals is relatively small through generations (Figure 7). The reason is that in this case agents who obtained poison have low fitness values by the equation (4), which results in the evolution that agents are directed away from signals. When the field contains both food and poison, the number of signals randomly oscillates. Such a chaotic behavior is due to the mixture of signals

number of signals

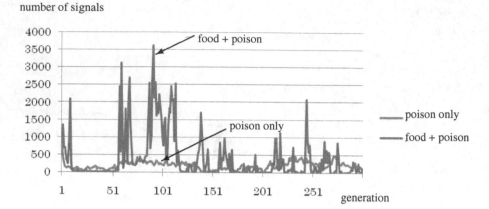

Fig. 7. Number of signals by obtaining poison

value

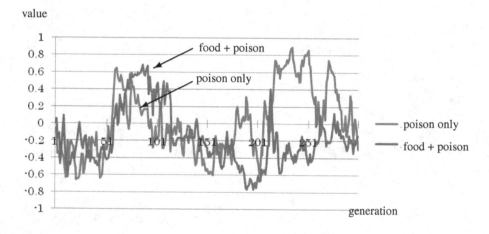

Fig. 8. Evolution of SW_{21}

from food and poison. The average synaptic weight SW_{21} also oscillates between positive and negative values (Figure 8). This means that in case of poison stopping signals does not imply any advantage for an agent, which results in no particular evolution of the synaptic weight SW_{21}.

3.3 Movement

We finally observe how each agent develops genes controlling its movement in generations. Figure 9 shows the evolution of average synaptic weights SW_{x2} and SW_{x3} ($x = 3, 4, 5, 6$) which control output neurons O_2 and O_3, respectively. In the figure, (a) shows the case of food only, (b) shows the case of poison only, and (c) shows the case

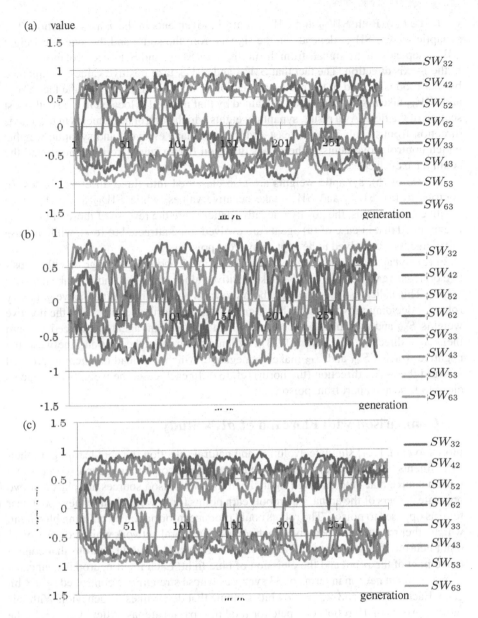

Fig. 9. Evolution of SW_{x2} and SW_{x3}

of food plus poison. Observing (a), synaptic weights are classified into three different classes. The first class takes positive values, the second class takes negative values, and the third class oscillates between positive and negative values. More precisely, SW_{42} and SW_{63} take positive values, while SW_{32} and SW_{53} take negative values. Others oscillate between positive and negative values. The synaptic weight SW_{42} depends on the signals from the east and the synaptic weight SW_{32} depends on the signals from the west (see (1)). Both SW_{42} and SW_{32} control movements in the x-axis direction. The synaptic weight SW_{53} depends on the signals from the south and the synaptic weight SW_{63} depends on the signals from the north. Both SW_{53} and SW_{63} control movements in the y-axis direction. The fact that SW_{42} and SW_{63} take positive values is explained by that food is located in the north-east section of the field. By contrast, the fact SW_{32} and SW_{53} take negative values is explained by that no food is located in the south-west section of the field. As a result, synaptic weights which control movements in the x-axis direction from signals from the east and the west have evolved, and synaptic weights which control movements in the y-axis direction from signals from the north and the south have evolved.

In case of (b), synaptic weights are also classified into three different classes. In contrast to (a), SW_{42} and SW_{63} take negative values, while SW_{32} and SW_{53} take positive values. Thus, the positive-negative patterns are the reverse of those of (a). This is explained that in case of (a) agents are evolved to be attracted to food, while in case of (b) agents are evolved to be kept away from poison.

In (c), synaptic weights, which are oscillated in (a) and (b), converge on either positive or negative values. S_{52} and S_{62} take positive values while S_{33} and S_{43} take negative values. This indicates when there are both food and poison, those synaptic weights play roles for deciding movements. When signals from the north and the south, the positive weights S_{52} and S_{62} make O_2 positive, which will lead an agent to proceed forward the x-axis direction (the east) (cf. (3)). When signals from the east and the west, the negative weights S_{33} and S_{43} make O_3 negative, which will lead an agent to proceed forward the y-axis direction (the north) (cf. (3)). In each case, the weights lead agents close to food and apart from poison.

4   Comparison with Floreano *et al.*'s Study

Floreano *et al.* [2,4] simulate evolution and natural selection in robot learning. In their experiments, robots are randomly placed in an arena containing a food source and a poison source that both emit red light. The food and poison sources are placed at two opposite corners of the arena. The robots earn points for how much time they spent near food as opposed to poison. The robots could produce information by emitting blue light, which other robots could perceive. Each robot has a neural network which consists of 11 input neurons that are connected to a robot's sensors and 3 output neurons that control movement of the robot and the emission of blue light. Each input neuron is connected to every output neuron in terms of 33 synapses whose strength are controlled by a 8-bit gene. Each robot has $33 \times 8 = 264$ bits genome that determines its behavior. With this setting, groups of 10 robots compete for food in separate arenas. After 100 rounds, the robots with the highest scores are selected for the next round. As robots become more efficient at finding and remaining near the food, the concentration of blue light near food also increases. Thus, blue light plays an inadvertent cue providing information on the food location. However, spacial constraints around the food source allow a maximum of 8 robots of 10 to feed simultaneously and result in higher robots density and increased competition and interference near the food. By the 50th generation, robots are selected to decrease the rate of blue light emission. Thus, selection is acting toward suppressing information on the food location.

As addressed in the introduction, our experimental setup is similar to [2,4], while there are some important differences as follows. First, in their experiment robots emit blue light randomly while it provides inadvertent social information on the food location. Once robots evolve the ability to find food and stay nearby, their increasing density near the food source translates into higher blue density near the food and a source of information for other robots in the arena. On the other hand, in our experiment, agents are initially set to send signals when they find food or poison. Floreano *et al.* observe how unintentional communication develops useful information, which generates self-centered behaviors of robots. By contrast, we observe that agents who behave cooperatively at first also turn to become self-interested in evolution. Second, neural nets used in [2,4] have 11 input neurons and 3 output neurons.[2] One of the input neurons is devoted to the sensing of food and another one is to the sensing of poison. Other 8 neurons are used for encoding the 360° visual input image, which is divided into four sections of 90° each. For each section, one neuron is used for perceiving blue light, and the other neuron is used for perceiving red light (food or poison). The activation of output neurons is computed as the sum of all inputs multiplied by the 8-bit synaptic weight of the connection and passed through the continuous hyperbolic function. Two of the 3 output neurons are used for controlling the two tracks, where the output value of each neuron gave the direction of rotation and velocity of one of the two tracks. The third output neuron determines whether to emit blue light. Thus the total length of the genetic string of an individual is: (8 bits) × (11 input neurons) × (3 output neurons) = 264 bits. In our multiagent model, each agent has a simpler neural network: 6 input neurons and 3 output neurons through 10 synaptic weights, and a 80-bit genetic string in total. Thus, an agent has less than one-third of genes compared with a robot of [2,4]. With this simplified neurons, we demonstrate an evolution similar to [2,4].

Thirdly, in the experimental setup, Floreano *et al.* consider the single environment in which both food and poison are located. By contrast, we set up three different environments: the first one contains food only, the second one contains poison only, and the third one contains both food and poison. We observe that such different settings affect the results of evolution of agents. Moreover, we analyze the evolution of synaptic weights in generations which are not reported in [2,4].

5 Summary

We experimentally realized a multiagent system to observe the evolution of self-interested agents to survive in a resource-restrictive environment. The results show that at the initial stage agents act altruistically to inform others of the location of food, while the increased population around the food results in the increase of self-interested agents who act egoistically to hide food information. Agents react to signals by other agents to obtain useful information, while once they successfully obtain food they evolve into agents who do not always send signals cooperatively. The evolution of self-interested nature of agents from simple action rules would explain a reason for the emergence of selfish behaviors of animals in resource-limited environments in nature.

---

[2] More precisely, 10 input neurons are used in [2] while 11 input neurons are used in [4]. The role of the additional input neuron is not clearly stated, however.

In this study, self-interested agents appear around food, while further evolution might generate dishonest agents who intentionally send "false" signals to other agents in order to keep them away from food (or even lead them to poison). Such deceptive signals exist in nature [3]. Further refinement of social models is needed to realize the evolution of agents who may act dishonestly.

References

1. Dawkins, R.: The Selfish Gene. Oxford University Press (1976)
2. Floreano, D., Mitri, S., Magnenat, S., Keller, L.: Evolutionary conditions for the emergence of communication in robots. Current Biology 17(6), 514–519 (2007)
3. Hasson, O.: Cheating Signals. Journal of Theoretical Biology 167(3), 223–238 (1994)
4. Mitri, S., Floreano, D., Keller, L.: The evolution of information suppression in communicating robots with conflicting interests. Proceedings of National Academy of Sciences 106(37), 15786–15790 (2009)
5. Rowell, J.T., Ellner, S.P., Reeve, H.K.: Why animals lie: how dishonesty and belief can coexist in a signaling system. American Naturalist 168(6), E180–E204 (2006)
6. Searcy, W.A., Nowicki, S.: The evolution of animal communication: reliability and deception in signaling systems. Princeton University Press, Princeton (2005)
7. Sober, E.: The primacy of truth-telling and the evolution of lying. In: From a Biological Point of View: Essays in Evolutionary Philosophy, Cambridge Studies in Philosophy and Biology, pp. 71–92 (1994)
8. Wade, M.J., Breden, F.: The evolution of cheating and selfish behavior. Behavioral Ecology and Sociobiology 7, 167–172 (1980)

Author Index